CHICAGO
MADE

HISTORICAL STUDIES OF URBAN AMERICA

Edited by Timothy J. Gilfoyle, James R. Grossman, and Becky M. Nicolaides

ALSO IN THE SERIES:

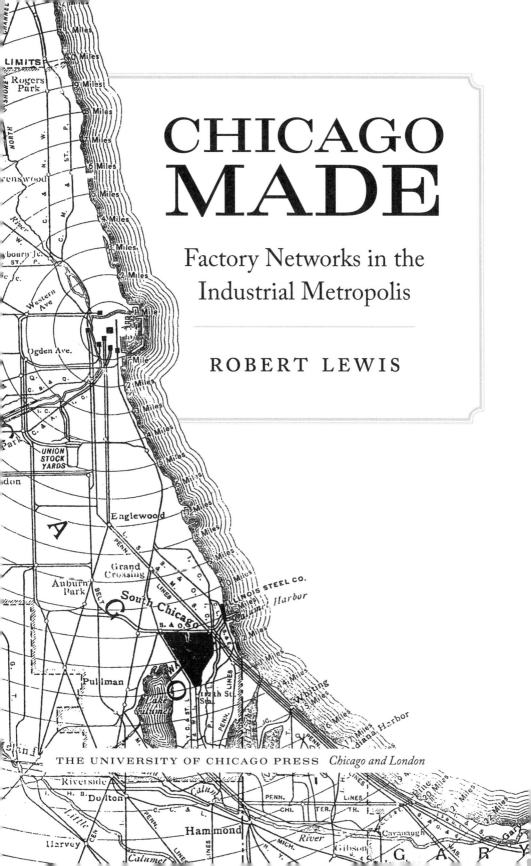

CHICAGO MADE

Factory Networks in the
Industrial Metropolis

ROBERT LEWIS

THE UNIVERSITY OF CHICAGO PRESS *Chicago and London*

Robert Lewis is professor of geography at the
University of Toronto. He is the author of
*Manufacturing Montreal: The Making of an Industrial
Landscape, 1850–1930* and the editor of
*The Manufacturing Suburb: Building Work and
Home on the Metropolitan Fringe.*

The University of Chicago Press, Chicago 60637
The University of Chicago Press, Ltd., London
© 2008 by The University of Chicago
All rights reserved. Published 2008
Printed in the United States of America

17 16 15 14 13 12 11 10 09 08   1 2 3 4 5

ISBN-13: 978-0-226-47701-5 (cloth)
ISBN-10: 0-226-47701-0 (cloth)

Library of Congress Cataloging-in-Publication Data

Lewis, Robert D., 1954–
Chicago made: factory networks in the industrial metropolis
/ Robert Lewis.
p. cm. — (Historical studies of urban America)
Includes bibliographical references and index.
ISBN-13: 978-0-226-47701-5 (cloth: alk. paper)
ISBN-10: 0-226-47701-0 (cloth: alk. paper)
1. Industrial districts—Illinois—Chicago—History.
2. Industrialization—Illinois—Chicago—History.
3. Urbanization—Illinois—Chicago—History. I. Title.
HC108.C4L49   2008
338.977311009'034—dc22
2008005801

CONTENTS

ILLUSTRATIONS

TABLES

ACKNOWLEDGMENTS

Any book is heavily indebted to many people. This one is no different. Those occupying the chair's office of the Department of Geography at the University of Toronto—Carl Amrhein, Larry Bourne, and Joe Desloges—provided me with terrific institutional support in all sorts of ways, both large and small. This is equally true of the department's staff. I would like to thank Ayesha Alli, Susan Calanza, Donna Jeynes, Marika Maslej, and Carla Vitoria for their help over the years. From discussions in the hallway and the classroom to conversation around a table at Rowers and elsewhere, my Toronto colleagues—Alana Boland, John Britton, Rick Difrancesco, Emily Gilbert, Jason Hackworth, Paul Hess, Debby Leslie, Scott Prudham, Katharine Rankin, and Virginia Maclaren—have given me intellectual sustenance, have been great drinking buddies, and have been tough billiard and foosball opponents. Other colleagues in various venues have contributed in all sorts of ways. In particular, conversations with David Meyer and Ted Muller over the years have been rewarding and greatly appreciated, while co-authorship with Richard Harris and Dick Walker has been stimulating and enlightening.

Funding institutions have been critical to bringing this book into being. A teaching release, courtesy of a Connaught Fellowship (2002–2003) from the University of Toronto, gave me time to write. A Social Science and Humanities Research Council grant (1997–2001) financed the weeks spent burrowed in the belly of the archival beast in Chicago and elsewhere.

I have been fortunate in having the research assistance of a stellar group of students. Norma Rantisi and March Burchfield made the initial forays into the archives, doing much of the early legwork looking for documents and alerting me to the archival delights of the city, which at that time I did not know very well. Their work helping me collect and then understand the bankruptcy case files is greatly appreciated. Jason Cooke and Charlie Strazzeri did excellent work collecting and compiling materials that were the foundation of the book. Similarly, I would like to thank Bryon Moldofsy, Mariange Beaudry, and Jane Davie of the University of Toronto's Cartography Laboratory for the wonderful maps they created from my mass and mess of materials.

The staff at various archives and libraries have played an important role in the making of the book. Two in particular have been important. Jane Lynch and staff at the University of Toronto's interlibrary loan office have greatly reduced the burden of being a Canadian studying the United States. I greatly appreciate Jane's dogged optimism in the face of my frequent request for fugitive and far-flung materials. The staff at the Chicago branch of the National Archives and Records Administration, especially Martin Touhy, have shown nothing but the highest standards of support, dedication, and friendliness. I owe them a great deal.

I am grateful to the following journals (and publishers) for permission to publish material in this book: *Journal of Planning History* (Sage), *Planning Perspectives* (Taylor and Francis), and the *Business History Review.*

The help and support of various people at the University of Chicago Press have been invaluable. The book's reviewers made trenchant and very useful comments, helping make the book a better thing. Series editor Tim Gilfoyle gave the manuscript a close read, made a large number of compelling suggestions, and stripped away a lot of the redundancies. In the process, he helped me write a better book. The editorial staff at the Press—most notably Robert Devens, Emilie Sandoz, and Erin DeWitt—have been a delight to work with. Their professionalism, friendliness, good humor, and patience are greatly appreciated.

I would like to extend a special thanks to three people. Jason Cooke read the entire first draft of the book as part of an undergraduate reading course he took with me. His cogent, generous, and intelligent comments were not only extremely helpful but also reflect the breadth of his knowledge and the maturity of his thinking. Gunter Gad has been the best of friends and colleagues. His commitment to teaching and historical research is inspiring. Richard Harris has been a friend, collaborator, and supporter. Not only did make valuable comments on the manuscript, but he also offered wise counsel during the rough patches. Thank you.

Lisa, Yonah, and Lev—what is there to say? You are the sunshine of my life. Finally, I would like to acknowledge my parents-in-law, Rita Weintraub and Marvin Weintraub. Although we have always been separated by thousands of miles, you have been a tremendous emotional and intellectual support to me over twenty-six years. For this I owe you so much.

The Metropolitan Production System

In July 1911 a group of Chicago's business elite took an industrial inspection tour through the city's suburban districts. The leader of the trip was A. R. Barnes, owner of one of the city's largest printing enterprises that manufactured a range of printing products in the heart of the city's printing district. He was also the chair of the Chicago Association of Commerce's Civic Industrial Committee, an important position in the region's most powerful and influential business organization. As the train pulled out of the Twelfth Street Station, Barnes told the committee's guests they were "to learn about this immense industrial territory that surrounds the city." The 40 business executives were going to be shown the industrial districts of Chicago's southwest side (fig. 1). Travelers first saw the familiar tenements, warehouses, and factories of the West Side's bustling immigrant and working-class districts. After a few miles they turned north, stopping to look over the Western Avenue and Cragin train yards of the Northwestern and Western Indiana Railroad companies. Next they crossed the Illinois and Michigan Canal and made their way through mile after mile of working-class residential and factory districts to the day's main attraction, the railroad yards at suburban Clearing and Argo. Traversing the Illinois Central Railroad's yards and shops, the committee's guests were regaled with the wonders of its connections to the Indiana Harbor Belt

Railroad and the Chicago Junction Railway. After a hearty lunch and visits to the Union Stock Yard and the industrial suburb of Harvey, the delegates were taken back to Twelfth Street Station.[1]

Barnes led the tour because he feared that the Chicago business classes increasingly had little knowledge of the resources of the vast territory outside of the downtown districts. "If one simply travels to and from his house and office," Barnes argued, "he sees very little of the extent of this territory." Trapped in daily routines, people saw little of the vast metropolis in which they lived. In his luncheon speech, Barnes rationalized why he and other members of the committee took the time and expense to lead a large number of guests on a daylong trip through the industrial territory of metropolitan Chicago. The assembled city officials, railroad leaders, merchants, manufacturers, and assorted others were told that the changing geography of manufacturing demanded a reformulation of how businesspeople understood the metropolitan area. "Years ago, the factories were mostly within what is known as the loop district," stated Barnes, but things had changed; "the city has grown, it has been necessary to push out and so this great territory has developed."[2]

Barnes raised a central concern, one directly touching on the interests of the audience. Along with the need to highlight the "industrial conditions at the various places where railroad connections favor the location of factories," the trip's main purpose—and Barnes was quite adamant about this—was to acquaint local decision makers with the need to cooperate in imposing the rules of local urban-industrial growth. In his mind, the trip demonstrated what the Chicago Association of Commerce's Civic Industrial Committee "stand for, and are trying to do." As a city booster, Barnes was adamant that rational metropolitan growth was a necessity, and as such, growth was possible only through the cooperation of Chicago's business classes. Metropolitan Chicago, as he told his audience over their complimentary food and drinks, "must expand in an orderly way—it must according to rules, limitations, and regulations."[3] An unstated purpose of the industrial trip was that such "rules, limitations, and regulations" had to be created and implemented by the right sort of people.

Barnes and many of his contemporaries knew that urban industrialization required planned control. Such control could only be achieved as long as businesspeople promoted an "orderly way" of building metropolitan districts. Industrial inspection tours—there were at least 15 others that covered Chicago's fringe between 1911 and 1914—reflect the push by Chicago's business and political elites to build a modern industrial landscape in the unsettled parts of the metropolitan fringe (fig. 2). To do this, the

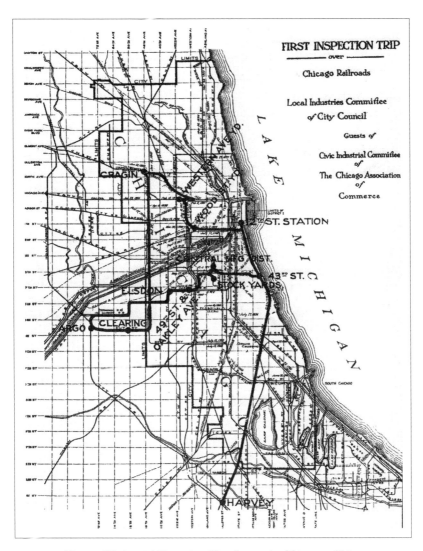

FIGURE 1. Route of Industrial Excursion. First Inspection Trip over Chicago Railroads, July 1911. Excursions such as the July 1911 trip to various parts of the fringe established the industrial possibilities of various parts of the city and the surrounding districts. Source: *Chicago Commerce* (July 21, 1911): 10.

metropolitan fabric had to be stretched and reshaped to accommodate industry's changing needs. The ability to create and manage the urban economy centered on business networks and contacts. For Barnes and his ilk, the development of factory districts could only be achieved through the building of industrial linkages across the wider metropolis.

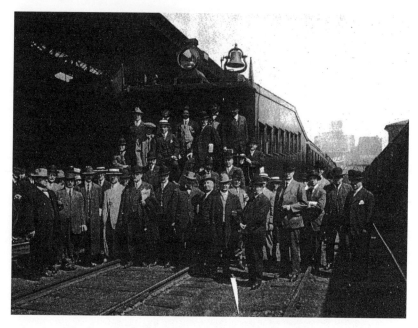

FIGURE 2. Men Who Know Their City and Its Industrial Necessities, June 1913. Chicago's bourgeoisie set out to establish the territorial parameters of the industry in the metropolitan area. Source: *Chicago Commerce* (June 13, 1913): 21.

Factory districts were a fundamental component of the American city-building process. Classic studies by Sam Bass Warner and Robert Fogelson and more recent work link urban growth with factory district formation throughout the metropolitan area. For the most part, however, central-city factory districts and their attendant external and transactional economies remain the focus of attention. Central-city economies offered manufacturers a multi-skilled labor force, a large set of commercial and financial services, an extensive transportation system, specialized input suppliers, and firms with which they could work cooperatively. The primary advantage for centrally located manufacturers was their ability to choose selectively from these to replace the elements they were unable to generate internally. Growing rapidly after 1850, central factory districts are considered the primary manufacturing employment node in metropolitan areas before World War II.[4]

Factory district formation was linked to central-city social, labor, and residential issues. Factory districts contained a wide range of residential, retailing, and associational functions alongside manufacturing ones. Dense city neighborhoods, such as Boston's North End or Chicago's Packingtown, housed distinctive class, racial, and ethnic family economies and social ecologies. As the primary reception area for new waves of immigrants, they featured a range of ethnic associations and retailers, labor organizations, and religious establishments. Central-city working-class political life was rooted in various strands of liberal, union, and socialist thought emerging out of resistance to the capitalist cauldron of work. Divided by race, ethnicity, and gender, the lifeways of the working class were forged in the factory district's sweatshops, factories, and shops; in the tenements and houses of the packed and diverse neighborhoods; and on the streets and public spaces. The development of highly divided labor markets along racial and ethnic lines generated segregated residential worlds. The combination of nativism, racism, and differences in income, job security, and skills structured access to housing and kept people apart.[5]

In contrast to the focus on central-city factory districts, few writers have considered suburban industrial areas before World War II to be of great importance. The massive surge in suburban studies over the past 40 years has produced numerous residential histories that range from wealthy suburban enclaves to owner-built, working-class suburbs.[6] Most writers have shown little interest in the manufacturing world of the suburbs. There are exceptions, of course. Studies ranging from Chicago's Pullman, South Chicago, and Hammond to Philadelphia's textile suburbs or Montreal's assorted industrial suburbs provide rich evidence of the suburban factory's importance in the urban-building process and the making of working-class life. Nevertheless, these studies have been overshadowed by the sheer volume of suburban residential histories. One reason for the neglect of suburban manufacturing is the belief that few firms were found outside of the central districts before the 1880s. Even then, suburban growth is considered to have been a trickle until the 1920s, only turning into a flood after World War II. Furthermore, most suburban firms are seen as large, self-contained, and unconnected to either central-city or other suburban economies. The suburban firm's relatively independent character has led historians to focus on the firm's internal organization, technologies, and industrial relations, and to view them as apart from the larger metropolitan world.[7]

This was not the case. In Chicago and elsewhere, a quite different story developed. First, American metropolitan areas experienced significant manufacturing suburbanization beginning in the late eighteenth century. Various enterprises sought out the urban fringe. From Philadelphia's textile suburbs and Pittsburgh's steel towns to the automotive-related districts of suburban Detroit and the assorted industrial suburbs of San Francisco,

Baltimore, Toronto, Montreal, and Los Angeles, firms of all kinds and sizes looked to the built-up edge as home. Chicago was no different. From the factory districts hugging the Chicago River in the 1860s to the large industrial suburbs developed in the Calumet district in the late nineteenth and early twentieth centuries, factory districts were an important part of the metropolitan fringe.[8]

Second, manufacturing suburbanization was a critical part of the metropolitan-building process. Manufacturers did not simply locate in a place; they made the place. The very act of setting down buildings, wharves, machinery, and necessary factory paraphernalia created a district and added to the built environment. Large-scale firms often functioned as magnets to smaller ones, offering semi-processed inputs, subcontracts, and links to other businesses. Workers' wages generated multiplier effects as dollars flowed into local housing, retail, and service sectors. Local government and private corporations supplied suburban businesses with their basic infrastructures, from the streets leading up to the factory gates to the electricity lines, water pipes, and railroad spurs that crossed the line separating the firm from the rest of the district. Finally, distinctive lifeworlds developed around the industrial relations of local companies, the specific ethnic character of the labor force, and the power firms had to shape the area's political climate.[9]

Third, city and suburban factory districts formed an integrated metropolitan economy. Most large suburban firms were not isolated, autonomous entities. They were part of social and economic tissues connected to other local companies. Different forms of urban hardware, from streets and streetcars to railroads and utilities, spanned municipal boundaries and allowed manufacturers to move goods from place to place. Company owners and managers operated as members of business associations whose interests extended across the metropolitan district. A forum for sorting out common differences, business associations allowed executives to maintain control of their firm within the broader ambit of the metropolitan economy. Suburban firms were linked to the wider economy through their reliance on central-city wholesalers, financial institutions, and service providers such as lawyers and accountants for many of their daily and infrequent transactions. As the focus and central coordinating node for the rest of the metropolis, the central district had extensive links to other local factory districts. Businesses also depended on the way in which the local economy was linked to regional and the national economies. They depended on the various components brought together within the metropolitan area's spatial boundaries.

A case study of Chicago demonstrates the complex interconnecting of local manufacturers and the making of the metropolitan landscape. While William Cronon and others persuasively show that the sociopolitical dynamics connecting the farmer, rural merchant, urban factory worker, corporate manager, and consumer were vital to urban and rural economic relations, it is equally important to reassert the metropolitan scale. The flows of capital, materials, goods, ideas, information, and people between Chicago and its region were driven, controlled, and coordinated for the most part from the city. Chicago industries such as meatpacking and furniture were rooted in the American rural economy, but the metropolis was the control center. Moreover, despite the intimate networks between city and countryside, the vast majority of metropolitan Chicago's networks and interactions were generated within the metropolis itself or with other industrial cities. The countryside mattered, but it is necessary to reassert the importance of the metropolis for understanding city building and urban expansion.[10]

The story is set out in two sections. The first provides an overview of Chicago's industrial geography between 1860 and 1940 and demonstrates how factory districts functioned as part of the city-building process in this period. No synthetic examination of this story exists for Chicago. The little known about Chicago's industrial geography is fragmented. Numerous individual monographs focus on specific neighborhoods and industries.[11] While studies by Mayer and Wade and Keating go some way to providing a comprehensive history, more needs to be done.[12] I frame metropolitan development within the process of post–Civil War industrial decentralization. Manufacturers, attempting to escape the problems they faced in the city's center, were attracted by the advantages offered at suburban greenfield sites. These factory districts, however, did not just appear. They were assembled and constructed by local alliances that profited by redevelopment of the urban fringe. In the process, thanks to the likes of A. R. Barnes, the Chicago Association of Commerce, real estate developers, and the railroads, this new industrial territory established the basis for Chicago's development as a sprawling multi-nodal metropolis and shaped the metropolitan area's internal structure.

The second section shows how the networks that manufacturers forged with each other, and with other agents, shaped metropolitan Chicago's industrial landscape. Inter-firm relations were essential to the operations of Chicago manufacturing and the manner in which the metropolis was built. The examination of key metropolitan firms, industries, and factory districts in this section highlights the importance of local business linkages for Chicago's metropolitan and industrial development from the end of the Civil War to the beginning of World War II.

METROPOLITAN FACTORY DISTRICTS

Factory districts were a key element of the modern industrial metropolis. This is demonstrably true for Chicago, where several studies reveal the fascinating history of the metropolitan area's industry development, from the establishment of metal and lumber firms in Bridgeport and the Lower West Side before the Civil War to the formation of steel districts in the Calumet district after 1900. Factory districts led the expansion of the metropolitan area's boundaries and shaped its social composition. Scholars have also given important insights into the planning, social, labor, and industrial histories of several districts, including Pullman, South Chicago, Packingtown, and Bridgeport. A great deal has been written about Chicago's city and suburban factory districts since the mid-nineteenth century.[13]

What is less understood is how individual industrial neighborhoods, suburbs, firms, and industries related to one another and, in general, how they functioned within the wider metropolitan area. How were individual factory districts linked to one another and to the metropolis itself? How were firms and districts integrated into the regional and national manufacturing economies? While we have a strong sense of the importance of industry growth to metropolitan, regional, and national growth, the business networks that held these manufacturing systems remain unknown. This is unfortunate, as factory districts were central to the development of the capitalist metropolis.

Our understanding of metropolitan factory district formation has been framed by a dualistic understanding of the relationship between industrial structure and the spatial organization of the metropolis. Following Alfred Chandler's formulation of the rise of the multi-unit corporation at the end of the nineteenth century, the industrial structure of American manufacturing is considered to be composed of two major types, the large corporation and the small proprietorial factory.[14] For many, this dualistic industrial structure established a polarized metropolitan spatial structure. David Gordon blends a class analysis with Chandler's outline of the rise of big business. He argues that industrial suburbanization occurred from the 1880s as the integrated corporation sought out the advantages of transportation and production innovations, such as the truck and the assembly line. In particular, the corporation moved to suburban sites for cheaper land on which they built their large, sprawling plants and escaped an increasing hostile labor force. Taking Gordon's ideas a step further, Allen Scott argues that small, labor-intensive industries such as printing and clothing remained central because they depended on external economies. In contrast, a few modern, vertically integrated corporations moved to the metropolitan fringe after 1900 because they were no longer reliant on external economies and could take advantage of large industrial sites.[15]

The polarized industrial structure and geography of the metropolis were linked to a particular set of business relations. According to the Gordon and Scott accounts, the large suburban business was a relatively autonomous entity, with most of its business done with national and international suppliers and customers. For all intents and purposes, the large firm was divorced from the main lines of the metropolitan area's networks. Suburban corporations in the steel, chemical, and automotive industries were characterized by arm's-length links to other firms, thus relying upon a few distant suppliers, a hierarchical internal structure, and control over most aspects of a suburb's political and social life. This contrasted sharply with the multitudinous inter-firm relations forged by small enterprises in the central city. Labor-intensive firms in the clothing, printing, and jewelry industries were tied to the central city through an extensive set of external economies. The result was the development of a polarized spatial structure, featuring a few large, capital-intensive corporations in the isolated industrial suburb and many small, labor-intensive firms in the central city.

While there is some truth to this position, the industrial structure of the metropolis does not fall so neatly into the large-small, suburban-central divide. The notion that the shift from the small competitive firm (industrial capitalism) to the large corporation (corporate capitalism) is the typical industrial trajectory was only one aspect of growth. Forsaking the teleology of Chandler's approach, Philip Scranton, John Ingham, and others point to the coexistence of different types of manufacturing dynamics, and highlight the variety of strategies that entrepreneurs assumed in their search for markets and profits.[16] There were obvious differences between businesses in the character of their raw materials, production process, semi-processed inputs, markets, and labor force. There were also contrasts between enterprises in any industry. The differences between vertically integrated giants such as United States Steel, the large number of large- and medium-size rolling mills and foundries, and the multitude of metalworking and machine shops do not do justice to the variety of the manufacturing strategies in the iron and steel industry. They certainly cannot be squeezed into the narrow confines of a dualistic structure. If nothing else, firms and industries were characterized by an array of production formats.

These production formats produced and depended upon different locational patterns. Gordon and Scott demonstrate that a few large corporations sought out the metropolitan fringe, while many small manufacturers were content to stay in the central city. But the industrial geography of the metropolis after 1850 was far more complicated than this picture allows. Some large, capital-intensive firms remained centrally located. Others moved parts of their operations out to the suburbs, while keeping some functions in their original location. At the same time, many small enterprises clustered on the metropolitan fringe, sometimes to service the demands of the local population, sometimes to be close to the large corporation. The combined effect of this collection of firms was the formation of diverse factory districts and the rise of collective economies throughout the metropolitan region.

METROPOLITAN MANUFACTURING LINKAGES

One of the most intriguing elements to emerge from recent studies of industrial America is the persistence of strong intra-metropolitan linkages. Particularly important in this respect are the relations forged between manufacturers, merchants, and suppliers; the coordination of local, regional, and international transactions by local institutions; and the development of a localized learning environment through the interaction of manufacturing firms and a specialized and multi-skilled labor force. In his study of industrial capitalism in the eastern seaboard between 1790 and 1860, for example, David Meyer shows how industrial districts centered on the machine, construction, metalworking (hardware, munitions, etc.), and consumer-goods (furniture, apparel, etc.) industries formed strong metropolitan linkages. Similarly, the textile suburbs of Philadelphia before the Civil War coalesced around a distinctive set of nonreplicable intra-metropolitan business relations.[17]

Evidence for the post–Civil War period suggests that central-city manufacturers were not the only ones making use of cost advantages obtained from sources external to the firm. Non-central businesses did the same. In many cases, external economies were available to manufacturers at a distance. Relational proximity—the ability to acquire services and material outputs from afar through inter-firm networks and linkages—was just as important as spatial proximity. Suburban firms had access to downtown financial houses, stock exchanges, and specialized machinery suppliers. At the same time, suburban areas generated their own external economies, from extensive labor pools to legal services and raw material supplies. Large suburban corporations generated a set of relations requiring spatial proximity to suppliers and customers. In Pittsburgh's steel, glass, and railroad equipment satellite towns, and in Los Angeles's resource-based suburbs, for example, the large firm paved the way for a hierarchically and territorially based firm complex on the urban fringe. Even in pre–World War I Detroit, the exemplar of mass production, large suburban automotive companies were surrounded by a multitude of small machine tool and fabricating metal firms. In other words, suburban businesses operated within a web of intra-metropolitan relations based on spatial and relational proximity.[18]

While these studies provide suggestive evidence about intra-metropolitan manufacturing linkages, very little is known about the economic relations of urban business. What we think we know is typically taken from anecdotes or histories of individual companies. Moreover, next to nothing is known about the range of business linkages that functioned across metropolitan Chicago. Did manufacturers receive inputs from local or non-local suppliers? How commonly did firms subcontract work out to other firms? Did firms share resources such as workers and machines? One reason for these lacunae is due in great part to the absence of suitable evidence. Furthermore, historians have been more interested in telling stories of the rise of individual business rather than their economic and urban milieu. The result is that a great deal is known about the history of firms and industries, but little about industry linkages and micro-geographies.

Existing studies emphasize three important aspects of metropolitan business life. First, they hint at the significance of local inter-firm networks for both small and large companies in the city and suburbs, and for the integrative character of business linkages at the metropolitan level. Second, the studies point to the importance of the building of bonds of cooperation by local agents. Although cooperation furthered the interests of those involved, it also refashioned the industrial geography of urban areas. Third, they highlight the critical relationship between manufacturing development and the social and political worlds of the worker. The building of factory districts throughout the metropolis framed the everyday life of the worker through their work and their home.

These histories also suggest conceptual similarities with the literature on present-day European industrial districts, where strong inter-firm networks, industry-specific expertise, specialized inputs and institutions, and embedded social relations underpinned the clustering of smaller manufacturers and a set of regional and metropolitan-wide interactions. The work on industrial districts—a staple of research on European areas as diverse as the Third (or northeastern) Italy, Baden-Württemberg, and Paris—emphasizes "the interdependence of firms, flexible firm boundaries, co-operative competition and the importance of trust in reproducing sustained collaboration among economic actors within the districts." Even though several types of industrial districts have been identified, all of them share basic principles: an array of firms operating within an intricate set of supporting institutions and a specialized division of labor. Not simply agglomerations of enterprises in one place, industrial districts function as local production systems consisting of place-based economic and social interdependencies.[19]

Unfortunately, the North American industrial district remains understudied. One reason is that the North American experience is interpreted as different from the European. Certainly, the actual presence of European-style industrial districts in the United States over the last few decades is spotty at best. The best-known examples are the development of the electronic industry in Silicon Valley and Boston's Route 128, or big-firm-led industrial districts such as Seattle, where Boeing and Microsoft created webs of inter-firm linkages at the metropolitan scale. Despite our understanding of the historical roots and contemporary interactions of these contemporary places, the historical geography of inter-firm business linkages in American metropolitan industrial districts is still unknown. This is an important gap, for, as Charles Sabel and Jonathan Zeitlin argue, the classic industrial district may have been a fundamental feature of American urban development.[20]

A related point is confusion as to what constitutes an industrial district. In some cases, the district is within a metropolitan area, such as Silicon Valley in San Francisco, while in others it encompasses an entire region composed of several towns and cities, such as those making up the northern part of Italy (or the Third Italy). Most historical research on links between industrial districts and inter-firm business privileges the development of regional economies, while downplaying those at the metropolitan level. In several important studies, David Meyer, William Cronon, and Brian Page and Richard Walker argue that increased industrial specialization after the Civil War led to greater flows of manufactures between regional industrial systems. More recently, Gordon Winder demonstrates the importance of regional, national, and international inter-firm linkages for metalworking firms located in small specialized industrial centers.[21]

For these writers, regions are considered to be reservoirs of social and business interaction that feed the growth of an integrated production system, often across vast distances. Metropolitan districts function as black boxes linked to the wider regional world. Urban dynamics are considered to be secondary to the full regional sweep of capital, labor, commodity, and information flows. A regional interrelationship of firms and cities account for the emerging oligopolistic world of mass markets, cheap and rapid transportation, and a segmented labor force.

The precise nature of the regional linkages has been debated. Page and Walker argue that agro-industrialization created and sustained long-term regional growth. Meyer, on the other hand, emphasizes regional subsystems that center around metropolitan districts. Despite the difference in emphasis, writers agree that regional growth was driven by the infusion of capital, labor, information, and commodities from outside the region. This not only added to internally generated factors of production, but provoked new productive capacity over and beyond their addition to the regional stock. The importance of business linkages within metropolitan industrial districts is minimized in these studies. This is exemplified by Winder, who downplays local sets of relations in favor of "networks, regulatory and governance structures at the scale of the [manufacturing] belt."[22]

Studies of Philadelphia, Pittsburgh, Los Angeles, Detroit, and Montreal suggest that the metropolitan scale was critical to the development of industrial capitalism after the Civil War. What they fail to provide is a detailed demonstration of the rich set of inter-firm relations and linkages operating at the metropolitan scale. But several writers, blending historical evidence with notions of industrial districts, external economies, and the division of labor, have devised some useful frameworks for probing this issue. Most useful is Ted Muller's notion of the metropolis as a localized production system. This concept—the territorially rooted set of firms, industries, and institutions forming around a metropolitan area's division of labor, strong inter-firm dependencies, external economies, and flows of goods, capital, labor, and professional personnel—establishes the industry-place dynamics necessary for metropolitan development.[23]

METROPOLITAN PRODUCTION SYSTEM

Muller's localized production system has important implications for the study of the development of the industrial metropolis. Metropolitan areas such as Chicago were politically and socially integrated, and this integration was built upon a well-developed set of linkages between local alliances and economic agents that spanned municipal boundaries. Despite varying quantities and strengths of non-local relations and industrial mobility, Chicago's industrial territorial endowments were centered around a rich set of place-based inter-firm and institutional relationships. Chicago was a metropolitan production system consisting of numerous factory districts linked by assorted inter-firm relations. This system consists of four parts: production chains, the division of labor, external economies, and locational assets.

The production (or supply) chain refers to the flow of goods from the raw material stage through manufacture to the marketing and distribution of the finished goods to the final consumer.[24] It has important implications for understanding the operation of the metropolitan economy, framing as it does the numerous relationships that firms labor under in the making of a commodity, while not privileging any one production stage (raw material extraction, manufacture, marketing, and distribution). Manufacturing would be difficult, if not impossible, if any part of the chain functioned poorly. Rather than focusing on only one stage of manufacture, the supply chain calls attention to the need to look as widely as possible at the entire manufacturing process. While most business history emphasizes the technological and corporate history of individual firms and industries, the production chain encourages the researcher to focus upon the functions of production (and distribution), rather than the specific companies that at any one point of time happen to perform only one or more of these functions. This allows us to examine the often varied relationships between firms and locations that are involved in the supply chain, and emphasizes production's networked character. All firms to varying degrees rely on other businesses—both as suppliers and customers—for inputs and outputs.

Metropolitan production chains are related to the industrial division of labor. Under capitalist industrialization, manufacturers deploy different labor forces, technologies, marketing schemes, and investment strategies. By the late nineteenth century, the division of labor formed four types of industrial groupings. Mass and flow producers (steel rail, petroleum, cigarettes) concentrated on the construction of complex technical and capital-intensive systems for making standardized goods in large volumes under exacting market and cost conditions. Bulk producers (belting, yarns, window glass) also produced large quantities of goods within difficult market and price situations, but in much less elaborate production systems. In contrast, batch producers (machinery, furniture, magazines) made products in varied lots on the basis of aggregated orders, while custom producers (turbines, linotype, jewelry) built specialty products for individual buyers. Each of these basic formats had different locational needs, business networks, industrial relations, and relations with the state. Outside the workplace, each of them was associated with different types of neighborhoods and social worlds.[25]

External economies operating at various spatial levels underlay the functional integration of the industrial metropolis. These economies relied on the operation of those components of commodity manufacture that were not encompassed within a company's boundaries. Many labor processes and direct material links were not integrated into a firm's production system. The optimal scales for different parts of the production chain were different, forcing businesses to seek outside support for manufacturing itself or services such as finance, legal advice, utilities, and transportation. For specialized goods, input costs were frequently lower when they were acquired from downstream producers. For many routinized and specialty companies, the uncertainty and instability of output markets forced them to subcontract their work. Finally, knowledge opportunities involved the need for technical, managerial, and labor know-how not readily available in the firm, forcing them to seek contacts and contracts outside the firm's boundaries. The playing out of these dynamics shaped the metropolitan manufacturing world and local labor-capital relations.[26]

Metropolitan areas such as Chicago formed around a distinctive set of place-based, material locational assets. These are locally rooted packages of infrastructures, institutions, and practices that generate and sustain growth over a generation or more. They are held together by the actions of a diverse set of actors who take advantage of the opportunities of industrial capitalism.[27] Builders of urban networks, such as utility companies, operated across municipal lines. Metropolitan and municipal associations worked as alliances to service local business interests. Planners quartered and zoned the metropolis. Railroad companies with their tracks, belt lines, switches, yards, depots, and stations forged an integrative metropolitan transportation system. Industrial managers worked hand in hand with transportation companies, local associations, and land developers.[28] Land developers, perhaps the most promiscuous urban actors, combined capital, land, and political support from diverse groups to build housing and factories at greater distances from the central city, linking their new subdivisions with the metropolitan world of work and consumption.[29] Metropolitan success depended on the ability of city promoters to coherently tie the various pieces together. All of this depended on the functioning of appropriate locational assets within an interactive, elaborate, and effective industrial territory.

The participants in the industrial inspection tours of metropolitan Chicago's manufacturing districts before World War I argued that what was good for them was good for Chicago. This was not a new idea, of course; city building from time immemorial has been controlled by the elite. As with medieval merchants or colonial administrators, Chicago's businesspeople, politicians, developers, and builders assumed that they spoke for all elements of Chicago's society. The tours themselves, traversing the factory districts of the older West Side and South Side areas to the new industrial suburbs of Clearing and Argo, and the matters they raised and discussed along the way and later on in their clubs, boardrooms, city hall offices, and homes were not self-involved acts of petty profit seeking and aggrandizement. These were not pleasant summertime junkets; rather, they were vital to the city's industrial, population, and income growth. As Alderman Albert Fisher told the group that visited Dolton, Indiana Harbor, Blue Island, and Elsdon at the end of July 1911, they were "preaching the gospel of one for all and all for one." Their success was Chicago's success.[30]

This book examines one critical aspect of that success: the relationship between manufacturing linkages and the construction of metropolitan Chicago. Metropolitan industrial development was built on a set of factory districts consisting of networks of firms that created and operated within a dense array of inter-firm linkages. These networks formed a distinctive metropolitan industrial district. Chicago's manufacturing geography before World War II was an assemblage of different pieces, from a densely built-up central city to a diversity of industrial suburbs stretching along major transportation networks. Commonly, large-scale firms, regardless of their location, were nodes around which formed networks of outsourcing and supply linkages, ever-widening metropolitan-wide labor markets, infrastructure improvements, community institutions, and land-development practices. Just as frequently, numerous smaller firms clustered together to take advantage of linkages with each other and the large firms. Suburban factory districts were not isolated from the central city, but maintained strong relationships with it. Information flows, markets of all kinds (capital, labor, wholesale, and retail), and inter-firm supply relations spanned the political boundaries of the suburbs and the central city and created a functional metropolitan economy.

These firms did not just appear on the metropolitan landscape. Chicago's manufacturers were connected to a bewildering array of agents along the production chain who were instrumental in creating place. Class-based alliances composed of land developers, real estate agents, transportation companies, lawyers, mechanical engineers, architects, and building inspectors furnished industrial land, factories, legal aid, and transportation. They also built working-class residential districts, frequently next to the factory areas. Family, friends, banks, and businesses supplied capital in various guises, from venture capital for purchasing machinery to family savings that paid the rent for a factory or workshop. A profusion of service, manufacturing, and wholesaling firms offered inputs to be converted into various outputs by the city's manufacturers. Once manufactured, products were wheeled around the corner to a local factory or transported thousands of miles away to a plant in New York City or Los Angeles. An assortment of marketing agencies, advertising companies, business organizations, and wholesalers greased the distribution of the hundreds upon hundreds of products that poured out of the metropolitan area's mills, workshops, sweatshops, and factories. These class coalitions made and remade metropolitan Chicago's production system. In the process, they built an industrial district.

SECTION I

BUILDING
the Industrial
Metropolis

CHAPTER I

Chicago, the Mighty City

Chicago shot to industrial preeminence almost overnight, becoming America's second largest industrial metropolis by 1900. As one early twentieth-century historian noted, from "the condition of mere frontier life with but a handful of men collected in a knot at the mouth of the Chicago River, surrounded by an almost untouched wilderness and having no industries but barter with the Indian and the collection of furs, we have seen Chicago and Vicinity, in 50 years, become one of the most populous, wealthy, and industrious centers of like area in the whole United States."[1] While Chicago never matched New York, the city's industrial growth was nevertheless breathtaking in its extent and speed. In a very short time, an extremely large and diverse industrial base—as measured by the number of firms and employees, the amount of capital investment and industrial output, and the range of industries and productive strategies—was created on the shores of Lake Michigan. As the promoters of one industrial development intoned, "Chicago, the mighty city which the fates have determined shall be the London and the Manchester of America rolled into one."[2]

Chicago, America's Manchester, exemplified the transformation of a small mid-nineteenth-century city into a large twentieth-century industrial metropolis by the process of geographical industrialization.[3] Capital accumulation drove industrialization and urbanization, refashioning existing urban form and producing new territorial areas on the urban fringe (fig. 3). Before the mid-nineteenth century, urban restructuring occurred

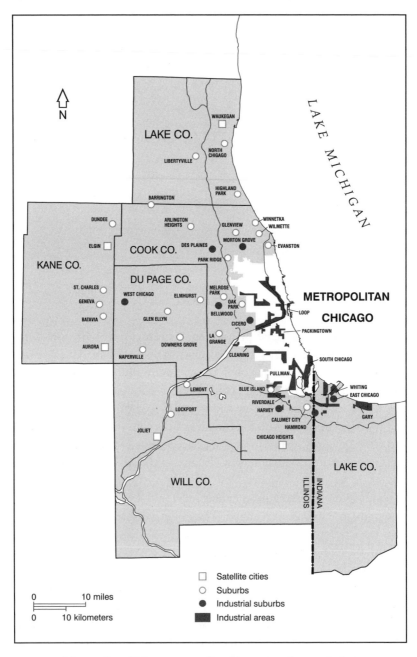

FIGURE 3. Metropolitan Chicago, c. 1925. Based in part on the map in James Grossman, Ann Durkin Keating, and Janice Reiff, eds., *The Encyclopedia of Chicago* (Chicago: University of Chicago Press, 2004), 258.

within the confines of a tightly bounded, monocentric form. Territorial growth was restricted as long as accumulation was based on mercantile exchange. With full-scale industrialization after 1850, however, urban areas became more than simple single-centered places as capitalism centralized industrial capital; created stronger webs of economic, political, and ideological coordination and control; extended the geographic reach of the urban bourgeoisie; and spun off an assortment of settlements on the periphery of rapidly growing industrial centers.[4]

MANUFACTURING CHICAGO:
BEHEMOTHS AND WORKSHOPS

Metropolitan Chicago's industrial growth occurred with remarkable speed. From its beginnings as a small outpost in the 1830s, an industrial center the equal of almost of any other was created on the shoes of Lake Michigan (table 1). Fewer than 7,000 workers were employed in metropolitan Chicago's factories, workshops, and mills in 1860. By 1929 more than half a million labored in the industrial establishments of the six-county metropolis. The bulk of these were located in the city. From a handful of workers in 1860, the number working in the city's factories grew to more than 400,000 by 1929. Only the massive losses during the Great Depression dented the long-term growth of the city's manufacturing base.[5] Chicago was second only to New York as an industrial emporium.[6] By any measure, Chicago's growth compared to other metropolitan areas was phenomenal. Although a small frontier settlement and inconsequential manufacturing town before the Civil War, by 1880 Chicago only lagged behind the great eastern seaboard centers of New York, Philadelphia, and Boston. Even though dwarfed by these three, Chicago had overtaken older eastern cities as well as its Midwest competitors, Cincinnati, Pittsburgh, and St. Louis. Chicago made its mark on the nation's economic scene a little over a decade after the end of the Civil War.

Industrial growth in the suburbs was also impressive. Despite the city's importance to the metropolis, a significant part of the periphery was associated in one way or another with manufacturing; the economic base of 31 metropolitan fringe towns incorporated between 1841 and 1900 was either agricultural processing or manufacturing. The manufacturing economy of the suburbs grew more rapidly than the city's between 1900 and 1940. Significant manufacturing employment in the metropolitan suburbs was led by the growth of a few large industrial suburbs, such as Gary, Cicero,

TABLE I. Metropolitan Chicago's Industry, 1860–1937

Year	City of Chicago		Metropolitan Chicago[a]	
	Firms (No.)	Workers (No.)	Firms (No.)	Workers (No.)
1860[b]	747	6,695	762	6,832
1870[b]	1,411	30,483	2,074	36,481
1880	3,519	79,414	4,595	88,841
1890	9,977	190,621	11,125	228,331
1900	19,203	262,621	21,891	297,738
1909[c]	9,656	293,977	10,202	393,859
1919	10,537	403,942	12,399	520,133
1929	10,201	405,399	11,774	550,903
1937	7,737	391,185	9,019	538,775

SOURCE: *Manufactures of the United States in 1860* (Washington, DC: Government Printing Office, 1865); *The Statistics of the Wealth and Industry of the United States*, vol. 3 (Washington, DC: Government Printing Office, 1872); U.S. Census, *Compendium of the Tenth Census (June 1, 1880)*, vol. 2 (Washington, DC: Government Printing Office, 1883); U.S. Census, *Report on Manufacturing Industries in the United States at the Eleventh Census: 1890*, Part II (Washington, DC: Government Printing Office, 1895); U.S. Census, *Thirteenth Census of the United States Taken in the Year 1910: Manufactures, 1909*, vol. 9 (Washington, DC: Government Printing Office, 1895); Department of Commerce, Bureau of the Census, *Fourteenth Census of the United States 1920: Manufactures, 1919*, vol. 9 (Washington, DC: Government Printing Office, 1923); Department of Commerce, Bureau of the Census, *Fifteenth Census of the United States, 1930: Manufactures*, vol. 1 (Washington, DC: Government Printing Office, 1933); Bureau of the Census, *Biennial Census of Manufactures: 1937*, part 1 (Washington, DC: Government Printing Office, 1939).

[a] Metropolitan Chicago consists of the following counties: Cook, Du Page, Kane, Lake, and Will, all in Illinois, and Lake (Indiana).

[b] City totals are estimates based on taking 98 percent of the Cook County totals. This percentage was derived from the fact that the city of Chicago's share of Cook County for the two variables (firms and hands) in 1880 and 1890 ran from 94 to 99 percent.

[c] Metropolitan Chicago is smaller than in the other years, and includes only the city of Chicago, parts of the rest of Cook County, and Lake County (Indiana). It was used because no information is given for counties in the 1909 census.

and East Chicago; the continued development of satellite cities, such as Joliet and Elgin; and the profusion of numerous smaller manufacturing centers. By 1937, despite the city's annexation of large chunks of suburban industrial territory, the metropolitan area's suburbs were home to 1,200 firms and more than 147,000 manufacturing workers, a number greater than the entire industrial workforce of Milwaukee, Bridgeport, and Cincinnati. In the first decades of the twentieth century, Chicago's suburbs and satellite cities formed a sprawling, multi-nodal industrial cluster that rivaled America's leading industrial centers.[7]

By the end of the nineteenth century, Chicago featured a few large corporations in industries such as meat, steel, and railroad equipment. In contrast, New York, Philadelphia, and Boston contained a large number of smaller proprietorial firms in labor-intensive, vertically disintegrated sectors. Chicago, while not a single-industry town, was controlled by a handful of large corporations. There was some truth to this picture. As early as 1873, 5 companies employed 1,000 or more workers and another 16 enterprises employed 275 or more workers. In other words, the large firm was securely implanted on the industrial landscape in the early post–Civil War period. This was even more so 50 years later. In 1924, according to the Illinois Manufacturers' Association's industrial directory, the largest 5 percent of Chicago firms, which totaled 315, employed 470,000 workers, almost two-thirds of the metropolitan area's industrial workforce. Western Electric, Chicago's largest, employed more workers than many midsize industrial cities. The next 4 largest—Illinois Steel, U.S. Steel, Wilson, and Crane—were substantial. Another 15 employed more than 4,000 workers each. All of these companies were part of large multi-unit corporations serving national and international markets.[8]

These large firms represented the core industries that developed in Chicago before the Great Depression (table 2). By 1880 many of the core sectors that would dominate Chicago for several generations were already in place. Fueled by more than 8,400 workers in men's clothing, the garment trade was the largest. This was followed by metalworking (primary metal, fabricated metal, machinery), meatpacking, woodworking (furniture and lumber), printing, and leather. Over the next 45 years, there was a great deal of stability as these sectors, with the exception of lumber and leather, remained central to the local economy. By 1924 the metalworking sector employed more than 215,000 workers. Similarly, clothing, printing, furniture, and meatpacking continued to hire a significant share of the metropolitan manufacturing labor force.

Despite stability at the top end, changes took place in metropolitan Chicago's industrial structure. Some sectors declined in importance or disappeared altogether. Along with the relative decline of leather and lumber, and the diminishing importance of beverage and tobacco making, craft-based industries rooted in pre-twentieth-century technologies and markets such as carriage making and blacksmithing faded from the local economy. In contrast, others grew. The most notable was the electrical equipment sector, which employed more than 61,000 workers in 1924. But there were others. The railroad equipment and petroleum sectors developed between 1880 and 1905, and an extensive automotive industry was in place by the end of World War I. Alongside the appearance of new industries, nineteenth-century fledgling sectors, notably scientific instruments

TABLE 2. The Industrial Structure of the City of Chicago, 1880, and Metropolitan Chicago, 1924

Sector	City of Chicago, 1880			
	Workers (No.)	Firms (No.)	Capital ($000)	Mean Firm
Primary metal	8,400	155	8,772	54.2
Fabricated metal	4,005	402	2,619	10.0
Electrical equipment	152	3	271	50.7
Clothing, hats and furs	12,111	183	7,716	66.2
Meatpacking	8,455	70	7,478	106.9
Furniture	6,464	236	3,420	27.4
Machinery	1,220	17	3,247	71.8
Printing and publishing	4,767	206	3,945	23.1
Railroad equipment	647	10	468	64.4
Food processing	1,405	80	3,008	17.6
Chemicals	1,471	72	3,614	20.4
Automotive	0	0	0	0
Lumber and wood	5,813	204	3,215	28.5
Baked goods	1,481	142	966	10.4
Textiles	1,719	82	484	21.0
Leather	3,870	259	3,787	14.9
Paper goods	380	18	94	21.1
Instruments	83	55	68	1.5
Petroleum	0	0	0	0
Construction materials	3,791	105	1,160	36.1
Beverage	1,844	38	4,691	48.5
Tobacco	1,978	291	826	6.8
Carriage making	1,810	161	1,575	11.2
Blacksmithing	346	146	110	2.4
Total	75,625	3,022	67,585	25.0

SOURCE: U.S. Census, *Compendium of the Tenth Census (June 1, 1880)*, vol. 2 (Washington, DC: Government Printing Office, 1883); *Directory of Illinois Manufacturers, 1924–1925* (Chicago: Illinois Manufacturers' Association, 1924).

NOTE: The number of firms with employees in 1924 is not the same as the total number of firms.

and paper goods, experienced moderate growth. Even though these sectors were never large, they built a respectable niche in the local economy before 1940.

Both new and old industries powered Chicago to manufacturing preeminence. After the Civil War, Chicago became a leading producer in several industries. Most famously perhaps, the city became America's Porkopolis.

Rank by Workers		Metropolitan Chicago, 1924			
1880	1924	Workers (No.)	Firms (No.)	Mean Firm	Median Firm
3	1	104,974	221	552.5	46
7	2	83,333	1,053	90.7	20
21	3	61,393	280	248.6	20
1	4	52,729	667	96.4	20
2	5	52,697	92	627.4	65
4	6	38,300	434	102.2	26
17	7	37,686	479	90.3	20
6	8	37,109	704	57.1	16
18	9	32,146	53	684.0	163
16	10	29,703	281	123.2	30
15	11	23,806	523	54.5	12
23	12	22,731	281	97.8	25
5	13	21,987	323	79.7	32
14	14	18,922	169	135.2	36
13	15	14,913	220	78.1	22
8	16	14,212	198	84.6	21
20	17	13,768	191	82.9	40
22	18	9,378	128	83.7	14
23	19	8,853	40	260.4	23
9	20	8,394	166	59.1	23
11	21	1,404	65	21.6	9
10	22	1,287	27	64.4	18
12	23	0	0	0	0
19	23	0	0	0	0
		727,303	7,316	115.4	22

Dominated by the local Big Three—Swift, Wilson, and Morris—the city remained the nation's biggest meat producer into the interwar period. Similarly, Chicago became one of the nation's, if not the world's, largest steel producers. By the 1920s the metropolitan steel industry made more finished rolled steel than Pittsburgh. By 1930 U.S. Steel—the country's largest steel corporation and originally from Pittsburgh—made more steel ingots in its Chicago mills than in those of its Pennsylvania competitor. In 1937, among other things, Chicago industrial-area manufacturers produced more food products (confectionery, chocolate, meat, and canned food), metal products (steel, machinery, foundry goods), chemicals, furniture, and radios than its major competitor, the New York industrial area. Chicago, with the exception of a few European industrial cities and New York City, of course, was the greatest industrial metropolis in the world.[9]

Chicago's preeminence rested on a wide diverse economy composed of industries established at different times. Tracing their origins back to the Civil War were the rolling mills, packinghouses, clothing sweatshops, furniture factories, and machine shops. But as the profusion of electrical equipment, radio, instrument, automotive, and petroleum workplaces suggests, the local economy stimulated new industries. In this respect, Chicago differed from the majority of industrial centers, such as Pittsburgh, Detroit, Akron, and Grand Rapids, with their specialized economies. Chicago's was much more diverse.[10] An important contribution to Chicago's industrial success was the coexistence of a strong manufacturing base rooted in a long industrial history with an industrial structure that reproduced itself through the development of new and rapidly growing and nationally important industries. Along with New York and Philadelphia, Chicago had the largest and most diversified economy in the country.

Metropolitan Chicago's heterogeneous industrial structure was paralleled by a range of production strategies. Business historians and economic geographers have pointed to the great variations between and within sectors. Differences in the material of the commodity, financial markets, the labor force, and distribution networks fashioned tremendous variety in a place's industrial structure.[11] The primary metal sector consisted of high-volume, vertically integrated steel mills covering hundreds of acres and employing thousands of workers, such as U.S. Steel and Inland Steel. These giants coexisted with smaller steel, aluminum, or brass foundries, such as Chicago Heights Brass Foundry and the city's Alloys and Brass Foundry, which made castings to order with a handful of workers. A similar story was evident for many other sectors. Printing was broken into several industries—printing (job, book, and magazine), publishing, lithographing, bookbinding—all of which differed greatly from one another in the way that they organized production. Large printing and newspaper publishing companies such as Donnelley and Sons and Illinois Publishing with thousands of employees rubbed shoulders with hand-to-mouth printers working a single machine with no workers. Similarly, furniture was divided along the lines of quality (ornate and rough), markets (office, restaurant, home) and material (wood and metal). Big firms with a large factory and workforce, such as Naperville's Kroehler Manufacturing (furniture) and Brunswick-Balke-Collender (billiard tables), operated in the same sector as smaller enterprises, such as Schreiber Window Shade and Victory Store Fixture with one or two employees.

Metropolitan Chicago was characterized by firms of all scales. Most sectors included a handful of corporations that accounted for a substantial share of their jobs. In some cases, a small number dominated some sectors. While 3 percent of metropolitan Chicago firms in 1924 employed almost 58 percent of workers, some sectors were more concentrated. Almost 90 percent of the primary metal workforce labored in 22 very large workplaces. The huge blast furnaces, open hearths, and Bessemer ovens of U.S. Steel and its subsidiaries dominated the industry, as well as commentators' vision of industrial Chicago. Similarly, only 2 percent of electrical equipment firms employed three-quarters of its 61,000 workers, many of whom worked for Western Electric. Likewise, 87 percent of meatpacking employment was found in Omaha Packing, Hammond, Armour, Allied Packers, Swift, and Wilson. A small number of large primary metal, electrical equipment, and meatpacking firms dominated their sectors and were decisive players in Chicago's employment history.[12]

Firms of other scales coexisted with the behemoths. The steel mills, packinghouses, and electrical factories were not alone on the industrial landscape. Alongside the blast furnaces, steel ovens, and pipe mills of the South Chicago and Gary steel complexes were a multitude of small to medium-size fabricated metal and machinery factories. At the furthest extreme and in direct contrast to the large multi-unit steel or metalworking corporation serving national markets was the small machine shop operated by a proprietor who employed one or two employees and serviced neighborhood needs. Some sectors had a relatively large share of its employment in smaller workplaces. Printing and leather—and to a lesser extent clothing, furniture, and fabricated metal—had more than their share of workers concentrated in firms employing fewer than 50 workers. In other words, while Chicago may have had more workers, capital, and value added per firm than New York, it resembled Philadelphia, Boston, and Cincinnati, which were renowned for their small proprietorial businesses, and lagged behind other centers, such as Pittsburgh, Providence, Cleveland, and Detroit, in terms of capital-intensity and corporate control.

Chicago's industrial specialization, diversity, and success were built upon a unique set of regional and national factors. Midwest farming with its family farms, widespread mechanization, growing productivity, and increasing disposable income laid the basis for the flow of accumulated capital and an ever-growing agricultural surplus into the city. Chicago became the home of several industries—most notably lumber, leather, food, and metal—that processed resources pulled in from the surrounding region. Agricultural goods, minerals, and timber provided cheap food and raw materials for Chicago's population and industrial base, stimulated the development of industries that serviced the resource sector, and spawned extensive local processing industries. A central part of the local economy was the farm implement industry, which, from McCormick's mid-nineteenth-century reaper factory to the large, multi-unit International Harvester Company at the beginning of the new century, propelled Chicago's industrial development. Similarly, producer durable firms—making mining machinery, edge tools, locomotives, stoves and engines for mining, timber, and agricultural purposes—fueled local growth.[13]

The stimulus for Chicago's growth came from forces outside the region. Chicago was the recipient of technology, knowledge, and capital from the eastern states and Europe. Eastern financiers invested capital to underwrite both industrial and non-industrial developments. Chicago was a magnet for firms and entrepreneurs who sought economic opportunities west of the Atlantic seaboard. The majority of nineteenth-century Chicago's manufacturing bourgeoisie and proprietors migrated from the eastern seaboard or the eastern section of the Great Lakes area. William Wrigley of chewing gum fame arrived in Chicago in 1891 from Philadelphia, while other important manufacturers and wholesalers such as Philip Armour (meatpacking), George Pullman (railroad cars), and Joseph Ryerson (steel products) made their way to Chicago from Stockbridge and, Brockton, New York, and Chester, Pennsylvania, respectively. Others came from Europe, sometimes directly, sometimes via the eastern seaboard. For example, Mathias Klein, the founder of the large toolmaking company, came to Chicago in 1855 from Philadelphia but was born in Worms, Germany, while Swiss-born John Brunswick of the billiard table firm came to Chicago by way of Cincinnati. Still others, such as Richard Warren Sears and Alvah Curtis Roebuck, the founders of Sears, Roebuck and Co., came from the Midwest.[14] Chicago's business class was drawn from a wide radius that spanned the region, continent, and the Atlantic Ocean.

Taking advantage of a window of opportunity associated with the emergence of new industries, native-born Americans and European immigrants moved to Chicago to escape the confines of the older eastern cities, which began their transition to industrial centers a generation or more before Chicago. In 1920 almost 30 percent of the city's population was foreign born, while another 42 percent had foreign parents. The suburban areas tended to have smaller numbers of immigrants, although more than a third of those living in industrial towns such as Whiting, East Chicago, and Cicero were born outside the country. By 1900 the flow of immigrants from the British Isles, Scandinavia, Germany, and other parts of northwest Europe that had dominated since the city's founding had been replaced by new arrivals from southern and eastern Europe. Chicago had been and continued to be an immigrant metropolis before the Great Depression.[15]

Local capitalists exploited government-funded infrastructural monies and created a set of locational assets that fueled industrialization in several ways. First, industrial development took place in a built environment, made up of harbors, canals, railroads, and other elements of the burgeoning metropolitan district that depended on state investment. This was supported by various governments, which invested in the region and metropolitan area's railroad, canal, and harbor systems, and coordinated urban space through such means as the formation of the Sanitary District of Chicago in 1889 by the Illinois General Assembly. Second, local capitalists developed institutional networks that formed the basis for the metropolitanization of industrial space. From the late nineteenth century, for example, the Sanitary District of Chicago and the Chicago Switching Districts coordinated the region's sewage and railroad traffic, respectively. Functioning at a metropolitan level, a range of infrastructures spatially segmented and channeled flows of capital, labor, information, and products. In the process, this contributed to the construction of niche factory spaces and functioned as anchors around which industries developed. Finally, the building of these infrastructures fed the growth of industries, such as railroad-car shops and construction trades, and attracted a skilled labor force.[16] Once in place, new firms created their own initial advantage, attracting new firms in existing industries and spinning off ancillary industries, thereby reproducing the trajectory of large-scale industrial growth in Chicago.

LOOPING THE LOOP AND PUSHING THE LIMITS

The awareness of the importance of the relationship between industrial development and metropolitan growth was commonplace in Chicago. The major vehicle for transmitting this was *Chicago Commerce*, the city's premier business journal. In issue after issue, the journal's editors, reporters, and business backers pushed the importance of manufacturing growth and Chicago's place in the regional and national industrial world. Many of the journal's writers recognized that the metropolitan area's spatial arrangement was critical to the success of its manufacturing base. As one reporter noted in 1923, the metropolis featured both an industrial downtown district and new factory districts outside of the built-up city:

> One feature of the rapid industrial growth of Chicago has been the establishing of districts, offering special facilities, in which many manufacturing plants are grouped. Formerly, the majority of manufacturing plants chose locations close to the downtown district, but in recent years they have found it advisable to move farther out and group themselves where the rail facilities are better and where there are other advantages.[17]

The report recognized that separate groupings of industries had district advantages and that industrial decentralization was a relatively new phenomenon. The writer acknowledged that specialized districts within the metropolis offered benefits that were otherwise not available. However, industrial decentralization was not a new phenomenon. As several writers have noted, most notably Ann Durkin Keating, suburban industrial districts characterized Chicago's development from its earliest days. Similarly, Graham Taylor in his work on satellite cities makes the case for the importance of manufacturing suburbanization in the Chicago area since the 1880s. Indeed, as with other metropolitan districts, such as Pittsburgh, San Francisco, and Montreal, each cycle of Chicago's urban growth from the mid-nineteenth century involved using the suburbs as a solution to the contradictions laid down in the early phases.[18]

The first ring of peripheral factory districts was established in the cycle of economic and urban expansion before the end of the 1870s. A new industrial geography emerged as Chicago manufacturing and associated firms relocated out of the settlement's core, the present-day Loop, to new greenfield sites on the urban fringe. The Chicago River and its branches became home to reaper works, tanneries, distilleries, flour mills, boiler works, as well as numerous grain elevators and warehouses. Foundries, sash and door mills, and flour mills moved to the West Side in large numbers and crossed the river to locate on the north bank. Several large firms, including the McCormick Reaper Works, settled on the north side close to the mouth of the river. Lumber firms relocated for half a mile on both sides of the South Branch. Grain elevators were built on the city's southern fringe near the railroad terminals.[19]

By the 1880s industrial firms formed an extremely compact pattern, with a central core and a set of suburban industrial districts anchored by the waterways and railroad lines pushing out of the center (table 3).[20] With 1,253 firms (45 percent of the city total), the Loop enjoyed the largest share of the city's manufacturing in 1881. The downtown zone, which surrounded the Loop and lined the banks of the Chicago River in all directions, accounted for 35 percent of all of Chicago's manufacturing firms. Chicago's industrial geography remained relatively compact as 80 percent of firms lay within three miles of the intersection of State and Lake. Outside of this tight cluster, a set of industrial districts up to five to ten miles from the Loop developed and became the basis for the geographic dispersal of Chicago's industry over several generations. In some cases, these industrial districts formed around earlier satellite clusters, while in others they were new nodes and signaled new patterns that developed in the following decades. Together they accounted for almost 350 firms (20 percent) in 1881.[21]

Manufacturing firms were not randomly scattered about the metropolis. Areas became associated with specific clusters of industries. Even though specialization can be identified from the earliest days of Chicago's industrial history, by 1881 a distinct division of labor was in place and industrial constellations by district had developed. Clothing, printing, and leather-working were heavily clustered in the Loop. The downtown districts specialized in metal goods as well as furniture and carriages. Woodworking firms (planing and sash mills, lumberyards, and cooperages) and a meatpacking cluster formed in the more distant, river-based districts to the north. Several major concentrations appeared on the South Side: meatpacking in Packingtown and an assortment of boiler, iron, and wood shops in Bridgeport and the Lower West Side. New districts extended out of earlier ones and belts of industry reached south. Areas such as Cleaverville with a soap factory and worker housing, which is now part of the Oakland community between Thirty-fifth and Thirty-ninth streets, became magnets for growth in the southern part of the expanding metropolis. Farther west, 14 miles of docks along the canal anchored some of Chicago's largest firms (the McCormick Reaper Works, American Bridge, the Union Rolling Mill, and Joliet Iron and Steel). These newer factory districts and the suburban clusters anchored the metropolitan area's future industrial growth.[22]

Another burst of growth from the 1880s through to the 1920s added to both the existing factory areas and the industrial base on the expanding urban fringe. By 1924 the pre-1881 city remained a massive industrial district (table 4). More than 5,000 firms accounted for more than two-thirds of the metropolitan area's manufacturing stock. With less than half of its

TABLE 3. The Geography of Chicago's Industry, 1881

			Selected Location Quotients[a]					
Zone	Firms (No.)	Five leading industries ranked by number of firms	Printing	Clothing	Leather	Metal	Wood	Meat
Loop	1,253	Printing, metal, clothing, food, leather	1.8	1.6	1.4	.9	.5	.1
Downtown Ring	971	Metal, furniture, wood, carriage, food	.3	.7	.5	1.3	1.0	.7
South Side	265	Wood, metal, food, carriage, furniture	.2	.1	.4	.6	3.1	5.3
North Side	243	Wood, metal, food, leather, furniture	.3	.5	1.9	1.0	1.3	2.2
West Side	39	Metal, carriage, wood	.8	0	0	1.1	.9	0
Total	2,771	Metal, wood, printing, furniture, clothing	1.0	1.0	1.0	1.0	1.0	1.0

SOURCE: *Lakeside Annual Business Directory for 1881* (Chicago, 1881).

[a]The location quotient (LQ) measures the degree to which an activity or a group is concentrated in a specific area compared to all activities. An LQ of 1 means that the activity is geographically distributed to the same extent as all other activities. A number greater than 1 means that the activity is concentrated, while less than 1 means that it is not. Typically, a score of more than 0.25 on either side of 1 is significant.

TABLE 4. Chicago's Industrial Geography by Stage of Development, 1924

Belt	Firm	Employ	Firms with Employ	Firm Size	Metro Share (%) Firms	Employ
Pre-1881 City	5,049	324,451	4,308	75.3	69.0	44.6
1881–1924 City	1,490	198,907	1,296	153.5	20.4	27.3
Suburbs	777	203,945	701	290.9	10.6	28.1
Metropolitan Chicago	7,316	727,303	6,305	115.4	100.0	100.0

SOURCE: *Directory of Illinois Manufacturers, 1924–1925* (Chicago: Illinois Manufacturers' Association, 1924).

workers, however, these centrally located firms were small compared to firms in the rest of the metropolis. Significant manufacturing development took place in new industrial belts of the expanding metropolitan fringe. By 1924 the city territory built up since 1881 had more than a quarter of the area's workers. Patches of industrial development coalesced outside the built-up perimeter as new outlying districts formed farther to the south around car shops, the Cornell Watch Factory at Seventh-sixth and Ellis, and, most importantly, at South Chicago and Pullman. To the west, the move of Western Electric and Crane stimulated the industrial settlement on both sides of the city boundary. The merging of railroad and manufacturing interests established the planned industrial districts of Clearing and the Central Manufacturing District.[23]

A third belt of manufacturing spread through the independent suburbs. Industrial suburbs and satellite cities at a distance of up to 40 miles from the Loop were home to more than 200,000 workers. Older railroad agricultural and commercial settlements such as Blue Island made the transition from market town to industrial suburb, while new places such as Harvey, Gary, and East Chicago were developed as manufacturing nodes. Even farther out, the satellite towns of Joliet and Aurora extended the metropolitan area's manufacturing base. Together, the post-1881 industrial perimeter of the city and the industrial suburban districts accounted for a third of all firms and more than 55 percent of all employment. Once again, Chicago had decanted its industry.[24]

The shifting city-suburban balance of manufacturing employment was associated with annexation. All of the land that was part of the 1881–1924 city was incorporated into Chicago at various times after the 1850s, and this contributed immediately to the city's industry and population, and established the parameters for extensive growth in the following years.

Annexed territory contained vast tracts of non-urbanized land that city promoters, such as those taking part in the industrial inspection tours before World War I, turned into new locational assets, industrial lands, and working-class residential areas. This process of territorial control through city political control formalized what was already in place. Through flows of information, labor, capital, and materials, these older suburban districts were effectively embedded into the city's political and economic ambit.[25]

Several factors anchored this industrial development. Large companies pulled businesses in their wake. The development of the Packingtown area in the 1860s drew most of the local packinghouses to the southern part of the city. Several ancillary industries including glue making and feathers flocked around the southern packing node. A generation later in 1912, the Crane Company, the country's largest valve maker, moved from a central location to the city fringe and in the process was responsible for the development of a cluster of firms that took advantage of the economies Crane had established on the urban edge. Similarly, groups of small firms, such as those found in the planned industrial districts to the southwest at Clearing and Kenwood, formed the nucleus of other districts. Annexed suburban districts, such as South Chicago, Pullman, and Riverdale, became magnets of industrial development.[26]

These annexed industrial areas were ringed by extensive factory construction in the new industrial suburbs. Suburban firms made up just 10 percent of the metropolitan area's total but were much larger than city counterparts and accounted for more than a quarter of all employees. Satellite cities (Aurora, Joliet, and Elgin) and new industrial communities (Gary) grew rapidly and became important components of metropolitan Chicago's industrial machine. Closer in, new suburbs developed. The move of Western Electric from Chicago in 1903 spurred development in Cicero, while steel manufacture production formed the economic base of many Calumet suburbs. In other cases, single-firm localities such as Summit, the home of Corn Products Refining Company, formed on the urban fringe. These large firms, and the multi-functional towns and suburbs up to 40 miles from downtown Chicago, anchored further industrial growth.[27]

By the 1920s, the city's industry's tightly bound character was undone and factory districts and suburbs were dispersed throughout the metropolitan area in all directions (see fig. 3). This in itself was not new, but the interwar pattern was scattered to a degree unknown before. Responsible for scouting out new locations and carving out industrial sites from the lands surrounding the 1881 built-up area were manufacturing, transportation, and real estate interests. Despite their different structure, organization, and focus, land developers and construction firms all had one central objective, the making of profits through the development of the metropolis. From the West Chicago Land Company, which linked residential building with railroad car shop development in Cicero, to the Maywood Company, which opened up the western suburb of the same name, the Calumet and Chicago Canal and Dock Company, which was responsible for industrial development in the southern reaches of the city, Samuel Gross's numerous subdivisions throughout the metropolitan area, and the Gary Land Company, which built, seemingly overnight, the "Magic City" of Gary—these local promoters channeled Chicago and non-local capital into ever more elaborate and more widely dispersed manufacturing sites.[28] They remade "empty" territory into "productive" industrial land. The seeds of the 1924 industrial geography were already planted by 1881. From the densely built-up core and the industrial belts running along the waterways and railroads traversing the metropolis to the suburban districts such as the Calumet district and satellite towns such as Joliet and Elgin, the basic framework could be discerned.

Metropolitan Chicago's dispersed industrial geography nevertheless remained highly specialized in 1924 (table 5). Light, consumer-based industries remained in the Loop and the downtown. For the most part, workers labored in small multi-storied workshops, factories, and sweatshops making furniture, clothing, hats, shoes, printed products, and fabricated metal goods. The central areas' importance rested on their array of locational assets: excellent transportation links, access to a large and multi-skilled labor force, and a wide range of building types. However, the seeds of industrial decline within the Loop were apparent by the 1920s as their share of Chicago's firms and workers had dwindled in relative importance since the nineteenth century. The number of manufacturing companies in the Loop declined between 1881 and 1924.

Specialization characterized other city districts. One quarter of metropolitan Chicago's workers labored away in the Lower West Side planing mills, Bridgeport foundries, Packingtown packinghouses, Pullman railroad car works, and South Chicago steel mills. The West Side district specialized in fabricated metal goods, furniture, lumber, and electrical appliances, while workers employed in the North Side factories and shops produced lumber, footwear, and car parts. Some suburban industrial districts continued to specialize. After the greater Pittsburgh region, the Calumet district was America's leading steel-producing center by the mid-1920s. The region was also home to an assortment of other industries, especially petroleum, metalworking, and transportation equipment. Not only did the

TABLE 5. Industrial Structure of Metropolitan Chicago's Districts, 1924

Districts	Firms (No.)	Workers (No.)	Firm Size[a]	Metro Area Share (%) Firms	Metro Area Share (%) Workers	Four Leading Industries Ranked by the Location Quotient
Loop	913	56,079	74.8	12.5	7.7	Clothing, printing, furniture, hats
Downtown Ring	2,298	140,838	71.4	31.4	19.4	Footwear, printing, machinery, clothing
South Side	1,108	128,310	134.8	15.1	17.6	Meat, lumber, furniture, chemicals
South Chicago	107	53,992	586.9	1.5	7.4	Railroad equipment, steel, vehicles, machinery
West Side	886	71,066	92.2	12.1	9.8	Fabricated metal, furniture, electrical, lumber
North Side	1,224	73,052	68.7	16.7	10.0	Vehicles, lumber, footwear, machinery
Calumet	234	105,797	494.4	3.2	14.5	Petroleum, steel, railroad equipment, chemicals
West Suburbs	403	84,118	235.0	5.5	11.6	Electrical equipment, furniture, steel, food
North Suburbs	140	14,032	108.8	1.9	1.9	Instruments, foundries, fabricated metal, paper
Suburbs	777	203,945	290.9	10.6	28.1	Steel, electrical equipment, fabricated metal, petroleum
City	6,539	523,358	93.4	89.4	71.9	Clothing, meat, printing, machinery
Metropolitan Area	7,316	727,303	115.4	100.0	100.0	—

SOURCE: *Directory of Illinois Manufacturers, 1924–1925* (Chicago: Illinois Manufacturers' Association, 1924).

[a] Calculated only from firms (n = 6,305) that provide employment.

suburban districts have many of the newest factories in fast-growth industries (electrical appliance, steel, petroleum, and instruments); they also had
a substantial share of the metropolitan area's largest firms.

Chicago's industrial frontier moved several times between the end
of the Civil War and World War II. Tentacles of industrial settlement
stretching one mile from city hall in the 1850s reached up to 40 miles in
all directions by the Great Depression. Specialization typified Chicago's
factory districts, as new lumps of industrial capital deposited firms from
select industries in the suburbs. While many manufacturers were pulled
to the central districts by their need to be close to the external economies,
others were pulled to the fringe. At all times factory owners and managers
sought out the advantages that were to be found on the metropolitan edge.
Following waterway and railroad corridors, executives of Western Electric,
Crane, U.S. Steel, Pullman, and many others sought cheaper and larger
plots of land where they could establish new production processes and
large factories, and the labor force that was moving to the new working-
class neighborhoods spread across the outer zones of the built-up area.

THE RESIDENTIAL GEOGRAPHY OF THE METROPOLIS

Chicago's explosive manufacturing development after the Civil War had
two major effects on Chicago's residential geography. First, it drove large-
scale population expansion and became the economic anchor of a range
of social areas within the city and numerous suburban districts spread out
across the expanding urban edge. Chicago's rise as a mighty industrial
center triggered the expansion of the six-county metropolitan area's population from a mere quarter of a million in 1860 to 2 million in 1900 and
more than 4.8 million in 1940. The city of Chicago was responsible for a
substantial share of the metropolitan area's population growth. Between
1860 and 1900, the city's population increased from little over 112,000 to almost 1.7 million. Growth was translated into a complex social ecology. The
downtown was home to an assortment of social areas, from the densely
built-up, working-class and immigrant districts surrounding the factory
districts to the mansions of Chicago's wealthy oligarches in the Prairie
Avenue district. Farther out, closer to the city's perimeter, were residential
districts ranging from the sylvan, low-density, middle-class areas to those
where workers' cottages intermingled with factories.

Even though the rest of the metropolitan's district grew slowly over
this period, this was a formative period for the areas on the metropolitan
edge. Before the Civil War, the city was surrounded by several small and

independent rural-based settlements. A generation later, many areas on the edge of metropolitan Chicago were suburban places directly attached to the central city or satellite cities at a distance. Although it is difficult to know precisely how many suburban places lay on the city's nineteenth-century fringe, Homer Hoyt estimated that there were 60 suburban towns and villages outside the city in 1880. Of the more than 230 settlements established in the nineteenth and early twentieth centuries identified by historian Ann Durkin Keating, close to three-quarters started as farm centers or industrial towns, while the others first developed as commuter-rail suburbs or recreational-institutional centers.[29] What is certain is that industrialization in the form of specialized manufacturing or general food processing outside the central city propelled the growth of an assortment of suburbs, towns, and satellite cities.

Annexation shaped the city and suburban population balance. As elsewhere, the city of Chicago's growth before 1900 was fueled by the addition of large tracts of adjacent areas. This began at an early date. By the time of the 1871 Great Fire, Chicago had extended its incorporation boundary. By the late 1860s, several residential and industrial suburbs, such as Packingtown and Bridgeport, were incorporated into the city. The most significant annexation occurred in 1889 when Chicago acquired 125 square miles and a population of more than 200,000. With its meatpacking plants, lumberyards, steelmaking facilities, and array of metalworking firms, the annexed sections contributed greatly to the city's industrial base. After 1889, annexation slowed, with very little new population or industrial additions to the city.[30]

In the 40 years after 1900, however, the city-suburb balance changed. Tremendous population growth took place in the suburbs, but the city remained dominant in terms of sheer numbers. Chicago's population doubled, reaching 3.4 million by 1940, and the range of social areas continued to be wide. Despite this increase, the city's relative growth slowed compared to the rest of the metropolitan area. In 1900 the city accounted for 81 percent of the metropolitan area's population, but only 70 percent 40 years later. The accelerated growth of the fringe is particularly evident in the number of large suburban towns and cities located across the outer ring. By the beginning of War World II, Gary had more than 100,000 people, while Cicero, East Chicago, Hammond, Oak Park, and Evanston were home to more than 50,000 residents each. The older satellite cities and suburban towns such as Joliet, Elgin, Aurora, Blue Island, and Whiting had sizable populations. By the early 1930s, suburban municipalities numbered in excess of 200, many of which were dependent on manufacturing. This collection of suburban places accounted for more than 1.5 million people on the eve of World War II and was tied to manufacturing suburbanization. As Helen Monchow points out in her study of Chicago's real estate development, "The suburban trend [could not] have reached it recent proportions had not considerable manufacturing expansion occurred to supply the economic base for at least a sizeable proportion of this suburban living."[31]

The second major effect of manufacturing growth on Chicago's residential geography was to reinforce social inequalities and accentuate social exclusion. High rates of geographic segregation within the city were matched in the suburban ring. Industrial capitalist growth ordered the metropolis by social status and geographic area. Chicago's social hierarchy manifested itself through a division of labor that forced immigrants and workers to reside in a graduated set of neighborhoods sorted by income, ethnicity, and race. The very worst, typically populated by the newest round of immigrants, were located in the older factory districts spread out along the railroad and water corridors. Subject to the surveys and reports of Chicago's social reformers including Sophonisba Breckinridge, Edith Abbott, and Robert Hunter, working-class districts such as the West Side, Pilsen, and Packingtown contained wretched housing, terrible environmental conditions, inadequate services, poor working conditions, and low wages. While many of Chicago's huge and multifarious working class had few options, housing conditions for others were more encouraging. Chicago's Poles, for example, resided in the full range of housing areas, from the poorest parts of Packingtown to the newer and more salubrious sections close to the city limits.[32]

Many of these working-class residential districts were directly dependent on industrial employment. In some cases, most notably in Packingtown, Pullman, and South Chicago, workers depended on one industry, while in others, such as West End, Bridgeport, and the North Branch, workers found employment in an assortment of workplaces. At a distance from the main city, factory districts were the newer working-class areas. Spread across the city, these areas, with the major exception of the African American neighborhoods running south from the Loop, typically had newer (and better) housing and services.[33]

The same pattern appeared in Chicago's suburbs. Suburban manufacturing plants and ancillary industries such as railroad shops and yards attracted workers. Three waves of manufacturing growth created a range of working-class suburbs on the metropolitan fringe before the Great Depression. Beginning prior to the Civil War, suburban districts such as

Bridgeport and Packingtown drew large working-class populations. A generation later, extensive working-class suburbs appeared in the Calumet district and the western suburbs. The continued expansion of these earlier suburban districts and the development of new ones farther out after 1900 formed the basis for a large working-class population on the growing metropolitan fringe. In some cases, industrial and working-class suburbs, such as Gary, East Chicago, and Cicero, were new; in others, farm centers, such as Blue Island, Joliet, and Elgin, became industrial suburbs.[34]

The city and the suburban working-class districts had several common features. Most workers had little mobility and were forced to live close to their place of work. Germans, Irish, and Poles crowded around Packingtown's and Bridgeport's lumber mills, foundries, and meatpacking factories. Farther out, 90 percent of East Chicago's workers in 1926 lived in the city and adjoining towns. The housing options available to most working-class families were at best limited, at worst dismal. Chicago's housing markets—stratified as they were by class, income, ethnicity, and race—were geographically segregated. Although some workers, such as those in Hammond, found decent housing accommodation, many Chicago workers, encumbered by their place within an unfavorable labor market, had few choices. Workers created strong family ties that revolved around a gendered family economy. These family loyalties were constructed and maintained within a cohesive, typically ethnic, community centered on an assortment of neighborhood institutions, most notably the church, associations, retailers, and the tavern.[35]

The result was that metropolitan Chicago was highly segregated. While the city's wealthy and professionals resided in the city's Gold Coast, Hyde Park, and Prairie Avenue districts as well as the exclusive and expensive suburbs of Oak Park and Kenilworth, the vast majority were not so fortunate. Vast territories of working-class districts stretched through the city and the suburbs, typically abutting railroad lines, canals, rivers, and factories. These social class districts were internally differentiated. People of different ethnic and racial backgrounds tended to live apart from each other, functioning within their own self-enclosed social worlds. Each suburb took on its own particular identity, shaped as they were by the presence or absence of industry and the associated class, racial, and ethnic constellations. Suburbs were also divided internally. In suburbs as industrially diverse as Blue Island, Cicero, and Chicago Heights, a residential geography calibrated by ethnic background and class was imposed on the landscape: the laboring poor ringed the factories while the white-collar resided in the better-off housing subdivisions on the suburb's edge.[36]

Between the end of the Civil War and the beginning of World War II, metropolitan Chicago became one of the great industrial centers of the world. From a small commercial node hugging the banks of the Chicago River, a metropolitan industrial district developed characterized by a diversity of production practices, industries, and scales. The large ironworks and steelworks on the South Side were anchors for the growth of working-class housing areas, ethnic neighborhoods and associations, domestic and producer infrastructures, and other firms. Such locational assets were replicated through the expanding metropolis. Printing, clothing, and other labor-intensive, consumer-based industries clustered close to the central core, forming a social ecological niche with the older working-class and ethnic neighborhoods.

Industrialization drove the metropolitan-building process. Class-based alliances made up of industrialists, developers, politicians, and streetcar promoters created factory districts and working-class neighborhoods. Manufacturing and working-class population growth forced local elites to seek new types of industrial space. The specific needs of a number of industries—most notably iron and steel, lumber, and meatpacking—forced industrialists to seek new industrial land and to construct new districts outside of the built-up areas of the city after 1850. From the late nineteenth century, the changes generated by the second industrial revolution, the rapid growth of the metropolitan industrial base, and the increasing scale of firms ensured that the city's older spaces were increasingly unsuited for many industries. Old congested industrial sites in blighted districts were poor working spaces, susceptible to attack from commercial, transportation, and financial uses, and increasingly ill-placed with respect to local and non-local markets. While many firms in printing, clothing, and furniture continued to find advantages in downtown areas, many others sought new sites in the factory districts close to the city's perimeter or in the rapidly expanding suburbs. Some of these manufacturers were more than willing to move to the farthest reaches of the metropolitan fringe and bypass the benefits of central agglomerations.

Chicago's expanding manufacturing base shaped the metropolitan residential landscape. Reflecting the scale and specialized character of an area's manufacturing structure, working-class districts and suburbs differed in terms of their social composition, housing stock, occupational structure, and service provision. The large expanse of industry running along the waterways and the railroad corridors and reaching out to the suburban districts became the home to large working populations bound together by their dependence on industry for their livelihood and for their everyday life. Every working-class city district and industrial suburb was, to varying degrees, relatively self-contained. The metropolis, however, was more than a collection of its parts. Bound together by the imperatives of capitalist-industrial growth, the metropolitan area was an extremely mixed and differentiated social residential landscape that was greatly shaped by the location of manufacturing.

This alignment of manufacturing and population growth from the mid-nineteenth century controlled by different factions of the capitalist class pushed out the perimeter of metropolitan Chicago. In the process, an increasing number of municipal boundaries separating suburb from suburb and from the central city were created. This profusion of separate municipalities—to the extent that each had political jurisdiction over a well-defined space, and their own unique structure and identity—represented the social and economic fragmentation of Chicago. At the same time, however, several processes and actors ensured that the reverse was also the case. Capitalist industrialization required that the spatial units in which urbanization occurred operated in a coherent and practical manner. Despite its seemingly fragmented character, metropolitan Chicago functioned as a large integrated territorial system, an industrial district. This was made possible by the search for and construction of a suburban solution by Chicago's business and political elites.

CHAPTER 2

The Suburban Solution

For many, Chicago's transformation into a great industrial metropolis after the Civil War was made possible by its leaders' ability to mold the prairie into industrial territory. The conversion of natural resources into productive use by Chicago's entrepreneurs and politicians turned an "empty" but potentially extremely rich landscape into one of the greatest manufacturing centers in the world.[1] The desire to remake Chicago's urban fringe was certainly in the mind of Charles D. Richards, a member of the Chicago Association of Commerce's Civic Industrial Committee, when he told his audience of industrial elites that the reason for the study of the region's "industrial locations and available sites" is "simply to get into our minds what there is to be developed in Chicago."[2]

This is certainly how Albert W. Beilfuss viewed it during an industrial inspection tour of suburban districts organized by some of the city's business and real estate interests in July 1911. Born in Germany in 1854, he moved to Chicago, became a partner in the printing and lithographing firm of Severinghaus and Beilfuss, and quickly immersed himself in the city's social and cultural business. An alderman for 10 terms, he was active in several committees, including the Special Park Commission. A year after his death in 1914, Mayor Carter H. Harrison suggested that the Humboldt Park natatorium be named after him. The Special Park Commission agreed. Alongside his interest in the city's cultural affairs, Beilfuss was also actively involved in promoting the city's economic business. His interest as a participant of one of the industrial inspection tours that traversed the metropolitan area before World War I was to promote the city's economic development. However, looking out of the Pullman car window as the train crossed the city's fringe, he told the *Chicago Commerce* journalist that he did not "see any very interesting landscape" in the city's fringe. For Beilfuss, the territory was of no interest because he believed that the districts they traveled through—Dolton, Indiana Harbor, and Blue Island—were practically devoid of human activity. What struck Beilfuss's imagination, and doubtless many others, was the picture of "these vacant lands here by the river, on the railroads, in the future being filed [*sic*] with factories."[3]

But these were not empty lands. Even ignoring aboriginal use, European human activity was apparent. Blue Island, a market town, was settled in 1836 and slowly grew into an industrial suburb by World War I. Similarly, Indiana Harbor on Lake Michigan was transformed by the opening of the Inland Steel mill in 1905. In shutting their minds to the degree and kind of activity taking place in these areas, Beilfuss, Richards, and others rhetorically reconstructed them to fit their needs. This involved molding how these areas were viewed by investors, framing the territory's potential for even more development, and establishing the manner in which the urban space could be transformed. For the participants in the industrial tours, this land was an urban frontier that had to be modernized to fit a regime of urban-industrial capitalism. Existing industrial settlements as well as the small commercial or agricultural ones had to be converted to functions fitting a large industrial metropolis.[4]

Two obstacles stood in the way. First, conflict between Chicago's capitalist classes over goals and strategies inhibited the drive to transform the fringe into a tame, settled, and modernized part of the metropolis. Second, the urban elite's inability to predict and manage the cyclical and competitive nature of capitalist economic growth hindered the seamless exploitation of metropolitan territory. Despite these obstacles, Chicago's entrepreneurial classes built a coherent growth-based industrial economy. In the process they transformed the city's undeveloped territory, creating the nation's second largest industrial metropolis. Chicago's manufacturing might was predicated on the local political and economic bourgeoisie's ability to convert what was considered useless prairie into productive industrial space. Businesspeople, politicians, middle-class observers, and workers increasingly looked to the suburbs as spaces where the problems of the older areas could be avoided, and in the process they made new economic geographies. They sought a suburban solution. Even though this was riven by conflicts of all kinds and created through the action of thousands of actors, a modern suburban reality did emerge. Chicago's elites were able to juxtapose and merge in both symbolic and material ways, images of heroic man (the white settler and entrepreneur), an unruly and unproductive nature (the prairie and the American native), and the good of the common weal (a capitalist urban economy).

THE EMPTY METROPOLIS AND INDUSTRIAL EXPANSION

Richards and Beilfuss were not alone in thinking that the territory surrounding the built-up city was composed of empty lands needing to be filled with human activity. The making of metropolitan Chicago's manufacturing base from the very beginning was formulated around the transformation of prairie land on the urban fringe. In 1873 a booster of Chicago's suburbs proclaimed that "westward [from the city] there is unlimited space, bounded only by the swamps of the Calumet, the Mississippi, the British provinces, and the imagination. Some day the Queen's dominions will be annexed, and then there will be absolutely no limit to Chicago enterprise. At present the real-estate dealer's horizon is bounded by a semicircle radiating about thirty miles from the courthouse."[5] While Canada is no longer part of the queen's dominion and continues to be formally independent of the United States, the point underlying the writer's hyperbole held true. The so-defined empty lands of the Chicago region could be transformed if the will and the material conditions existed.

The editor of a Chicago yearbook made the same point in 1885. Chicago settlers were "far-sighted" and had "visions of its destiny." They tamed the "unproductive, Indian-haunted swamp . . . drained the marsh [and] pushed back the billows of the tossing lake." The capitalists who came to Chicago were fleeing the "less progressive cities" in the east and investing with "tremendous vigor" in the factories and mills filling up the booming metropolitan district's territory. A few years later the Harvey Land Association, the property development company charting the transition of the open prairie to the industrial suburb of Harvey, assured potential investors and clients that through its "ability, capital and energy," the land association was "capable of changing this vacant waste into a busy beehive of human industry."[6]

Of course it is easy to dissect writings such as these and find examples of purple prose and the appropriately revealing booster sentiment. Yet these sentiments represented the views of Chicago's commercial, manufacturing, real estate, and transportation interests desperate to exert control over the environment, to shape nature to urban-capitalist ends, and to modernize the prairie lands and market towns on the urban fringe. This, of course, was no simple matter. As historian William Cronon shows, the transformation of the prairie into industrial land took a great deal more work than the rhetorical projection of wrestling with and overcoming nature. The commodification of land on Chicago's expanding urbanized frontier required a set of mechanisms linking property investment on the frontier to the search for profits. Flows of capital from other sectors (especially manufacturing and finance) and places (especially the city of Chicago) occurred through various property development forms such as real estate corporations, improvement companies, land associations, and syndicates. From the small subdivisions laid out along the railroad lines to the corporate creation of places such as Pullman and Gary, the prairie was remade as urban-industrial land.[7]

Not surprisingly, those building urban infrastructures—electricity, railroad tracks, docks, streets, homes, and factories—were in the vanguard of defining what was metropolitan space. Local agents directly involved in the process of capital accumulation through urban development took the lead in coordinating the conversion of the prairie into manufacturing land. One group in particular, property developers, were instrumental in defining metropolitan Chicago's territory. South of the city, developers such as Marcus Towle opened up worker housing in Hammond, while syndicates managed by Chicago realtor Oliver Brooks and Chicago lumberman Turlington Harvey built up West Hammond and Harvey, respectively. To the west, the Riverside Improvement Company bought 1,600 acres, hired Frederick Law Olmsted, and built an elite suburb.[8] Not all developers were successful, but the cumulative effect was that they pushed the limits of urban settlement by laying out subdivisions and building infrastructures, homes, and factories. In the process of subdivision and the construction of built form, they redefined metropolitan boundaries. Institutions such as the Chicago Real Estate Board and the Cook County Real Estate Board coordinated and oversaw individual needs. They ensured that metropolitan development, from the constant rebuilding of the city core to new development on the expanding fringe, contributed to the definition of the city and the surrounding districts.[9]

In a few cases, industrial places were constructed by large corporations. While thousands of entrepreneurs worked mightily and long to build Chicago's industrial districts and suburbs, corporations such as Pullman built new towns in quick time. The most dramatic was Gary, "the Magic City," a creature of U.S. Steel. Founded in 1906, it was the largest company town ever built by private interests in the United States. The impetus was the decision by Eugene Gary, U.S. Steel's chairman of the board, that the corporation needed a new industrial center if it was to remain competitive. Working on Gary's directive, corporation officials under the leadership of Eugene Buffington, president of Illinois Steel, a company subsidiary, fixed on 9,000 acres among the sand dunes and swamps of Lake Michigan's Indiana shore. This site provided U.S. Steel with cheap land, ample space, excellent rail and water access, and a large local market and labor supply. U.S. Steel undertook two main actions. First, they built several large mills, plants, and finishing shops. Some of the largest of their kind in the United States, the town quickly became the corporation's major producing center. Second, the company, through the auspices of the company-owned Gary Land Company, laid out the town and constructed more than 800 dwellings, most of which were sold to workers, as well as various utilities and public buildings. By World War I, U.S. Steel in conjunction with private real estate promoters had built a new city on the empty shores of the lake (figs. 4 and 5). While Gary became a highly profitable and successful economic machine, the town, in the words of historians Raymond Mohl and Neil Betten, "proved a failure from a planning standpoint." A combination of poor planning, high rents, environmental degradation, and the search for profits ensured that Gary would, like so many other company towns, be terrible places for the local working population.[10]

Gary was only one model; land developers and land companies transformed the fringe in different ways. Samuel Gross and hundreds of others opened up working-class, middle-class, and elite residential suburbs. Similarly, Arend Van Vlissingen and Nelson Thomasson actively subdivided, promoted, sold, and leased factory sites throughout the metropolitan district. Hamilton Bogue, of Bogue and Hoyt, for example, specialized in industrial land in the southern suburbs. He first learned his trade after the Civil War through his association with the Chicago Land Company. After 13 years at that job, he became vice president and manager of one of the most important real estate and industrial developments in the southern suburbs, the Calumet and Chicago Canal and Dock Company. After some time he moved on to join business forces with the Hoyt family. Pursuing industrial development and land sales in the rapidly expanding southern industrial districts, he became involved in the Hegewisch trust, a syndicate controlling large landed interests at the forks of the Calumet River close to the large manufacturing shops of the United States Rolling Stock Company. Deploying a well-known strategy, Bogue attempted to market industrial land by linking his properties with the development to a large anchor firm.[11]

FIGURE 4. The Corner of Fifth Avenue and Broadway, Gary, 1906. The "empty land" of the prairie is being prepared for the building of the industrial city of Gary. Source: U.S. Steel Gary Works, Photograph Collection, 1906–1971, Calumet Regional Archives, CRA-42-100-021.

Land-development companies focused on different parts of the metropolis. Some such as Bogue and Hoyt specialized in industrial land to the south. Others such as the West Chicago Land Company cast a wider net, building up both the residential and industrial character of the metropolitan area's western limit. Regardless of their property focus and the districts they specialized in, all of them were linked to the dynamics of metropolitan-building. Haphazard and uncoordinated at the individual level, the overall impact of the thousands of individual actions, from the identification of the farms to be platted to the laying down of infrastructures, contributed to the incremental construction of metropolitan Chicago. Developers and builders laid out the land, roads, and sewers; built the homes; wrote up the restrictive covenants; and, to ensure the connection between the suburbs and the central city, worked hand in hand with railroad and intra-urban transportation companies to build the tracks and chart the schedules.[12]

FIGURE 5. The Corner of Fifth Avenue and Broadway, Gary, 1931. Twenty-five years later, the prairie at Fifth and Broadway has been transformed into a commercial center servicing the workers of the Gary steel mills. Source: U.S. Steel Gary Works, Photograph Collection, 1906–1971, Calumet Regional Archives, CRA-42-114-057.

Suburban-based booster groups coordinated many of these activities. In the 1880s, for example, the West Chicago Land Company ensured that street horse cars and commuter railroads reached out to its developments on the western city limits, lying between Madison and Kinzie streets, just west of Central Park. To the south in West Hammond, a syndicate of investors orchestrated by Chicago real estate agent Oliver Brooks subdivided 1,728 lots, subsidized a Catholic church, and, through the use of a Polish agent, turned what was empty land before 1890 into a vibrant Polish community. To the north in Evanston, developers combined with municipal officials to create an upper-middle-class suburb. Banning manufacturing establishments and alcohol, the town laid out extensive lots, streets, and a modern sewerage system in order to attract a prosperous clientele.[13]

Numerous methods financed property development. Real estate developers of well-to-do and working-class residential suburbs sold lots and houses by auction and monthly payment schedules. Improvement and manufacturing companies constructed manufacturing suburbs throughout the metropolitan area, especially on the South Side and the Calumet district. The coalitions of entrepreneurs who built up Harvey, Hammond, Riverside, Evanston, West Chicago, and other parts of the suburban ring provided capital for large-scale real estate development and residential, transportation, and manufacturing projects. Also important were coalitions of investors who supplied large amounts of capital, typically to promote industry in order to attract a population. Some land associations established the framework for and offered inducements to industrial development, while others became more directly involved in manufacturing.[14]

Utility companies were major players in the construction of manufacturing Chicago. A key agent here was Samuel Insull and his creation of an urban electrical realm. After consolidating control over the businesses and larger users of central Chicago, Insull went after the large number of small consumers spread throughout the city and the adjacent suburbs. Combining technological innovation, new markets, and political consensus, Insull rewired the geographical context of metropolitan energy consumption. Building an electrical load, constructing generation stations, and laying the lines, Insull extended the limits of what constituted the metropolitan region between 1900 and the Great Depression.[15]

The outward movement of railroad shops stretched the metropolitan boundary, facilitated the making of an industrial base, and contributed to the growth of a specialized metropolitan geography. After the late 1840s, a large number of railroad tracks crisscrossed Chicago and in the process shaped the locations of urban development. At least 233 settlements in the nineteenth and early twentieth centuries were established as railroad suburbs. A large share were commuter, recreational, institutional, and farm centers. More than 70, however, were industrial. Home to manufacturing establishments, they extended the territory in which goods and people could function and featured an array of residential neighborhoods and community institutions. Similarly, urban development coalesced around railroad yards on the urban fringe and established the perimeter of what were Chicago's metropolitan boundaries.[16]

The Chicago railroads also defined the metropolitan manufacturing geography by the creation of switching and terminal districts. Established in 1910, the Chicago Switching District was a territorial unit covering 400 square miles that was governed by a set of tariffs providing for certain uniform practices and rates. Shipments originating on belt lines cost the shipper no more than if they originated directly on the lines of the road-haul carrier by which they left the district. Industries were free from the disadvantage of having to pay additional freight charges on all traffic except that handled directly by the line-haul carrier. The Chicago terminal district defined a larger region than the switching district. By the 1920s a crescent-shaped district extending from Lake Michigan to the Outer Belt and from Waukegan in the north to Porter in the south, the terminal encompassed 1,750 square miles and 4 million people, and had 4,330 firms located along 7,726 miles of railroad track. More than 300 freight trains arrived and departed daily. These railway facilities framed one version of the territorial limits of metropolitan Chicago.[17]

Harbor facilities also localized industrial resources and built up what constituted the manufacturing metropolis. The Chicago Harbor and the Illinois and Michigan Canal were the center of Chicago's water-based metropolitan region in the nineteenth century. Fueled by federal appropriations and the establishment of large plants in the southern part of Cook County dependent upon waterborne supplies after 1870, the harbor's expansion signaled a shift in the scale and focus of the industrial metropolis. This became apparent with two events in the first decade of the twentieth century. The first was in 1906, when the traffic of the Calumet Harbor surpassed that of the Chicago Harbor. The second occurred three years later, with the publication of the 1909 Chicago Harbor Commission report. Not only did it emphasize the idea that the metropolitan area was an integral functional unit; it made several recommendations that acknowledged the superiority of the southern port and laid out the basis of the metropolitan district. The report recognized that the geographical limits of Chicago extended from Waukegan to Gary, constituting one great industrial community. State boundaries were irrelevant. Indiana Harbor and Gary, even though located in Indiana and having no formal political links to Chicago, formed part of the industrial community focused on the metropolitan region.[18]

The Calumet district illustrates the ability of growth coalitions to transform the prairie's apparently vacant and unlimited spaces into productive
industrial territory. In Calumet, as elsewhere, developers assembled the
locational assets necessary for suburban settlement. Local property alliances brought together into one bounded district a range of interlinked
infrastructures that connected firms to businesses both inside and outside
the district. Successful suburban manufacturing development depended
on bridging local initiatives with state and national infrastructure improvements. From the building of new harbors to the dredging of silted
rivers, the laying down of railroad tracks, and the wiring of a region, government intervention in the form of enabling legislation, surveys, and finance were instrumental to the creation of greenfield manufacturing sites
and the building of industrial suburban districts making up the Calumet
district.[19]

The origins of South Chicago, the first major steel center in the southern portion of the metropolis, originated with the Calumet and Chicago
Canal and Dock Company (CCCD), formed in 1869 by two major land
holders and several other capitalists. After making a fortune in downtown
Chicago, president Colonel James Bowen of the CCCD turned his attention to the empty land south of the city. Bowen (1822–1881), a self-made
man among other things, amassed a fortune in commerce and real estate,
helped George Pullman purchase land for his new industrial town on the
shore of Lake Calumet, and acted as a U.S. commissioner to the Paris Exhibition of 1867. Like many other boosters, Brown "pictured a great harbor,
an industrial empire and great resources as yet untapped," and, of course,
great profits. Bowen worked the typical promoter trick; he combined land
(6,000 acres) with new infrastructures (docks built through government
funding), railroad connections, and subdivisions for workers and industry.
By 1874 the company had established South Chicago as a vibrant industrial and residential district with inexpensive workers' frame houses and
enterprises such as the Illinois Steam Forge, Chicago Iron and Steel, and
some lumberyards and grain elevators.[20]

Further subdivisions' activity and manufacturing growth in the 1880s
laid the basis for the large-scale development that took place in the last
decades of the century. The most important single event was the move by
officials of the North Chicago Rolling Mill in 1880 to organize the North
Chicago Steel Company for the purposes of the building a plant on the
South Side. The older mill on the North Branch of the Chicago River was

closed, and the new South Chicago plant located at the meeting place of the Calumet River and Lake Michigan became the flagship mill of the company and the metropolitan area. The first blast furnaces were blown in June 1881. A year later the company opened a Bessemer steel converting plant, a rail mill for rolling heavy sections, and two other blast furnaces. The establishment of one of Chicago's most important workplaces triggered further residential development and population growth. Numerous new subdivisions along with the lots laid out by the CCCD became home to a large, mixed working-class population. From less than 2,000 residents in 1880, the area numbered 16,000 three years later and more than 26,000 in 1890.[21]

The development of South Chicago's manufacturing could not have occurred without infrastructure. Four railroad lines cut through the area before 1865 while several more were laid down between 1874 and 1895. Railroads were magnets for industry. Industrial belts were strung out along the railroads crisscrossing the district, while rail sidings serviced firms of all sizes. The building of the Chicago Outer Belt Line, a set of lines forming a crescent at a distance 25 to 40 miles from the Loop and bounded by Gary and Waukegan, further linked South Chicago to the city, the Midwest, and the nation. The Belt carried the district's steel and lumber to other Calumet firms and to markets elsewhere. Similarly, after the federal appropriation for the Calumet River harbor improvements in 1870, the river became a conduit for moving goods in and out of South Chicago's factories. By the early twentieth century, successive rounds of federal funds resulted in the dredging of a four-mile channel from the mouth of the river and the construction of what would be Chicago's leading harbor. The result, as Henry Lee breathlessly exclaimed in 1907, was "miles and miles of docks and harbor frontage unequaled in the whole world as manufacturing sites."[22]

South Chicago's development was appreciated by John Emerson, a Chicago alderman and a guest on a 1911 industrial inspection tour, who boasted that the Calumet district was "destined to be the greatest manufacturing center in the United States." He reminded his fellow guests that the area's potential for greatness was of recent origins: "It was not much more than twenty-five years ago that along the banks of the Calumet at South Chicago was a little fishing hamlet. In that small beginning has grown a manufacturing district that employs in the neighborhood of 12,000 men."[23] Many agreed with Emerson. According to one 1912 advertisement, South Chicago was home to the "best factory site in the United States" (fig. 6). The property was located on two railroad lines, close to

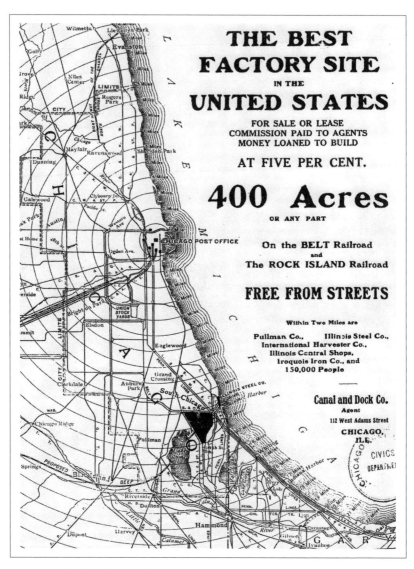

FIGURE 6. The Best Factory Site in the United States, 1912. Typical of industrial property schemes throughout the United States, the promoters of sites in South Chicago pointed to the locational advantages of the land for sale. The usual ingredients, in South Chicago as elsewhere, were proximity to transportation, markets, and labor. Source: *Chicago Commerce* (July 5, 1912): 44.

several industrial plants, and within reach of a large workforce. The agent, the Canal and Dock Company, also promised to invest in the site. The entrepreneur had successfully grappled with the unruly environment to produce a modern industrial landscape.

Through its property development, the CCCD laid the basis for industrial development and for the rise of local coalitions of farmers, real estate developers, industrialists, and commercial interests who shaped the new industrial environment. They bought and subdivided land, supplied the necessary urban hardware, provided manufacturers with subsidiaries and cheap land, and availed themselves of federal and state monies to finance the building of harbors, roads, railroads tracks, factories, stores, and homes. Even though frequently in conflict and fluctuating in composition, and caught in the ups and downs of the business cycle, South Chicago's local coalitions, in the words of one of Calumet's historians, became "closely intertwined" with Chicago and transformed "the heretofore worthless marshes and sand dunes along the Indiana shore of Lake Michigan."[24]

South Chicago was just one of the many districts carved out of the territory south of the city before World War I. At the same time as the CCCD worked to crystallize various entrepreneurial initiatives, George Pullman pulled up his Detroit railroad passenger car factory and constructed the planned industrial suburb of Pullman. While Pullman built a model town designed to produce individual profits and social harmony, another form of land development took place at West Pullman. In November 1891 Chicago and eastern investors formed the West Pullman Land Association and purchased 4,800 acres of an old farm. They laid out sewer, gas and water mains, cement sidewalks, and macadamized streets at a cost of $350,000 before the actual house building was done. They induced the railroads to extend their lines and build passenger and freight stations, and constructed a system of terminal tracks connecting these trunk lines and reaching all parts of the factory districts. They planned a strict separation of land uses: the factory district was situated far from the residential district, while the higher class of residences were placed on the higher and wooded portion of the development. Over the next few years, several factories were established and a manufacturing and residential district formed.[25]

As these examples indicate, property development institutions played a critical part in the making of the South Side's manufacturing base in several ways. They facilitated the formation of large amounts of land that were quickly subdivided for industry and residence. By bringing together different individuals and associates, they made large amounts of capital available to underwrite property acquisition, subdivision, development, and, sometimes, housing construction. In some cases, capital was transferred directly to manufacturing through the agency of a land association. In most cases, however, the effects of capital investment on manufacturing were indirect and not through direct ownership. Manufacturing firms obviously gained great benefits through the building of infrastructures, most notably railroads, waterways, and harbors, most of which were built by bridging local capital and expertise with state and national institutional support and funds. The provision of subsidies, such as free land or low taxes, were attractive enticements for some firms.[26]

The building of the Calumet district, however, was not unproblematic. The district experienced several false starts as developers went too far ahead of the urban frontier. Firms bought land, drew up factory designs, and planned production only to find their schemes thwarted and the capital for infrastructure improvements not forthcoming. This was partly because the building of local coalitions, consisting of public and private interests, was fraught with difficulty. Individuals had a different sense of how and where their capital should be used. In some cases, the infrastructures necessary for a business to operate productively were not accessible or adequate. In other cases, firms, especially those dependent on well-established external economies, were more than happy to remain closer to the city center.

Two examples illustrate this point. Silicon Steel purchased 26 acres of prairie in South Chicago in 1873. Foreseeing a great future, the company laid down streets and drew up plans for workers' homes and stores. But, with the exception of a slight flurry of land development in the surrounding area, nothing materialized as Chicago capitalists proved slow to invest. The 1870s economic downturn made investors more guarded with their capital, and significant industrial activity was delayed by a decade. Almost 40 years later, a similar event took place. In 1912 Ferdinand Schlesinger, a Milwaukee businessmen with interests in iron mines, bought 425 acres in Hammond, intending to build blast and open hearth furnaces and steel finishing mills to manufacture implement bars and special shapes. The project was viewed with some skepticism by local interests and never did come to fruition. One factor undermining the project was the failure to construct the Indiana Harbor ship canal for the use of the plant.[27]

By the 1880s South Side developers faced another problem: the frequent attempts by Chicago-based interests to restrict the flow of capital to the Calumet district. William Rowan, a South Side alderman and historian of the Calumet region, complained about the rivalry between Chicago and

Calumet interests over their respective rivers and harbor appropriation. "Chicago always feared the Calumet in those early days and whatever concessions were made were done so very grudgingly," Rowan remembered. "The powerful Chicago river clique was unwilling to talk of an industrial harbor at the Calumet or an inland harbor and there is no question that this attitude of the Big Brother had its influence upon many."[28]

Similarly, a determining element of the Chicago Real Estate Board was intra-metropolitan conflict. While various elements of the property industry had interests in common, conflicts developed based on the geography of the city. According to Everett Hughes, the board's historian, Chicago "was something to be nursed, and even to be protected from its own suburban offspring." City interests were extremely wary of competition from the suburbs. Various suburban plans—the Belt Line railway to facilitate freight transfer, the building of the Calumet River's dock, and even the 1893 World's Fair—were considered to be dangerous. Loop-based realtors looked to enhance the values of central districts, while the urban fringe was left to neighborhood agents.[29] The fighting over the size and location of capital funds was institutionally locked in place.

The Calumet district was not unique. Other new districts were carved out of the prairie. Cicero and Clearing became major industrial districts by the 1920s. Smaller in scale and built on the "vast prairie" where "bull-frogs, swamps and rag-weed abounded," Austin was home to assorted manufacturers before the Great Depression.[30] As city builders and boosters such as Richards, Beilfuss, and Emerson made very clear, Chicago grew from a small frontier outpost to a leading industrial metropolis in a very short period of time. Through their pronouncements and actions, they unquestionably celebrated the ability of Chicago and non-local entrepreneurs to transform an empty prairie into an industrial metropolis. To their mind, not only did they construct a temple of trade and manufacture, but, by the relentless pursuit of profits, property, and industrial modernization, they overcame the obstacles imposed by the prairie.

DIFFERENT VISIONS

Neither the rhetoric nor actions of those building metropolitan factory districts, however, were shared by all observers. More than one vision existed about how the city should develop. Visitors to the city, for example, noted a quite different geography than the one presented by boosters. Local promoters dwelt on the potential of the metropolitan fringe at the expense of the already-constructed Chicago. With a more distant and critical eye than local commercial, political, railroad, and manufacturing elites, however, visitors to the city probed the underbelly of Chicago's factory and working-class districts.

In 1894 the French novelist Paul Bourget toured the city. Traversing a different territory than that taken on the prewar industrial inspection excursions and using a different eye than the one used by the excursion members, he discovered a less appealing city. The city's landscape, he noted as he took a trip to the stockyards, was both hectic and well-developed. It was not empty, awaiting investment from industrialists. Taking a slower form of transportation than the railroad cars that freighted the industrial tour members out to Chicago's fringe districts, he highlighted a built landscape busy with work and activity. He also showed that Chicago was a place of great contrasts. Bourget's carriage passed "hotels which are palaces, and laborers' houses which are hovels. It skirts large plots of ground where market-gardeners are cultivating cabbages among heaps of refuse, and others which bear nothing but advertisements." As if this wasn't enough, the city seemed to go on forever: "The advertising fields give way to more houses, more railways, under a sky black with clouds, or smoke, one hardly knows which, and on both sides of the road begin to appear fenced enclosures, where cattle are penned by the hundred."[31]

Bourget was not alone in viewing industrial Chicago with a jaundiced eye. Two years after the French novelist's visit, the English journalist George Steevens took his own tour. Again, the city's contrasts were plain to see: "Away from the towering offices, lying off from the smiling parks, is a vast wilderness of shabby houses—a large and more desolate Whitechapel that can hardly have a parallel for sordid dreariness in the whole world. This is the home of labour, and of nothing else." Like Bourget, Steevens sensed the seemingly endlessness of the city: "Street stretches beyond street of little houses, mostly wooden, begrimed with soot, rotting, falling to pieces. The pathways are of rickety and worm-eaten planks, such as we should not tolerate a day in London as a temporary gangway where a house is being built. The streets are quagmires of black mud, and no attempt is made to repair them. They are miserably lighted, and nobody thinks of illuminating them." For Steevens, much of Chicago was a dismal place: "All these miles of unkempt slum and wilderness betray a disregard for human life which is more than half barbarous."[32]

While Chicago's boosters portrayed the prairie as pristine and empty, spaces to be worked on, the older built-up districts were worn out long before their time. While the fringe districts were ripe for new productive uses, the desolate central-city factory and residential districts showed signs of significant disrepair and diseconomies. While the metropolitan fringe was ready for investment, the miserable districts of the city faced disinvestment. The divergence between the city's working-class and factory districts as described by the French and English visitors and the empty lands of the territorial fringes promoted on the industrial tours was never more compelling. The contrast of Steevens's and Bourget's view between the images transmitted by Beilfuss, Harvey, Towle, and Brooks could not be more obvious.

Visitors were not the only ones to look askance at the contrasts between the promises induced by Chicago's industrial growth and the city's sordid built environment and social inequities. Numerous local commentators and activists agreed with Steevens's and Bourget's assessment of the city and excavated beneath the earnest facade of the industrial Chicago advertised by city boosters. A quite different picture was presented by the likes of Ernest Bicknell of the local Bureau of Charities; George Hooker, the City Club's secretary; Christian and professional social reformers Graham Taylor, William Stead, Sophonisba Breckinridge, Edith Abbott, and Jane Addams; philanthropic upper-class women groups, such as the Woman's City Club of Chicago; and the novelists Upton Sinclair, Henry Fuller, and Theodore Dreiser.[33]

These middle-class commentators took many opportunities to advertise the problems of housing, congestion, and poverty plaguing the city. Bourget's and Steevens's impressionistic depictions of Chicago were also supported by numerous studies. The best known were the housing reports by Breckinridge and Abbott, and the social-ecological examination of the city by the University of Chicago sociologists. Even some of those taking the prewar industrial inspection tours viewed the development of Chicago with jaundiced eyes. In 1911 one tour guest, Dr. G. Young, the city commissioner of health, echoed the concerns of local social reformers when he stated that "the sanitary conditions which govern the congested districts" posed serious problems. This line of thought was directly plugged into the coalitions of bourgeois women who raised and wrestled with issues of pollution and housing.[34] Together these voices contributed to visions competing with those put forward by city boosters. This in turn led to different assessments of the impacts of industrial growth and an array of solutions that needed to be implemented if the problems were to be dealt with.

Inner-city industrial districts such as the Near West Side became the focus of discussion for the city's elite and the basis for the contrasts with suburban greenfields. From the late nineteenth century, suburban promoters sought to escape what were perceived as deteriorating areas such as the Near West Side, Chicago's largest manufacturing district. Located immediately west of the river across from the Loop, it was bounded by railroad tracks and the South Branch of the Chicago River. A mix of manufacture, residence, commerce, and transportation, it was the city's most congested factory district. As the *Chicago Tribune* noted in 1887, this area "near the south branch of the canal draws houses, stores, machine shops, planing mills toward it as a magnet draws iron filings." By the time of the industrial inspection tours, however, its reputation as a problem area was firmly in place. With its heterogeneous population, numerous ethnic and race-based sub-areas, wide range of housing types, and large unskilled workforce with low and irregular incomes, it was the model of a run-down central district. For many manufacturers, a location in a new factory district on the metropolitan fringe was more preferable than one in blighted areas such as the West Side. For others, it was a place that needed to be repaired.[35]

Despite the proliferation of dissenting views about the impact of Chicago's rapid expansion, the vision put forward by city boosters and the industrial tour organizers proved the most influential. Despite the various voices raised about the social and geographic inequalities generated by capitalist urban-industrial growth, the program of the pro-growth business and political elites was the one that most determinedly drove metropolitan expansion. The drive to transform the empty prairie into an industrial landscape was not to be stopped. From Harvey, South Chicago, the Central Manufacturing District to Clearing and Austin, the practice of identifying fringe land as empty and converting it to productive uses dominated the urban agenda. New industrial territory fanned out along transportation routes in all directions from the central core as local alliances wrestled with and subdued the open prairie. The success of these local alliances, despite the forces arrayed against them, ensured that a large and dynamic manufacturing base was built on the urban fringe. New industrial landscapes were implanted on the prairie and were incorporated into an expanding metropolitan district.

CENTRAL-CITY REORGANIZATION AND THE SUBURBAN SOLUTION

As the different visions of the city suggest, solutions put forward to manage Chicago's industrial growth varied. Rapid industrial expansion forced city builders, industrialists, boosters, and others to search for ways to accommodate the desired steel mills, meatpacking houses, and electrical appliance factories. For many such as Beilfuss, the solution was obvious and uncomplicated. All that was necessary was for farsighted promoters, such as themselves, to appropriate the useless prairie and convert it into productive industrial space. For people such as Bourget, Steevens, Breckinridge, Abbott, and Young, however, the solutions to the dilemmas of large-scale industrial growth in Chicago were not so simple. While the former group favored market conditions and minimal government intervention, the latter one promoted a more proactive state. Out of the tension between those who sought untrammeled industrial development and those who wanted moderate and regulated economic growth developed a bipartite approach to urban form. Although there was general agreement that the city needed to be spatially remade, there was disagreement about how this was to be achieved. Two distinct spatial solutions or fixes were developed to deal with the geographic impacts of Chicago's large-scale geographic economic growth.

One solution was urban redevelopment. By the late nineteenth century, America's urban middle class and business elites realized that the problems inherited from the previous burst of urban-industrial growth demanded greater control and coordination of urban functions and spaces. A better urban society, however defined, was only possible if a new set of principles was imposed upon the built environment. For many, it was clear that escalating environmental and housing problems, growing ethnic and class segregation, and spreading labor dissent could only be solved if the existing city was remodeled. This required increased investment. America's urban middle class and elites stressed that redevelopment involved both the razing and repair of existing structures. This creative destruction was typically framed within a broader plan of renewal or reorganization. Some commentators called for the pulling down of the derelict parts of town and their replacement by new business, transportation, and residential functions. Others called for an upgrading of existing facilities. Regardless, the emphasis was upon the reconstruction and coordination of the city's existing built environment.[36]

Urban reformers understood that solutions to Chicago's many problems were vital to the city's welfare. One element that was viewed as a particular obstacle to better conditions was housing. Little was done, however, as the notion that housing was best left to the market was far too powerful to dislodge. However, spurred on by housing studies made by Robert Hunter, Breckinridge, and Abbott, the Chicago Association of Commerce's Housing Committee in the summer of 1913 tabled a brief on the city's housing conditions. The thin gruel of a report did not pose any important questions nor suggest any useful and practical solutions. The report stressed that Chicago's housing—from rents to supply and conditions—was dismal, and that many individual Chicagoans and city institutions—like the City Club, the Chicago Woman's Club, the City Planning Commission, and local churches and newspapers—did appreciate the problems faced by the city. It also noted that many other cities, such as its rival New York, were taking positive steps to remedy problems. The best that the committee could do, however, was to recommend undertaking another housing survey.[37] This impetus was soon lost to the war economy, and even in the face of the dismal conditions revealed by the existing studies, the association and the city did little to deal with housing problems.

The best-known attempt to reorganize the central city and deal with the city's problems came with Daniel Burnham and Edward Bennett's *Plan of Chicago* (1909). In the words of its authors, the *Plan* was a response to the growing realization that the "formless growth of the city is neither economical nor satisfactory; and that overcrowding and congestion of traffic paralyze the vital functions of the city." Along with city beautification, Burnham, Bennett, and their sponsors, the Chicago Commercial Club, wished to "bring order out of the chaos incident to rapid growth." The solution put forward by "practical men of affairs" was a plan "whereby the city may be made an efficient instrument." While they had regional interests that stretched outside the city limits, the *Plan*'s authors and its supporters were interested in producing "a well-ordered, convenient, and unified city."[38]

While Burnham, the Chicago Commercial Club, and the other local activists sought to address the city's ailments by highlighting how the central city could be changed, a less monumental but no less important solution was put into place on the fringes of the metropolitan area. In Chicago, as elsewhere, an easier, more effective, and, for some, profitable solution to urban problems was industrial and working-class decentralization. In the minds of many, the clustering of factories and industrial workers in the city's existing built-up districts was the major cause of the city's social and environmental problems. According to James Ford, writing in 1992, factories and the working class were "largely responsible for the unhealthy crowding of American cities." More than a decade earlier, the central theme of the Second National Conference on City Planning held at Rochester was urban congestion and its attendant evils. Similar issues were taken up by Graham Taylor and Edward Pratt in their studies of central-city congestion and population and industrial deconcentration.[39]

Many middle-class Chicagoans promoted industrial and working-class decentralization as a solution to the city's problems. In this view, suburbanization would foster orderly and scientific control over urban form and provide a healthier, more efficient, and loyal workforce. Accordingly, local governments, business leaders, and middle-class professionals attempted to forge the various metropolitan components into a coherent whole. The most eloquent voices were the New York Regional Plan's Robert Haig and Roswell McCrea, who equated metropolitan New York to "a piece of productive economic machinery competing with other metropolitan machines." Internally, each piece of the metropolitan machine had to have a specialized role: "The area of New York and its environs may be likened to the floor space of a factory." Rooted in industrial modernist notions of functionality and integration, manufacturing and working-class suburbs would contribute to a balanced urban structure and would allow for an orderly spatial arrangement of urban functions. The move of factories and people to the metropolitan fringe would reduce congestion, lower production costs, and provide workers with better-quality housing and living conditions. Manufacturing and working-class suburbs, it was believed, were essential to the development of a healthy and modern urban society.[40]

Many believed that decentralization could save the central city from itself by building new fringe districts. Once again, Ford was representative of this train of thought. Building on the points set out by the planners associated with the New York Regional Survey, he declared that "if industries could be removed from communities now crowded, to open land, if new industries could be induced to establish themselves in open country, and if the industrial population could be enticed from the city to these industrial villages, the characteristic housing ills of America would be largely reduced." Edward Pratt in his study of New York's congestion approved of the "distinct tendency among manufacturers to remove from Manhattan and to locate in the outlying suburbs." Some commentators stressed the importance of satellite cities—independent areas beyond the city fringe that were a functional part of the metropolitan districts—for industry. In his study of the satellite cities of St. Louis, Birmingham, Chicago, and Cincinnati, Graham Taylor, for example, points to the rewards that decentralization would have as long as it was properly managed.[41]

Influential Chicagoans recognized the need for careful planning of suburbanization. There was little doubt in the minds of the participants of Chicago's industrial inspection tours that the "useless" prairie surrounding the built-up area offered potential relief to the central city's congested conditions. The fact that they could make a great deal of money certainly added to their faith in decentralization. The same ideas were clearly delineated in George Hooker's discussion of Chicago, where he elaborated on the various forms of congestion and talked about the "positive forces" producing the "wide decentralization of industry in general in Chicago." Piecing together the various components of the expanding city into a functional whole, planners together with other professionals actively pursued the building of a new set of rational social and economic spaces.[42]

For others, the reasons were more self-serving. For the makers of the industrial landscape such as George Pullman, the CCCD, and the industrial tour promoters, the suburbs were a territorial solution to their economic problems. To some extent, this was driven by the need of industrial property boosters to profit from the urban fringe land they held in speculation. The search for suitable factory space on the metropolitan area's edge was also motivated by the fact that the older noisy and congested districts closer to the city core were becoming more expensive and inefficient. But it went beyond the simple calculations associated with the profits to be made from property or from the costs incurred from run-down areas that had experienced little long-term investment for a generation or more. Promoters were also looking to reproduce more effective industrial spaces.

Chicago's industrial interests sought to create a new industrial landscape that avoided the past landscape they and their predecessors had previously created. Industrial decentralization allowed them to escape the problems created by the neglect of older factory districts and to evade the costs of any future investment. This older landscape was already in place; it was fixed. The benefits of the upward curve of developing these industrial districts was already captured. At best, these older areas could now only provide small incremental benefits to those who had the entire metropolitan district within their purview. At worst, they were already problematic zones characterized by disinvestment and industrial decline. The larger question of metropolitan Chicago's growth was not the focus of attention. By building factory districts in the open prairie, they would start afresh as well as escape the past, even if they could not erase it.

This was not true for everyone. For owners and managers of manufacturing firms in older industrial areas such as the Near West Side, the character of the districts where their plants were located continued to be of vital importance. For the workers who built up family and kin economies and invested in homes in the neighborhoods surrounding the factories and warehouses where they were employed, the ongoing and everyday events of the locality were their chief concern. For the industrial tour participants, however, it was the future of metropolitan Chicago's industrial fabric and the state of their pocketbooks that were paramount. Members of the Chicago Association of Commerce's Civic Industrial Committee, city councillors, and the railroads believed that they had a larger vision, the building of a new Chicago.

Despite the mundaneness of the basic premise—converting non-urban land to industrial uses by developing vacant land on the urban fringe—the new Chicago, as envisaged by the city's power brokers, was transcendental. A set of development episodes continued to contribute, as Alderman Charles Reading noted, to the "greatness" of the city. The crusading tone was even more explicit in the words of the association's president, Harry Wheeler, who stated in July 1911 that "we have found a territory capable of industrial development which would make the present development of Chicago appear almost insignificant compared with the prospects for the future."[43] The historical geography of metropolitan Chicago was in the making.

Wheeler's claim, of course, was outrageous and was unfulfilled. But the attitude of Wheeler and other business and political elites initiated and shaped the vision of an ever-greater, more dynamic Chicago. The actions and beliefs of the capitalists taking the industrial inspection tours defined what the metropolis was and where it ended. For them, the city and its surrounding areas were lands to be changed through the civilizing impetus of urban-industrial growth. The prairie for Chicago's bourgeoisie was the urban frontier. The not very interesting landscape that Alderman Beilfuss saw from his train window in the summer of 1911 was empty and useless land. Transforming that land would allow city builders to construct the city's massive industrial base.

From the earliest days of Chicago's industrial development, class coalitions propagated images of Chicago that were instruments for their ends. To do this they denied the complexity of the present, provided a narrow interpretation of the past, and projected a heroic one for the future. The lack of complexity lay in the silence about issues of vital concern for the industrial landscape that was already created. This narrowness was rooted in a picture of industrial development that reified the past as a natural step toward progress. The heroic future was driven by a vision of Chicago that claimed an endless bounty of industrial growth if the city elites—and through them, the urban masses—would only grasp the opportunity. The ideological ruminations of the city's boosters reinforced their common images of the relationship between industry and the landscape.

Local alliance members shared a similar vision of what constituted the city-building process between the Civil War and World War II. Framed by a fast-growth mentality, city builders constructed suburb after suburb. Many of the suburbs were middle-class residential areas. Others, however, were industrial suburbs complete with manufacturing establishments, urban infrastructures, transportation systems, and workers' homes. Industrial growth both drove and shaped metropolitan growth. From the older Calumet steel areas and satellite towns to the newer industrial districts coming on stream after 1900, the making of new industrial space was essential to the formation of new urban space and the conversion of the city to a metropolis.

This was not unproblematic, however. Even though their aim may have been the same—transforming space to make a profit—city builders had different interests. Frequently these interests conflicted with each other. Manufacturers' interests often collided with those of the construction industry and the streetcar operators. Regardless, each group was forced to function as part of a local growth machine if their interests were to be met. To this end, real estate developers, members of the Chicago Association of Commerce, city councillors, manufacturers, utility suppliers, and the railroads laid out their common version of the past and present industrial landscape. In the process they established where the future industrial landscape would be built and what it would look like. Defining an image centered on the establishment of a frontier consisting of an "empty" landscape in the surrounding fringe areas of the city was critical to this undertaking.

CHAPTER 3

Four Factory Districts, 1860–1940

For many writers, the beginnings of industrial suburbanization in Chicago and other American cities date from the 1880s, when large modern corporations left the central city for suburban industrial areas. Some, such as Allen Scott and David Gordon, point to corporate vertical integration and the need to escape central-city workers as major factors driving the search for locations outside the built-up area. Historians, such as Stanley Buder and Graham Taylor, argue that suburbanization began in the 1880s with the development of Pullman and South Chicago. But, as Ann Durkin Keating and others recognize, factory districts appeared on Chicago's periphery long before the well-known 1880s industrial promotions taking place in the city's southern sections.[1] The emergence and development of four nineteenth-century factory districts—Bridgeport, the Lower West Side, Packingtown, and Milwaukee Avenue—illustrate the forces underpinning factory-district formation, the changing character of Chicago's industrial space, and metropolitan growth between 1860 and 1940.

FOUR FACTORY DISTRICTS

New factory districts on Chicago's fringe were built in each round of industrial growth after the Civil War and determined the location of new ones. Four important factory areas emerged outside the older central districts by the 1860s. Three of them—Bridgeport, the Lower West Side, and

Packingtown—spread out of Chicago's manufacturing zone running along the South Branch of the Chicago River. The other, Milwaukee Avenue, stretched along the river's North Branch and built on early developments north of the river's main course. Even though all of them were annexed by the city before 1889, their origins were as independent industrial districts outside the city limits. After annexation they remained home to nineteenth-century growth industries and anchored a new generation of urban fringe industrial growth by attracting both new enterprises and firms moving from elsewhere in Chicago.

These four industrial areas provided a solution to the problems of Chicago's nineteenth-century industrial growth. The contradictions thrown up by rapid industrialization forced industrialists, real estate developers, merchants, transportation companies, and the state to rethink the locational logic of manufacturing. This was not an easy matter, for many obstacles impeded the establishment of new industrial spaces. Nineteenth-century industrial sites, which frequently lacked basic infrastructures such as water supply and roads, were outside the main lines of intra-urban movements and were a distance from pools of labor. For business, the costs of searching for and moving to an appropriate site was extremely onerous, both in terms of time and money. New fringe developments, unlike central areas, did not have well-established external economies. Despite these obstacles, manufacturers settled in the four districts. In so doing, they created specialized suburban industrial areas, solved the locational problems of firms, rearranged the city's industrial geography, and established growth corridors that shaped Chicago's geography for decades to come.[2]

When the city of Chicago was incorporated, Bridgeport was a boggy area far outside the city's commercial and residential activity. Momentum for development came with the construction of the Illinois and Michigan Canal in 1836. The early locational assets of the canal and the railroad, large lots of industrial land, and a labor force established the basis for a suburban industrial district. In 1848 the canal was completed, and the town boomed as industries moved into the area. This process was further spurred when Samuel Walker built dock frontage by dredging out a series of slips on the river's north side. The basis for the district, which would eventually run north from the canal to Thirty-ninth Street and between State Street and Western Avenue, had been established.[3]

Bridgeport had a varied industrial base. An assortment of metalworking enterprises opened their doors before the end of the Civil War, some of which combined manufacture for the general trade with the making of specialized products. Weir and Craig, for example, made plumbers', steam, and gas-fitters' supplies as well as a specialty line of meatpacking machinery for local, national, and international customers. Breweries, slaughterhouses, brickyards, and railroads became major employers, while lumberyards, furniture factories, and planing mills also furnished work. Growth continued after annexation in 1863, and Bridgeport remained the home of an assortment of metalworking firms, some of which were new starts, while others were relocations. Clustered around firms such as Weir and Craig by the 1880s were packinghouses and associated industries, the lumberyards, and railroad yards, all of which continued to be important sources of employment for Bridgeport residents into the twentieth century (table 6).[4]

Like Bridgeport, the Lower West Side began as an industrial district outside the city's built-up area and municipal limits. First settled in the 1860s, the Lower West Side, which stretched between the canal, Burlington railroad tracks, and State Street and Kedzie Avenue, took off as an industrial district after the Chicago Fire of 1871, and by 1881 it had several industries, including metalworking and carriages. One of the most important was the McCormick Reaper Works, which moved to Blue Island and Western avenues in 1873 from the city's North Side. But, unlike Bridgeport, a very specialized industrial structure emerged there, centered

TABLE 6. The Industrial Structure of Four Factory Districts, 1881

	Firms									
	Lower West Side		Bridgeport		Packingtown		Milwaukee Avenue		Chicago	
Industry	No.	LQ	No.	LQ	No.	LQ	No.	LQ	No.	LQ
Clothing	1	.1	0	0	0	0	5	.6	165	1.0
Printing	0	0	0	0	0	0	3	.2	283	1.0
Furniture	6	.5	0	0	0	0	17	1.3	238	1.0
Leather	4	.5	1	.5	0	0	21	2.8	144	1.0
Food	6	.5	5	1.5	1	.4	13	1.1	228	1.0
Lumber	86	5.1	10	2.0	1	.3	29	1.7	327	1.0
Meat	3	1.1	3	3.9	20	39.2	8	3.0	51	1.0
Metal	18	.6	9	1.0	0	0	27	.9	579	1.0
Chemical	1	.2	1	.6	3	.9	4	.2	111	1.0
Carriages	10	1.2	2	.8	1	.6	6	.7	164	1.0
All	145	1.0	42	1.0	27	1.0	148	1.0	2,772	1.0

SOURCE: Author's calculations from the *Lakeside Annual Business Directory for 1881* (Chicago, 1881).

around the lumber industry. Some lumberyards had moved out of the Loop by the early 1850s and lined the west bank of the South Branch stretching as far south as Twenty-second Street. But by the late 1860s, the industry experienced decisive locational shifts and the Lower West Side became the city's lumber center. Firms first transferred their operations to the New Lumber District on the north side of the canal. More than four miles long and with 500 acres of lumberyards, the New Lumber District developed as the world's largest lumber market. By the mid-1880s, the industry shipped more than 3 billion feet of lumber and shingles to local and national markets. Growth was accommodated by the excavation of canals, slips, and a dock front of 12,000 feet by the South Branch Dock Company. Each lot had a dock, street front, and switch track.[5]

The lumber industry's move to the Lower West Side refashioned the industry's geography. In the mid-nineteenth-century, all of its functions (office, storage, processing) were to be found alongside the river in the Loop. In a couple of bursts, this unraveled as the trade moved away from a business district increasingly populated by commercial, legal, financial, retail, and entertainment functions. While many lumber offices remained in the Loop, storage and processing functions shifted to the Lower West Side. By 1871 the district extended as far west as Ashland Avenue, was connected with the main tracks and yards of the Chicago, Burlington and Quincy, and had an interchange with other Chicago railroads. Moreover, a related set of woodworking industries, such as planing and sash and door mills and carriage works, clustered around the lumberyards.[6]

After a generation of expansion, the problems of congestion, declining conditions, and high dockage rates forced firms to seek other locations. From the 1880s, firms migrated from the main lumber docks to new quarters in the vicinity of Thirty-fifth Street and the stockyards. By 1891 new lumber docks lay along the north bank of the Calumet River, while growth of the city toward the north led to the establishment of yards at several points along the North Branch. Despite these changes, the Lower West Side remained an important location for the lumber trade into the early twentieth century.[7]

The Lower West Side's lumber industry was intimately tied to Chicago's role as the Midwest's major distribution center and financial nexus. The lumberyards stretching along the South Branch were tightly interwoven with a production chain stretching out into the resource hinterland. Sawmill operators fell and cut logs from the great forests of Wisconsin, Minnesota, and Michigan and shipped them to Chicago. Once there, the business of selling the shiploads of unplaned, unsorted, and ungraded lumber took place at the downtown wharves. The lumber was consigned to commission brokers working out of the Loop who represented the mill owners and, for a fee, sold to lumber merchants and manufacturers. The new owners shipped the goods to the yards lining the riverbanks, where it was unloaded and placed into huge storage piles.[8]

The next step of the chain involved the selling of newly cut and semi-processed wood to other producers. While a great deal was shipped out of the city to regional consumers, a large and, from the 1880s, an increasing share remained within Chicago. The local industry had very strong and direct links to a variety of semi-intermediate producers, such as coopers, planed wood, sash, and box manufacturers, furniture makers, and builders. As one observer noted in the mid-1880s, "The lumber brought here was dried and dressed for flooring and sidings and made ready for use in building before it was shipped to other points. There were also many large establishments for making doors and sash and blinds; and others still were heavily engaged on the manufacture of every line and quality of furniture, and organs and pianos." In the Lower West Side, this assortment of industries, which functioned as a very specialized knot of businesses, found their major source of raw and semi-processed materials. Tying the local and regional lumber industry together were several trade organizations. The Board of Trade regulated lumber inspection between 1855–57. The wholesalers then formed the Lumberman's Board of Trade of Chicago in 1857 and the Lumberman's Exchange of Chicago in 1868. The lumber manufacturers organized the Lumber Manufacturers' Association in 1859.[9]

Even more specialized than the Lower West Side's industrial mix was the one-industry district of Packingtown, the area lying between Thirty-ninth and Fifty-fifth streets and State Street and Western Avenue (see table 6). One of the most infamous factory areas in the United States, Packingtown illustrates how the creation of a big-firm-led factory district established the parameters of the metropolitan area's social and economic features for several generations. The district started in 1864 when a Chicago entrepreneur, politician, editor, and speculator, John Wentworth, along with his father-in-law, Riley Loomis, sold 40 acres for $100,000, for which they had paid $74,000 in 1852, to the recently created Union Stock Yard and Transit Company as a site for the city's new and consolidated stockyard. Wentworth was a typical urban booster of the period, with interests in a range of city business, social, and political affairs. Born in 1815 in New Hampshire, he moved to Chicago in 1836, where he opened a law practice, worked for many years as a newspaper editor for the *Chicago Democrat*, supported the Democrat and Irish nationalist causes, and was a six-term congressman and two-term Chicago mayor. Working with various entrepreneurs, such as Stephen A. Douglas and William Ogden, he made a fortune through large land developments in the city and the suburbs, including the sale of the stockyard land. One of the city's foremost railroad promoters, he was responsible for bringing the Illinois Central and Michigan Central lines into the city. He died a millionaire in 1888.[10]

Wentworth's sale of the land in the suburban town of Lake in 1864 made possible the development of what became one of the metropolitan district's major industrial anchors. The suburban land, which became Packingtown, was not a piece of the built-up part of the city in the 1860s. Eight miles from the city's center, the area was, as one commentator noted, "a reedy, undrained marsh, remote from the inhabited part of the city, and not regarded as a quarter which would ever be suitable for residence or business purposes." The Union Stock Yard Company's developers quickly got to work, and when the stockyard opened in December 1865 not only was the marsh drained, but business on a scale never envisaged for Chicago was born. A seemingly unsuitable area was transformed into a leading industrial territory.[11]

The building of the stockyard on the urban fringe at the end of the Civil War was a response to the inadequacies of the existing distribution of the trade. Early slaughterhouses and packinghouses were centrally located along the Chicago River, facilitating the shipping of preserved meat and supplies, such as salt and barrel-making materials. The city's central area, however, became increasingly untenable. The land-extensive, dirty, and noisy trade clashed with the desire for a respectable central district, while the economic and geographic costs of centrality rose as the city grew and the trade became larger. Grazing and holding on the edge of the city became necessary as the city grew before the Civil War, and taverns on the highways leading out of Chicago began to offer pens and pastures. The opening of the Illinois and Michigan Canal in 1848 and the building of the railroads from the late 1840s led to the modernization of the trade and construction of more slaughterhouses and packinghouses along the river's South Branch. The modern system of packing—with commission men, market reports, hotels, scales, and rail connections—was put into place near each railroad terminal. Even though the packers were not highly clustered, they did share similar sorts of sites along the branches of the Chicago River and in the suburbs.[12]

Despite reorganization, the packing industry and animal stocking by 1860 presented an incoherent and irrational locational pattern. As one writer noted, Chicago's growing share of the expansion of demand for packed meat "proved that the many separate stockyards located on the various railroads were not adapted to efficient and economical trading and handling." Two reasons account for this state of affairs. First, as the city grew and the distances between yards became longer, it was increasingly more difficult and expensive to make comparisons of quality, volume, and price. Second, Chicago's population and economic growth led to congestion along the routes to the various yards. There was little room for railroad expansion, and the old stockyards became hemmed in by factories, stores, and homes. The ability to function effectively and competitively was severely compromised.[13]

The solution imposed by the railroads was the construction of one large, consolidated stockyard located outside the city yet accessible to the railroads and the packers. The building of a more effective industry involved redrawing the geography of animal holding and meat production. By the Civil War, according to one observer, "there was a general demand from all elements of the trade for a new large Union Stockyard where the supply of livestock could be concentrated and all sellers and buyers could meet in open competition." Commission men and packers wanted a centralized location so that they could better monitor changing market conditions. The railroad companies were interested because of the lucrativeness of the livestock trade and anxious to deliver and receive cars over a line that would connect all of their lines with the consolidated market to avoid excessive switching. Following the lead of the Chicago Pork Packers' Association, the railroads began negotiations for consolidation, and by the mid-1860s Chicago's packing associations and nine main railroads joined forces. They abandoned the separate yards on the separate lines and financed a new yard. Of the $1 million initial capitalization, $925,000 was from the railroads. The Union Stock Yard opened in 1865, and within a few days the other stockyards had closed.[14]

The new stockyard was gigantic. It accommodated 100,000 head of stock, while planked pens covered 120 acres, a network of streets crisscrossed the area, 15 miles of tracks—plus switches, turntables, water tanks and wood yards, unloading chutes traversed the districts, and 50 miles of sewers carried surplus to the Chicago River. In contrast to the older ones, the new stockyard was constructed in an orderly and rational manner. The yard was divided into four sections; three for receiving livestock and one for shipments east. Each railroad was assigned to a particular loading and unloading quadrant, although all of them shared the tracks.[15]

The opening of the stockyard did not have an immediate effect on the packers. Located two and half miles to the north on the river where they were well established, it was several years before they formed a distinct cluster around the yard. The packinghouses began moving out of Bridgeport and other Chicago locations to the stockyard district in the early 1870s. In order for this relocation to happen, additional land had to be purchased adjacent to the yard. The land was sold outright to the packers, but the Union Stock Yard Company retained control over the district through continued ownership of streets and railroads. Of the 40 Chicago-area packing firms at the end of the 1870s, about half remained in the city because they served local demand, had good and efficient plants, and had no trouble satisfying city and state sanitary regulations. The rest clustered around the yard.[16]

The locational switch was particularly advantageous to the large packing firm. The packers who moved to the Packers' Addition had ready access to raw materials, while saving on intra-urban switching and drayage costs. The stockyard was still outside city limits and thus outside city control, especially those laws prohibiting cattle in the streets and imposing restrictions on rendering plants. The district also provided cheap land and immediate access to the stockyard railroad tracks. Packers could take advantage of the expanse of suburban land to construct big packinghouses with the latest production processes, a task that was difficult to achieve in the cramped central city. Migrating firms accounted for the dramatic increase in the slaughter industry during the 1870s. As early as 1875, meatpacking was the most important industry in the city, and Packingtown—a one-industry, big-firm area—was its most important district.[17]

In contrast was the Milwaukee Avenue factory district, where local and non-local capital investment built a significant manufacturing district on the urban fringe. The origins of the district, which is commonly known today as the Clybourn Corridor, dates from 1857 when Captain Eber B. Ward established the Chicago Rolling Mill on the river's North Branch with the aid of three Boston capitalists. The Chicago area's first iron mill, the company grew over the next 30 years. By 1884 the firm operated several plants, had expanded the original mill, and employed 5,800 workers, many of whom continued to work in the North Side mill. The Milwaukee Avenue plant became a magnet for many other businesses attracted by the advantages of locating close to the mill. The Chicago Foundry, for example, was not only geographically close to the Chicago Rolling Mill, but provided it with heavy castings.[18]

A large factory district coalesced around the iron firm from the late 1860s. The area bounded by Chicago and Diversey avenues running parallel to Milwaukee Avenue and the North Branch of the Chicago River was home to a wide assortment of factories, mills, and workshops of all sizes and from an array of industries. Taking some of the spillover from the factory area to the south, Milwaukee Avenue benefited from Chicago's rapid post–Civil War commercial and industrial development. Even though the district lost several industries after the 1871 fire, others remained. By 1881 the district had 148 firms and a specialized industrial structure (see table 6). Other than the Loop, it was Chicago's major leather district, having most of the city's tanneries as well as numerous shoe, belting, and baggage firms. Several lumberyards, running along the river, supplied local furniture factories and builders with various grades and types of lumber, while packing and metalworking firms employed large numbers.[19]

Milwaukee Avenue's industrial growth and specialization took various forms. One was *in situ* growth. The Chicago Steel Works, for example, began as a small proprietorial car spring maker. Bought out and reorganized by four new investors in 1873, the firm built over time a sprawling plant consisting of a foundry, machine shop, blacksmith shop, and rolling mill. A second way was by attracting firms that switched from commercial to manufacturing activities. Pettibone Mulliken Company, for example, was organized in 1880 as a dealer in general railroad supplies. Six years later, sensing that profits were better on the manufacturing side, the general supply business was sold and the company moved into the manufacture of railway track supplies in a new district shop.[20]

Finally, the relocation of manufacturers from other districts fed Milwaukee's growth. While the district received few firms from the clothing, printing, and jewelry industries, it did attract those from industries lured by the district's locational assets. In the 1880s, for example, Crucible Steel Casting bought land and moved its manufacture plant from the South Side to larger premises three miles out of the Loop. Gaining more space was just one factor enticing the company to the northern reaches of the city. There were others. A railroad location provided quick transportation, especially as the company had a railroad switch and dockage on the Chicago River. A labor force with a range of skills had developed in the surrounding neighborhoods, and new metalworking enterprises drew on a well-trained German and Swedish workforce.[21]

Bridgeport, the Lower West Side, Packingtown, and Milwaukee Avenue were just four of several factory districts that grew up along the river in the nineteenth century. All of them were initially established outside Chicago's city limits, and all were transformed from empty prairie into important industrial areas. First established before or during the Civil War, these and other nineteenth-century factory districts shaped the character of Chicago's industrial geography for generations. They were part of the solution that the local bourgeoisie put into place in response to nineteenth-century industrialization. As the previously occupied industrial districts became inadequate, new ones on the metropolitan fringe became home to the growth industries of the day. Expansion continued after 1900 (table 7). With 138,000 workers employed in 1,100 firms in 1924, the four districts maintained their specializations: lumber and ancillary industries dominated the Lower West Side; Milwaukee and Bridgeport remained relatively mixed, although each had their own specialties; and Packing-town was still the city's and America's major meat-producing district.

CAPITAL SWITCHING AND FACTORY DISTRICTS

By the early twentieth century, a combination of industrial changes, competition from other factory districts, and worsening social and environmental conditions led to the restructuring of the four districts' place in Chicago's industrial geography. Traces of decline can be discerned in these districts even as they became the solution to the city's rapidly expanding industrial problems after the Civil War. By the interwar years, they at best had reached a plateau of growth, and at worst were experiencing the first tremors of deindustrialization. Without exception, all of them were transformed from the answer to the locational needs of the nineteenth century to the problem areas of the interwar period. All of them, although to differing degrees, were recast as outmoded, inadequate, and run-down factory districts. The transition from suburban solution to an inner-city problem did not happen overnight nor did it occur in a seamless linear pattern. But over the course of two to three generations, these old suburban districts became less attractive to those investing in the urban landscape.

One reason for the decline was the shifting nature of industrial capital. This was illustrated by the erratic but nonetheless ongoing disinvestment in Packingtown's meatpacking industry from the 1880s. The intense competition among packing companies alongside large increases in land prices made it virtually impossible to find affordable land. In addition, there was little ongoing investment in Packingtown's physical environment, and its old streets, roads, and railroad tracks became increasingly run-down. Adding to the problems were the growing production capacity and attractiveness of Chicago's major competitors, Kansas City, East St. Louis, and Omaha. From the 1880s, the receipts and number of slaughtered animals of these centers increased relative to Chicago's, which remained relatively

TABLE 7. The Industrial Structure of Four Factory Districts, 1924

Industry	Lower West Side			Bridgeport			Packingtown			Milwaukee Avenue		
	Firms	Employees		Firms	Employees		Firms	Employees		Firms	Employees	
	No.	No.	LQ	No.	No.	LQ	No.	No.	LQ	No.	No.	LQ
Lumber	46	4,097	4.5	10	1,428	2.0	12	692	.4	19	2,662	3.1
Furniture	24	5,268	3.7	17	1,700	1.5	4	54	.1	40	2,353	1.8
Textiles	13	1,473	2.6	7	181	.4	2	45	.1	16	313	.6
Transport equipment	2	3,400	2.4	1	53	.1	7	1,573	.6	0	0	0
Machinery	20	2,083	1.3	14	1,450	1.2	5	433	.2	30	1,548	1.0
Fabricated metal	51	3,937	1.1	29	1,803	.6	16	1,020	.2	42	3,207	1.0
Non-Metallic	9	389	1.0	3	86	.3	5	341	.5	10	590	1.6
Automotive	13	801	.9	17	742	1.0	4	278	.2	6	60	.1
Food	23	1,410	.7	17	2,825	1.7	13	1,594	.4	23	2,543	1.3
Paper	9	420	.7	10	3,250	7.1	4	112	.1	13	1,162	2.1
Meat	3	1,270	.6	22	4,866	2.8	25	43,361	11.0	2	657	.3
Chemicals	16	268	.4	25	1,470	2.6	15	841	.7	23	717	1.1
Leather	2	108	.4	1	0	0	1	15	.1	19	2,662	11.7
Foundries	16	1,434	.3	10	686	.2	6	604	.1	9	560	.1
Clothing	9	456	.2	4	87	.1	1	40	.1	60	3,136	1.7
Shoe	0	0	0	0	0	0	1	70	.1	8	1,044	3.7
Other	59	3,459	.4	39	3,572	.5	16	3,472	.2	115	5,369	.7
Total	315	30,273	1.0	226	24,199	1.0	137	54,595	1.0	425	28,583	1.0

SOURCE: *Directory of Illinois Manufacturers, 1924–1925* (Chicago: Illinois Manufacturers' Association, 1924).

stationary. Together these factors ensured that the district faced increasingly difficult times from the end of the century.[22]

Despite relative decline, Chicago remained America's major packing city. In 1914 the value of Chicago's meat industry still equaled the combined total of its three competing cities. Nevertheless, the shifting character of capital investment signaled that little future employment growth was likely in the stockyard and packinghouses. Hopes that new industries would develop in the neighborhood and provide work never materialized. In 1924 almost 80 percent of its 54,000 workers were employed in the packinghouses. Things had not changed much by 1940. In that year more than three-quarters of the district's workers labored in the packinghouses while another 5 to 10 percent were employed in ancillary firms. Packingtown's fortunes resolutely remained tied to its meatpacking industry.[23]

Packingtown's status within the industry was not helped by the frequent struggle over production space and profits. This was most evident when the meat packers decided in 1890 to move their works to the Calumet district. The packers considered moving because they were unable to gain control over the Union Stock Yard, which would allow them to increase profits by reducing terminal charges and underselling the small packers. In addition, growing discontent in Chicago about the district's smells and polluted waterways provided ammunition to those wanting to restrict the activities of the industry and to expel the packers from the city altogether.[24]

The situation became critical when the Vanderbilt family and certain English interests bought the stockyard. The big three packers were concerned that they would be strangled by their old railroad enemies. In 1890 they organized the Chicago and Calumet Stock Yards Company and announced their intention to move the livestock market to Tolleston, Indiana, where they had purchased several thousand acres. This forced the stockyard to negotiate a settlement. The packers not only got most of what they wanted in return for staying in Packingtown; they also retained the land in Indiana, most of which they sold to U.S. Steel to build Gary's steel mills.[25]

A similar set of processes played themselves out in the lumber district as Chicago's place in the regional lumber nexus changed after 1880. The local trade began to decline as regional sawmill operators no longer depended to the same degree on Chicago wholesalers. In response, lumber operators closed down their Chicago yards or cut back operations. The substitution of metal for wood also had it effects. As this was an incremental process, the effects on the lumber industry were slow, but, nevertheless, accumulative and felt by most industries. Chicago's cigar box manufacturers, in conjunction with some of the lumber wholesalers, for example, responded to the substitution of tin for wood after World War I by creating the Wood Cigar Box Boosters' Club. Despite these trends, the Lower West Side remained an important, even though declining, center for lumber. Chicago's lumber interests continued to service regional markets, but as the regional trade dropped, however, they shifted their focus to servicing Chicago's rapidly growing local demand.[26]

The vicissitudes and changing geography of capital also affected Bridgeport. One of the most important firms in that district was Acme Steel Goods. Starting in a small way in 1880 as a maker of barbed box straps and other shipping supplies, by the 1920s the company was a leader in cold rolled strip steel, barrels, and shipment packing. Between 1880 and 1918, its main plant was at Bridgeport, where output was limited to cold rolled strip steel and miscellaneous manufactured products. In 1918, however, the firm built a new plant at Riverdale because "its business had grown to such proportions that the need for a more dependable and reliable source of raw materials." Not only did it want to have closer access to Calumet's steel mills from which it received its main raw materials; it also desired larger premises and needed to expand its production lines to meet the increasing diversified demand for steel by the automobile, furniture, and hardware industries. This had important implications for Bridgeport. The new plant's large capacity produced important economies in raw material costs, thus ensuring that the ensuing economies of scale produced cheaper rolled steel. With the shift in investment away from Bridgeport to Riverdale, the old workplace was reduced in scale and importance by the 1920s. It was a barometer of the central factory district's economic decline.[27]

Further contributing to the old suburbs' relative decline before 1940 was that many of the industries driving earlier factory-district formation experienced little to no growth after 1900 in comparison to the new leading industries of the twentieth century, many of which were seeking locations in new spaces. The meatpacking and lumber industries, for example, were ranked second and fifth of all industries by number of production workers in 1880. By 1924 they were ranked fifth and thirteenth, respectively. In contrast, electrical equipment and vehicles rose from twenty-first to third and twenty-third to twelfth, respectively.[28] Little changed by the end of the Great Depression. In other words, the nineteenth-century industries that produced and sustained the old suburbs' factory districts over a generation or more were no longer important employment generators. Nor did these older factory districts become home to jobs in the early twentieth-century growth sectors.

Chicago industrial geography before 1940 was reconfigured as many new starts skirted older districts and old firms relocated to new districts. Before World War I, this was evident to astute observers. In 1910 City Club secretary George Hooker stated that "within the last few years a very marked outward movement of industrial plants has been taking place." His idea that Chicago had experienced a great deal of industrial decentralization was confirmed five years later by Graham Taylor in his study of satellite towns and industrial suburbs, where he pointed to the fleeing of firms from Chicago's central districts from the 1880s. This point was reinforced a generation later by William Mitchell in a study of Chicago's industry. In a systematic examination of Chicago's locational trends in the 1920s, he highlighted the displacement of industry from the central city as industrialists sought suburbs where they could find the "optimum combination of rural and urban locational advantage." Moreover, he showed that a "significant proportion" of the 249 relocating plants moved from the older parts of the city to the "suburbs immediately outside the city limits." This pattern of central-city factory-district relative decline continued through the 1930s and early postwar period.[29]

The territorial moves of one of Chicago's biggest firms and the nation's largest valve maker, the Crane Company, illustrates this reality. In 1913 the ground was broken on a huge tract of property lying at Kedzie and Thirty-ninth. The Chicago Works, the name given to the new complex, was the firm's latest attempt to establish a viable locational fix. With 49 buildings and a total floor space of 50 acres, more than twice the size of the old works, the Kedzie complex was a monument to the latest in factory building and territorial expansion. Implementing some of the latest factory design principles, the architect and engineers planned factory buildings that were laid out "to give an uninterrupted movement from the time the raw material is received until the finished product is shipped out." Not only did the design meet the firm's needs at the time; it also catered to the future: "The layout is so designed that buildings may be added to take care of the future growth of the business without changing the general scheme."[30]

The Chicago Works was the last in the line of factories that Crane had possessed since the 1850s. The company was established in a small way in the brass business in 1855. Over the next couple of years, several moves were made, ending at a small three-story frame building on Lake Street. The plant was enlarged in 1862 to accommodate a foundry for the manufacture of steam machinery. Two years later the company expanded by adding the first wrought-iron pipe mill west of Pittsburgh and a malleable iron foundry. Even this massive program of extensions was not enough to keep up with demand, and in 1865 the company built their South Jefferson plant. With several foundries (gray iron, brass, and malleable iron), an annealing department, various shops (machine work, blacksmithing, pattern making, steam fitting, brass finishing, and iron fittings), and the company offices, South Jefferson became the manufacturing heart of the expanding Crane corporation. So successful was the company that it made numerous additions between 1870 and 1874.[31]

Production and plant expansion continued apace. As early as 1880, the pipe mill erected in 1864 was too small. By the early 1880s, however, the company's expansion plans could no longer be accommodated by building new additions to the existing complex. After scouting out various sites, the company elected to build the new mill at the corner of Canal and Judd. By 1891 the valve and fittings business consisted of three separate divisions with their own group of buildings: malleable iron fittings, cast-iron fittings, and the brass department. Over the next 20 or so years, the company added and dropped lines, and opened and closed buildings at its Jefferson complex.[32]

By 1913, however, Crane was once again running into problems. With a sprawling set of buildings constructed at different times over the previous 50 years, Crane needed to rationalize its space. The firm was finding it increasingly difficult to coordinate a complex production process, supervise a growing and increasingly militant labor force, and accommodate a changing and bewildering array of product lines in its haphazardly designed set of buildings. Moreover, the older neighborhoods along the South Branch of the Chicago River became less suitable for large manufacturing concerns as wholesale, financial, and retail functions pushed up land prices and stable working-class residential areas were undermined by blight and slum development.

Under these conditions, the company had two options. One was to rationalize manufacturing operations as best they could within the existing buildings. The other was to rebuild in a location outside the existing built-up manufacturing district. They chose the second. The new Kedzie complex with its new design principles and location on the open prairie represented, as the Judd Street plant did in the 1880s, the merging of a firm's need for manufacturing space with the territorial ambitions of a metropolitan manufacturing bourgeoisie. Keeping a few old downtown properties and consolidating all office functions on South Michigan

Avenue, the company had built new factory complexes in different parts of the city during Chicago's different industrial cycles.[33]

As the Crane case suggests, newer metropolitan fringe industrial districts outpaced the older districts. A new set of suburbs emerging as major industrial centers between 1890 and 1930 contributed to the reconfiguration of the older districts as second-rate industrial areas. There had always been a locational hierarchy of place in Chicago. At any one time, some districts were more structurally favored than others. Chicago commentators after the turn of the century advocated industrial suburbanization as a panacea to the troubled central city. From the early twentieth century at the latest, the growth of new districts and the resulting reorganization of metropolitan Chicago's industrial geography reinforced the declining position of nineteenth-century factory districts such as Bridgeport and the Lower West Side within the metropolitan area's hierarchy. This was particularly the case as many new firms located in satellite towns and industrial suburbs, such as Whiting, Joliet, Elgin, and Cicero.

In Whiting the establishment of the Standard Oil refinery in 1889 was an important moment in Calumet's industrial development. Like much of the surrounding area, Whiting was uninviting for settlement. The area consisted of a mixture of swamp and sand dunes, and remained a sleepy hamlet until 1889 despite the railroads that opened up the area to eastern and western producers, the low land prices that attracted land speculators, and the growth of neighboring Chicago. By the 1880s, however, the lands south of the city were becoming increasingly attractive. For Standard Oil, Chicago—with excellent shipping and rail systems, abundant water supply, and a large market—was an advantageous location. Once the firm had decided on Chicago as the site for its new refinery, executives had to make a choice about where to actually build the plant. Standard chose Whiting, and not a more central factory district, because it could assemble a modern refinery on a site linked to Lake Michigan, the railroads, and the company's Lima–South Chicago pipeline. Over the next 40 years, the company expanded production, increased the size of the labor force, and, along with Sinclair Refining in neighboring East Chicago, attracted several chemical firms dependent on petroleum-based products for inputs to their production process.[34]

While Whiting's fortunes were dictated by the fortunes of one industry, other industrial developments were more mixed. Joliet's industrial development took off in the boom of the 1870s, and it continued to attract capital and firms for many years. Its industrial origins lay in the establishment of Joliet Iron and Steel and a few other factories. Over the next 50 years, growth continued in bursts, stimulated by the promotion and construction of the Outer Belt Line in the late 1880s, the Sanitary and Ship Canal after 1900, and the building of several inter-urban railways. Industrial development was fed by the growth of Joliet Steel (later absorbed by Illinois Steel) and other forms of industrial expansion, most notably improvements to local steel plants, the consolidation of several wire plants into American Steel and Wire in 1899, and the opening of a U.S. Steel by-product coke plant. These and other local firms had direct and indirect ties with Joliet Steel. By 1924 the town boasted 85 manufacturing firms and almost 14,000 workers.[35]

Elgin's industrial development took off with the establishment of Elgin National Watch after the Civil War. Originally established in Chicago in 1864, the firm moved to Elgin three years later. Small at first, the firm grew rapidly at the end of the century. Bursts of town expansion were related to additions to the watch firm, the move of Cook Publishing from Chicago, and the building of several substantial industrial concerns. A specialized industrial base developed. Firms related to Elgin National Watch (Illinois Watch Case, Elgin Clock, and Elgin Silver Plating) and ancillary firms (such as Western Casket Hardware and Elgin Metal Novelty) located in the town. In the process they created a strong industrial fabric around specialized metalworking and silver plating. Transportation developments added to Elgin's industrial attractiveness. Closely linked by railroad connections to the metropolis and the resource-rich Midwest, and tied by financial flows to Chicago banks, Elgin was a haven for twentieth-century entrepreneurs eager to invest their capital in safer and less run-down sites than those characterizing many Chicago factory districts.[36]

Cicero's development dates from the late nineteenth century when the Grant Locomotive Works established a plant in the suburb's northern section. Growth was slow, but by 1900 a small industrial base was in place, which was fed by the inexorable westward movement of Chicago's industry, especially Western Electric's move from Chicago to Cicero in 1903. Cicero, unlike the two satellite cities Joliet and Elgin, was largely a one-industry town. Electrical machinery, machinery, and steel product firms moved to Cicero to be close to Western Electric, while other firms migrated there for its labor force, lower taxes, cheaper land, minimal regulations, and excellent railroad facilities. In combination, these locational factors made Cicero an attractive place for firms wishing to leave central-city districts such as Milwaukee Avenue and the Lower West Side. By World War I, the western suburb was an important factory district and a competitor to the run-down inner-city factories.[37]

Not only factories became run-down. Older factory districts also were challenged to maintain decent housing and environmental conditions. By the end of the nineteenth century, second-generation immigrants moved out of Bridgeport or Milwaukee Avenue to better housing in the newer residential districts of the city and working-class suburbs such as Hammond. As they did, they left behind large swaths of cheap and run-down housing. Poorly constructed when first built, it did not age well. In addition, congestion increased as the original housing was replaced or divided up. In some cases, single-family, working-class frame cottages were torn down and replaced by multi-unit tenements. In other cases, as Edith Abbott made clear in her 1936 study of Bridgeport's Lithuanian section, "the old dwelling intended for one family has been requisitioned for two or three families that were old, more or less run down."[38]

Other environmental problems plagued the central factory districts. The districts' waterways, a major locational asset, were polluted and hazardous, while the garbage dumps west of Packingtown transmitted their malodorous and toxic effects. All four Chicago districts were near the bottom of the scale of most environmental and housing indexes as measured by infant mortality rates, education levels, and rooms per family. This was not lost on contemporaries. A map showing the geographical relation of industry to poor environmental conditions in Charles Bushnell's 1902 study of Packingtown made this very point. All four factory districts, along with the steel-producing district of South Chicago, had high rates of child mortality, overcrowding, and economic distress. For Bushnell, the "infinitely tragical struggle" of people living in these factory districts was "not only a menace to those finer dreams of a noble, joyous, and beautiful national life, but a threat even to the very essentials of a common and decent civilization itself." His sentiments were reinforced a year later by Ernest Bicknell, who pointed out that these neglected and dilapidated districts contained a large proportion of people living in poverty and run-down housing.[39]

The four districts' poor environmental and housing conditions were directly related to their inhabitants' low incomes and high rates of unemployment and underemployment. Packingtown, as one early 1920s study noted, was "a community of laborers, the majority of whom are scarcely able to provide their families with the barest necessities on the wages received. Many fall below the borderline when any complication such as accident, sickness or unemployment come [sic], for the meager earnings of these people do not allow for savings against the day of adversity."[40] Economic conditions did not get better during the 1920s despite the boom of that decade. Conditions only worsened during the 1930s.

The decline of old factory districts, as Bushnell and Bicknell pointed out, was related to their position at the losing end of the vicious cycle of neighborhood disinvestment. Industrial out-immigration, increasing competition from other metropolitan areas, poor housing, and declining environmental conditions contributed to the disdain toward the four districts felt by those with the capital to invest in housing or infrastructures. This in turn only further undermined the attractiveness of the districts for new capital investment in factory plant. Many Chicago commentators took this one step further by linking poverty and poor conditions with social values. With the nineteenth-century inner-suburb districts in mind, Charles Wacker, chair of the Chicago Plan Commission, pronounced in 1921, "Congestion breeds vice, crime, and disease. There is a direct relationship between crowding and disease and crime."[41]

He was not alone in equating poverty and poor living conditions with a population's social values, despite the work of some of his contemporaries to counter such fallacies. Many influential Chicagoans promoted this idea. As environmental and housing conditions remained bad or got worse, especially in comparison to new districts, the tendency to equate morality with the physical conditions of the districts became stronger. In the process, the discourse reinforced the trend in which these poor neighborhoods were viewed as run-down and not attractive to capital investment.[42] The growing body of professionals—notably social workers, planners, and housing reformers—reinforced the growing polarization between old and new industrial spaces. Indicative of the representations of the older space was Faith Miller, who described Packingtown in 1923 in this way: "Situated as it was at the very back door of the 'greatest slaughter houses in the world,' with obnoxious, disease breeding Bubbly Creek on one side and the city dump on the other, this congested community of foreigners presented a serious problem of health and morals."[43]

Despite the good intentions and efforts of reformers, politicians, and planners, metropolitan expansion of twentieth-century Chicago was not, as Miller's words made quite clear, orderly and scientific. Along with the social, ethnic, race, and class tensions that urban managers feared, urban growth proceeded hurly-burly as population and economic activities spilled haphazardly over existing city boundaries and set up homes, offices, and factories in surrounding independent suburban municipalities. In response to these social tensions and urban expansion, professionals reframed the image of urban districts by World War I if not earlier. While new suburbs were viewed as the panacea, older districts such as the Lower East Side and Milwaukee Avenue were represented as the problem. No longer the throbbing heart of the expanding suburban belt or the solution to central-city problems, they were inner-city neighborhoods where social pathologies were concentrated.

The four districts' changing economic and social conditions reinforced the idea that they were part of Chicago's belt of inner-city blighted areas. The response of some, like Edith Abbott, was to create employment, maintain a sense of local identity, and rebuild the built environment. The response by many, however, was to raze the areas. Again Wacker voiced the concerns and opinions of many other middle-class Chicagoans when he stated that "slums must be wiped out and in their stead there must be created districts made healthful by sunshine; made invigorating by fresh air; and made pleasant by places of recreation." Similarly, while his agenda may have differed from Wacker's, a meatpacking company's lawyer put it succinctly when he declared in 1918 that the only solution to Packingtown's housing conditions was "absolute destruction of the district. You should tear down the district, burn all the houses."[44]

The seeds of decline were already planted by the end of the nineteenth century. The combination of capital investment strategies that favored new sites with a host of central-city problems, including rising property values and growing congestion, forced many managers to seek new factory districts away from the built-up core. The housing, infrastructures, and environment of the older districts were poor when they were built. The absence of investment did little to make these districts competitive. As immigrant working-class districts, they received trivial amounts of public investment. At the same time, the combination of landlords seeking the highest returns and resident owners unable to reinvest heavily in the upkeep of their housing only reinforced the poor conditions of the older suburbs. By the end of the nineteenth century, the move of capital from old industrial suburbs such as Bridgeport and Packingtown, which were now inner-city districts, to the newer industrial suburbs such as Cicero and the expanding industrial satellites such as Aurora compounded the difficulties faced by the older neighborhoods.

The geographical processes underpinning their origins and subsequent histories of the older suburban districts were simplified. This simplification is related in part to the amnesia and disinterest of pragmatic urban stakeholders to the historical and geographical processes producing urban neighborhoods, both industrial and residential. Planners, politicians, and business elites were more concerned with the immediate problems that they faced. While understanding the origins and shifting geography of slum and blighted-area formation might be a useful academic exercise, it did not provide an immediate and clear solution. To build a modern city, the pressing problems of poor housing, poverty, environmental deterioration, and social despair had to be dealt with in a practical manner. Forgetting that industrial suburbs became slums over several generations was a convenient way to deal with the contradictions of capitalist urban growth.

The simplification is also related to the needs of reformers, businesspeople, and planners to redesign the meaning and function of urban space. To acknowledge that industrial suburbs were a solution to urban problems since the mid-nineteenth century would have raised serious questions about the modern twentieth-century city that they wished to build. It also meant taking responsibility for the decline of the past and the costs of the present. Investment in older working-class and factory districts was not considered to be the solution to the problems generated by capitalist urban-industrial growth. Rather, the solution was the creation of new spaces on the urban fringe. The result was the emergence of a new set of suburban factory districts on the metropolitan fringe by the interwar period. Waves of investment and disinvestment refashioned Chicago's industrial landscape between the Civil War and the Great Depression. A new suburban solution was implemented.

CHAPTER 4

The Shifting Geography of Metropolitan Employment

STARTS, ADDITIONS, & MOVES

In 1924 Sullivan Machinery Company, one of the country's largest makers of air compressors and mining equipment, moved its plant from Chicago to Michigan City, Indiana. From humble roots as a small repair shop and wholesaling outfit, the company built up a large competency after 1900 by purchasing the plant and business of the mine hoist maker M.C. Bullock Manufacturing. For the next 25 years, the company produced a variety of mining equipment at a downtown factory and developed into a large corporation, with plants in Chicago and New Hampshire, and branch offices in 14 American cities as well as seven in Europe, South America, and Australia. By the early 1920s, however, decisions had to be made about the aging Chicago plant. The company either had to undertake expensive rebuilding at the old site or seek a location elsewhere. It chose the latter. Unfortunately for its Chicago employees, the decision was made to move out of Chicago altogether. After an extensive search, the firm finally settled on a 120-acre site on the Pere Marquette Railroad in Michigan City. The new location had several advantages, including prompt delivery of casting, parts, and completed machines for stock from the Claremont, New Hampshire, works, and low industrial and residential land costs compared to Chicago.[1]

Questions of plant location and firm migration from cities were burning issues by the interwar period. City councils were concerned about their tax base. Insurance companies were worried about rising costs and company profits. Unions and churches were terrified about job loss and the breaking up of communities. A 1929 Metropolitan Life Insurance Company survey of 16,000 American and Canadian manufacturing firms estimated that "nine-tenths of all industries are improperly located. They have been placed where they are by rule-of-thumb methods, for personal reasons, or for other causes of unscientific character." Since World War I, however, many firms attempted "to arrive at a future location by scientific means." While such statements must be taken with a grain of salt, they nevertheless reflected concern with the changing geography of America's industry as evidenced by large-scale regional movements. The migration of the cotton industry from New England to the South and the steel industry west of Ohio resulted in the relative decline of the Boston and Pittsburgh areas. Large-scale regional change was a matter of serious vexation for local social, economic, and political interests. Similarly, the shifting geography of metropolitan manufacturing firms and employment was high on the agenda of most elements of urban society. The geography of jobs mattered.[2]

Metropolitan Life's findings were replicated in studies of interwar Chicago. In 1933 William Mitchell found that manufacturing employment in the city of Chicago in the 1920s declined relative to the rest of the metropolitan district. A large part of this industrial displacement was because industrial suburbs such as Clearing, Hammond, and Cicero attracted plants from other locales, especially the city of Chicago. Although the city's decline was far from dramatic—its share of manufacturing employment fell from 77 to 73 percent between 1919 and 1929—the city-suburban job shift signaled the important changes occurring to the metropolitan geography of manufacturing employment. This point was picked up a decade later by the Chicago Plan Commission. In an overview of Chicago's industrial geography, the authors stated that "perhaps the most portentous industrial development in [the interwar] period was the decentralization trend." While they tried to make the best of this—by asserting that most jobs were going to Chicago's industrial suburbs—they were concerned about the central city's industrial base.[3]

As these and other studies suggest, the interwar period featured extensive regional and metropolitan plant relocation.[4] This chapter explores this issue through a study of the locational outcomes of 1,020 fixed capital actions made by Chicago companies in the 1920s.[5] These decisions are divided into four main types: the building of new plants, making additions to existing factories, the leasing of work space by firms new to Chicago, and moving production to another site in the metropolitan area (table 8).[6]

TABLE 8. Plant Changes, Chicago, 1923 and 1929

	No. of Firms		Total	
Type	1923	1929	No.	%
New Plant	98	69	167	16.4
Additions	126	238	368	36.1
New Starts	141	52	193	18.9
Moves	168	102	270	26.5
Other	10	12	22	2.1
Total	543	477	1,020	100.0

SOURCE: Material compiled by the author from "Industrial Buildings Sets Great Record," *Chicago Commerce* (January 5, 1924): 13; "Big Industrial Buildings of 1929," *Chicago Commerce* (January 11, 1930): 122; "Industrial Building Figures Record Number of Small Structures," *Chicago Commerce* (January 18, 1930): 10, 29–31; and the "Machinery News" section of *Iron Age* for the years 1922–1924, 1928–1930.

The study of these locational decisions brings out three significant findings. First, firm relocation was a complex process involving an array of movements. Companies did not simply move from the city center to the suburbs. Second, moving was not the only choice that manufacturers made. Companies committed a great deal of energy and investment to plant additions and to existing neighborhoods. Third, the industrial landscape underwent incremental not dramatic change. Despite the apparent large scale of industrial change, the manufacturing landscape evolved slowly and through tens of thousands of separate but yet interlinked decisions.

STARTS AND ADDITIONS

Manufacturers do not move location very often. Firms tend to be stuck to place despite increased capital mobility associated with the rise of the large corporation, the expansion of markets, and the generalization of cheap transportation. To be sure, firms could be highly mobile, especially those in new industries. In this case, the opportunities for movement were great because companies had little prior history in any particular place and were less impeded by existing labor pools, institutional rules, raw material supplies, and inter-firm networks. This is apparent from the development of nineteenth-century Chicago's meatpacking industry, early twentieth-century Detroit's automotive industry, and the postwar California computer industry. In all three cases, new growth industries located in new places. This was also evident in the older industries that experienced large-scale regional movement in response to wholesale reorganization or the development of new production structures.[7]

For most businesses, however, the historical endowment of a place creates the spatial envelope from which they operate. For entrepreneurs, the economic, social, political, and institutional assets constructed over time by numerous interacting individuals and groups maintain existing investment in place. Place-bound attributes function as transmission mechanisms, channeling industrial and social capital into particular urban places. Similarly, within metropolitan agglomerations, specific factory districts develop an array of ubiquitous and specialized locational assets, establish path dependency through their endowments, and become magnets for future growth. Firms, either by building new establishments or investing in existing plant, both feed off and contribute to these assets. In the process, existing factory districts become attractive to entrepreneurs because they have some combination of the requisite contacts, institutional structures, knowledge channels, labor pool, and transportation links.[8]

Many entrepreneurs are committed to local assets. People do not make decisions about capital investment in a geographic vacuum. The world in which they inhabit is spatially bounded by the contacts and contracts they make, the areas they know, and the resources they command. Chicago's institutions, particularly those coalescing around the property industry, mediated the spatial embeddedness of the manufacturing industry (fig. 7). Not only did industrial real estate companies such as Hodge, Nicolson and Porter have the "pleasure and privilege of serving" industrial concerns; they were also very active in creating the networks in which firms operated. As exemplified by the prewar inspection tours through Chicago's industrial areas, the commodification of land and the making of industrial spaces were formulated through the machinations of the local real estate industry, local business associations, financial institutions, and manufacturers. Helping solve the problems of, but sometimes reinforcing, the imperfect knowledge that individual manufacturers had of the local property market, institutions were responsible for placing boundaries on the search activity of industrial firms.

In other words, place can be extremely sticky and there is strong pressure for manufacturers to settle in existing firm clusters, many of which are plugged into well-developed local networks. This is evident from the results of the 1929 Metropolitan Life survey. According to the report, more than 80 percent of all new factory starts were made by local entrepreneurs. Only 9 percent were branch plants of multi-unit firms, and the remainder were formed by people from elsewhere. Entrepreneurs were overwhelmingly

FIGURE 7. Serving Industry in the Chicago District, 1928. Real estate firms that specialized in industrial property such as Hodge, Nicolson and Porter, Inc., promoted metropolitan expansion and industrial development by delivering land to manufacturers and bringing together a wide range of actors responsible for land conversion. Source: *Chicago Commerce* (May 5, 1928): 3.

local and built their industrial world upon known circuits of capital and knowledge. Similarly, the vast majority of enterprises (82 percent) lost to urban communities were companies that went out of business, while only 18 percent moved away. Firms were born and died at home.[9]

While not strictly comparable, the 1923 and 1929 data on Chicago's additions, moves, and starts suggest a similar picture to the Metropolitan Life survey. Of the 1,020 actions, only 193 (or 19 percent) were new starts—companies with no prior history in the city. The other 827 actions were undertaken by enterprises already located in the city. It is impossible to determine in any systematic way whether or not new starts were initiated by local or non-local capital. Nevertheless, the evidence shows that only a handful were established by entrepreneurs with non-local capital and resources. Only nine can be identified that moved from other cities, mainly from the eastern Great Lakes area and Illinois. There can be little doubt, despite the difficulty of identifying new firms' origins, that most of the 1923 and 1929 new starts had local roots.[10]

The importance of local knowledge for new starts can be clearly observed in the case of spin-offs. These were businesses begun by people who transferred industrial, organizational, and managerial skills and knowledge acquired as employees in another enterprise to a new start-up firm. Two metalworking examples illustrate this point. Albert Cochrane, president of American Tool and Manufacturing, was previously the sales manager of another tube company. Robert Hofstafer, the firm's industrial engineer, worked for another local machine-tool firm. Cochrane's establishment of his own firm may not have been entirely voluntary. Four months after opening American Tool, his old employer was bought by Youngstown Sheet and Tube. Youngstown immediately initiated a restructuring of the new subsidiary, which involved closing down "isolated properties" in Kalamazoo, Michigan, and Mayville, Wisconsin, and reducing operating costs at the company's other plants. The blast furnaces and steel mill at the two Chicago mills (South Chicago and Indiana Harbor) were retained, but Cochrane may have sensed that times were changing, and maybe not in his best interests. Whether this was the case or not, he probably knew about the impending buyout and fancied his chances operating on his own.[11]

Similarly, the Halligan Pipe and Foundry was founded in 1923 by George Halligan to assemble industrial piping. Halligan's previous experience included work with various Chicago pipe dealers, and the new firm's sales manager, Harry Shoff, had been a former vice president of another local pipe dealer.[12] For both Cochrane and Halligan, local know-how was critical to their knowledge, among other things, of finance, credit and labor, national markets and prices, and the strategies of competing firms. Moreover, they knew enough about the local pipe, tool, and machinery scene to bring with them people with industrial and managerial skills. Hofsafer, the industrial engineer, was a linchpin of American Tool's development of new products and the re-jigging of existing products. Shoff, the experienced salesman, provided Halligan with the knowledge and the networks necessary to distribute pipe products of the new jobbing and manufacturing firm.

While new starts were important, the most significant capital investments made in 1923 and 1929, at least in terms of numbers, were enlarging existing plants. More than a third of all actions consisted of new investment to existing plant facilities (table 8). Most of this was through additions, that is, the construction of an adjacent structure, the addition to existing plant, and, in a few cases, the remodeling of existing factories. The significance of additions lies in a manufacturer's commitment to place. In some situations, of course, firms had little choice—making an addition, at least in the short term, was cheaper and less time-consuming and disruptive than building in a new location. In others, however, it demonstrated the importance of the existing location and the web of links that a company had with its neighborhood.

Size varied greatly. Some enterprises were small. The smallest was a $10,000 two-story addition made to a printing plant in 1923 and the construction of a $6,000 steel tube warehouse six years later. At the other extreme, a few firms poured huge amounts of capital into their plants. Western Electric directed $4 million into a new warehouse and production complex at their Hawthorne plant in 1923. Six years later it spent more than $2 million on five different additions, including a cable shop and a garage. Other very large expenditures were made by Lever Brothers ($5 million at its Hammond soap factory), James Kirk ($2 million at its Chicago soap factory), Grisgby-Grunow ($1.6 million for a new radio tube addition), and Crane ($1 million at its sprawling Kedzie Avenue plant). Regardless of the value of investment, the energy and money spent was a commitment to the existing factory and reflected the company's history and the neighborhood's attractiveness. The Western Electric and Crane investments in the 1920s built up the industrial complexes they constructed between 1905 and 1913. For these and other enterprises, the fixed capital embedded in the factory and the links made with surrounding working-class communities, transportation facilities, and, in the case of industrial suburbs, local governments combined to form locational assets that were difficult to ignore. For Western Electric, Crane, and others, plowing capital into an existing factory was a more rational strategy than investing in a new one elsewhere in the metropolitan area or in another locality.

The most common reason companies built additions was to accommodate expansion. Goodman Manufacturing, makers of mining machinery and locomotives, illustrate this point. The additions made to the company's turn-of-the-century plant were not particularly large ($60,000 and $90,000), but they did signify the importance the firm placed on the site. Located in the heart of a South Side factory district, the firm found access to metalworking and machinery equipment producers, a large labor pool, and transportation facilities, particularly important given its production strategy (unique products made under batch conditions), large workforce (more than 1,000 workers in 1924), and close relationship with a neighboring firm (Link Belt).[13] For companies such as Goodman, place mattered enough to warrant considerable investments in their old buildings over long periods of time.

Another view of the choices that manufacturers made about plant additions can be gleaned from the additions made to the Calumet steel mills in the 1920s. According to an annual listing of new capacity by *Iron Age* there were 50 major additions to the Calumet mills between 1920 and 1929. Some firms were more active than others. Small mills tended to make only one addition: Columbia Steel of Chicago Heights added two electric furnaces in 1926, while Highland Iron and Steel built a new eight-inch rolling mill to roll small bars and shapes at its Pullman plant three years later. Although corporations such as Wisconsin Steel and Inland Steel continued to invest in their plants, the number of major changes that they made was relatively small. Even though Inland built new blast furnaces, open hearth ovens, and a splice bar mill, its main focus was on vertical acquisitions—a Milwaukee rolled-back sheet mill and a Chicago Heights steel post firm—rather than horizontal expansion through existing capacity.[14]

In other cases, large-scale expansion was part of the firm's strategy. For example, Youngstown Sheet, after the ending of the Pittsburgh-plus pricing system in 1923, bought the furnace and mills of the South Chicago and Indiana Harbor plants of the Sheet and Tube Company of America. The Chicago firms gave Youngstown low-cost plants with excellent transportation facilities in the expanding Chicago and Midwest markets and access to coal and iron ore mines in Kentucky and West Virginia. At the same time, the buyout linked the Calumet plants to a parent company with an elaborate set of interests along the steelmaking production chain: extensive raw material reserves, several subsidiary firms making spring steel, rod, and wire ore reserves, and, centered at its main Youngstown mill complex, the third largest steel-producing capacity in the county. Between 1924 and 1929, the company invested heavily in the two Chicago plants, adding a new blast furnace, a coke plant with 70 Koppers ovens, a continuous 21-inch skelp mill, a creosote plant, three 100-ton open hearth furnaces, a metallurgical laboratory, two butt weld pipe mills, and tin mills.[15]

The expansion activity in Chicago was part of Youngstown's strategy to meet the competition of other steel firms, especially U.S. Steel. Youngstown's decision to move into Chicago was spurred by the construction in 1923 of a large tube plant at the Gary plant of the National Tube Company, a U.S. Steel subsidiary. In order to meet the growing capacity of U.S. Steel, the first addition made by Youngstown at its Chicago plants was pipe mills. New capacity served various purposes: it made Youngstown a presence and major competitor in Chicago, added to the fixed capital of the company's Calumet mills, and consolidated Youngstown's presence in the national market. Other Calumet steel mills—most notably Acme Steel, Illinois Steel, and Interstate Iron and Steel—also built extensive additions to their factory stock over the course of the 1920s.[16]

In a few cases, the expansion of production facilities took place outside the walls of the existing plant but within the city itself. The decision to build elsewhere in Chicago confirms the importance of the local environment for firms such as the architectural iron maker Hansell-Elcock that were experiencing increased production. Rather than adding in place or building an entirely new, self-contained plant elsewhere, the firm built an addition, consisting of a gray iron foundry and an architectural iron shop, five miles west of its Bridgeport plant. Internalizing two aspects of the company's production chain but at a different location, the foundry supplied all of the castings required in the old shops and furnished many of the new iron shop's inputs. The new plant gave the firm the best of two worlds. By supplying goods and complementing existing facilities, it ensured that Hansell-Elcock did not have to invest in new premises. The new shops were designed with modern production layout in mind. Here the intent was to ensure "flexibility of output" and "diversification of production" under conditions specific to their unique production requirements, the foundry's need to produce bulky and heavy work on short notice.[17]

As the Hansell-Elcock case suggests, additions were associated with changes to production processes and product lines. Production process changes varied greatly, from replacing old machinery to the overhaul of the way that production was undertaken. Manufacturers frequently combined the decision to build an addition with production changes. Not only were these changes the driving force behind the need to build, but the addition gave the firm the opportunity to implement the changes without being confined by the older plant layout. Indeed, by the 1920s new factory spaces were increasingly being constructed around the production process. A new modernist factory replaced the old nineteenth-century mill structure. The refashioning of the W. A. Jones Foundry and Machine Company in 1929, for example, was contingent upon the building of a 100-foot extension costing $18,000. The firm's ability to lower costs and speed up production was made through a combination of better-coordinated through-flow and a new layout for machinery, both old and new. This would not have been possible without the new addition, which provided uninterrupted space for better through-flow and machinery rearrangement.[18]

A final type of fixed capital investment was the construction of new factories. A total of 167 new plants were erected in 1923 and 1929. As figure 8 shows, the vast majority of the new plants were small and medium in scale (valued at less than $500,000) and were scattered across the industrial landscape of metropolitan Chicago. At the lower end of the scale was the downtown factory valued at $9,800 and taken over by the furniture polish maker Straub and Yorke. Larger were the new $100,000 acetylene plant constructed in the suburb of Summit for Compressed Industrial Gases and the Borden dairy in Forest Park that cost $350,000. The map of newly built plants also shows that 16 large plants (constructed at more than $500,000) were built in the 1920s. The largest valued at $4 million was the paper mill built for Federal Paper at California Avenue and Thirty-first Street. Other multimillion-dollar manufacturing plants were built for Pines Winterfront (automotive parts), United Autograph Register (office machines), and Illinois Anthracite (carbonized coal) (fig 9). There was also great variation in the types of industries building new plants; dairy products and baked goods, chemicals, metal goods, furniture, radios, and a host of other products. Some of these were new starts, while others were built by firms moving from other locations in the city and in some cases from other parts of the country.

One firm combines many aspects of new plant construction. In 1929 the car battery maker Vesta Battery signed a 17-year lease with the Clearing Industrial District for the building of a new plant. Vesta began in Chicago in 1897, opening a small plant in the Loop. The company made several moves over the next 16 years, ending up at a four-story factory in the heart of the Near South Side's automotive district. With rapid expansion in the postwar period, companies such as Vesta built up a large competency by

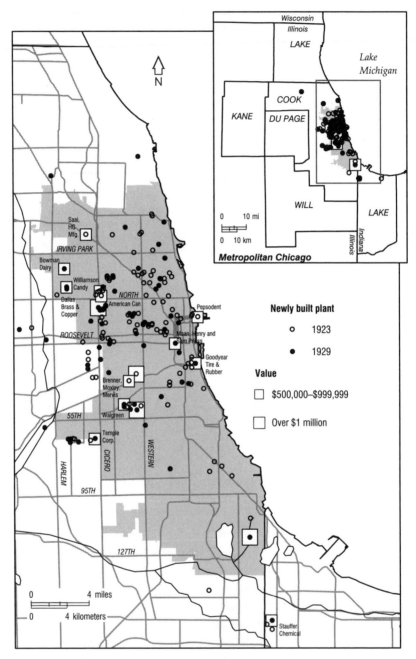

FIGURE 8. The Geography of Newly Built Plant in Chicago, 1923 and 1929. Source: "Industrial Building Sets Great Record," *Chicago Commerce* (January 5, 1924): 13; "Big Industrial Buildings of 1930," *Chicago Commerce* (January 11, 1930): 122; "Industrial Building Figures Record Number of Small Structures," *Chicago Commerce* (January 18, 1930): 10, 29–31.

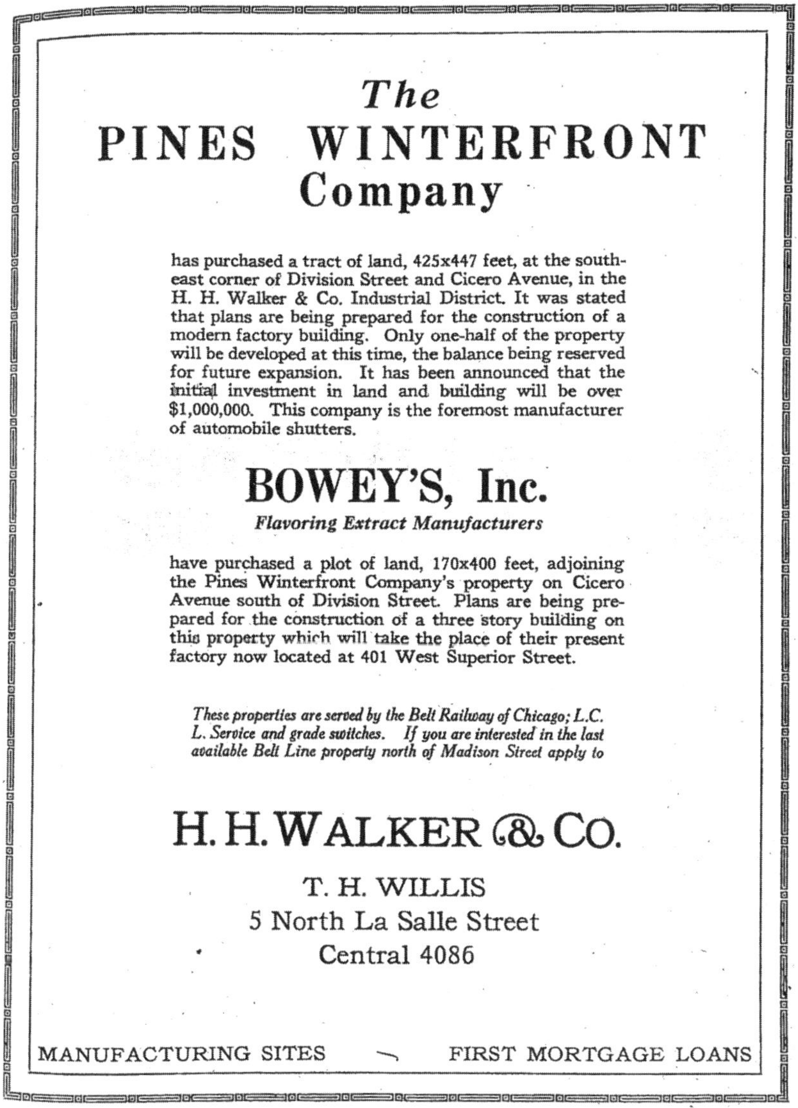

FIGURE 9. The Pines Waterfront Company and Bowey's, Inc., 1928. Loop real estate
firms such as H.H. Walker and Co. sold "manufacturing sites" on the fringe of the
city and arranged "first mortgage loans" for Chicago firms. Source: *Chicago Commerce*
(November 17, 1928): 19.

making the correct production decisions. One important element of a successful production strategy was location. In many cases, this involved the construction of new factories at new sites. While many firms remained in the old automotive districts ringing the central zone, others moved to the city edge. Vesta was one of these. Clearing, which became home to Vesta's new production facility, was a rapidly growing factory area in the first part of the twentieth century. Located on the southwestern fringe of the city, the planned industrial district was developed by a business alliance led by the railroads. Vesta's one-story Clearing factory was almost double the space of its old plant, and the lease contained an option to purchase an additional 80,000 square feet for future expansion. Operating in an expanding industry, Vesta produced light automotive parts in a specially designed factory.[19]

Additions, new starts, and new plants were not randomly distributed across space; there was a distinct geographic pattern to fixed capital investment in metropolitan Chicago. One feature, as reflected in the concentration index, was the absence of activity in the center zone (the Loop and the surrounding downtown area) (table 9).[20] This was particularly pronounced in the Loop and the areas immediately to the south and the north, which suffered the greatest competition from financial, legal, commercial, administrative, and cultural land uses. As figure 10 shows, new starts were concentrated in the center area of the metropolis. Clustering close to the river, new firms such as Federal Gauge, which leased a fifth-floor factory in a building at Adams and Canal to manufacture electric gauges, and E.C. Mith Manufacturing, which rented a floor in a multi-storied building at Superior Street and Sedgwick Street to produce bathroom accessories, illustrate the continued importance of small, downtown working spaces in old multi-storied mill structures available at affordable rents and for short leases. While firms new to the city enjoyed the advantages of the center, some local firms built additions to existing plant (fig. 11). In some cases, firms such as the flavoring extract maker Horine and Bovey added an extra floor to their plant. In others, new firms bought an existing plant and made additions, such as the Connecticut brass goods producer Scoville Manufacturing, which built an addition to the premises at Washington and Roosevelt. Small firms still sought out the range of existing, flexible, and cheap working spaces that were so common there by making additions rather than moving elsewhere. They also would have been able to tap into the hundreds of thousands of workers of all kinds and skills that lived within easy walking distance of the central area's factory districts.

TABLE 9. The Geography of Additions, New Starts, and New Plants, 1923 and 1929

Zone	Additions			New Plants			New Starts			Firms 1924	
	No.	%	CI	No.	%	CI	No.	%	CI	No.	%
Center	58	18.0	.4	25	15.0	.3	76	39.4	.9	3,211	43.9
South	83	25.7	1.6	24	14.4	.9	34	17.6	1.1	1,215	16.6
West	46	14.2	2.2	33	19.8	3.0	24	12.4	1.9	470	6.4
North	98	30.3	1.4	66	39.5	1.8	49	25.4	1.1	1,640	22.5
Suburbs	38	11.8	1.1	19	11.4	1.1	10	5.2	.5	777	10.6
Total	323	100	1.0	167	100	1.0	193	100	1.0	7,313	100

SOURCE: Material compiled by the author from "Industrial Building Sets Great Record," *Chicago Commerce* (January 5, 1924): 13; "Big Industrial Buildings of 1929," *Chicago Commerce* (January 11, 1930): 122; "Industrial Building Figures Record Number of Small Structures," *Chicago Commerce* (January 18, 1930): 10, 29–31; and the "Machinery News" section of *Iron Age* for the years 1922–1924, 1928–1930.

NOTE: CI = concentration index. The CI is obtained by taking the percent of additions or new plant or new starts by zone and dividing by the percent of number of firms in that zone according to the 1924 directory.

Center = Loop, Northwest Side, Near South Side, Near North Side, West Side.

South = Bridgeport, Central Manufacturing District, Douglas, Englewood, Lower West Side, Packingtown, Pullman, South Chicago.

West = Archer Heights, Austin, Clearing, Garfield Park.

North = Humboldt Park, Lincoln Park, Lincoln Square, Milwaukee, Portage Park, Uptown.

The other major feature of the metropolitan geography of fixed capital was the clustering of investment in the West and North districts. With the exception of new starts in the North, these two districts had concentrations of additions, new starts, and new plants. Firms, ranging from large metal fabricators such as Crane and Hansell-Elcock to a host of small firms, continued to invest in these zones, either in existing plants or by constructing entirely new buildings. In some cases, fixed capital expenditures reinforced existing factory districts, while in others it heralded the development of new ones. This is most evident in the case of the unsettled fringes of the city at Portage Park, Lincoln Square, and Clearing. These areas attracted investment on a scale not commensurate with their existing business number. They also attracted the appropriate locational assets. While in some cases firms located or made additions to plants in areas containing the essential infrastructures and working-class residential neighborhoods, in others this was not the case. A coalition consisting of local politicians, developers, builders, and transportation and utility companies typically followed in the wake of the settling firms and constructed the necessary sewers, roads, streetcar lines, houses, schools, and churches.

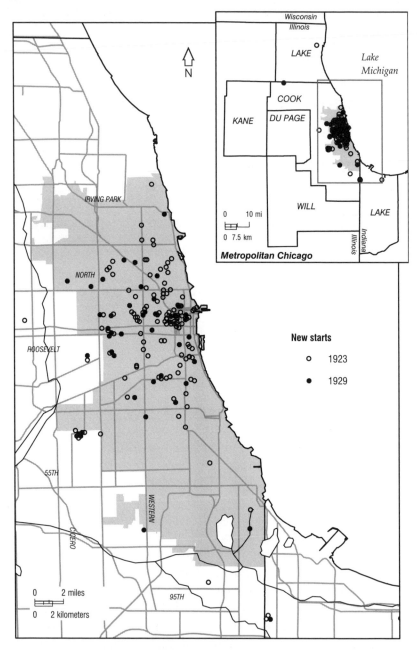

FIGURE 10. The Geography of New Starts in Chicago, 1923 and 1929. Source: "Industrial Building Sets Great Record," *Chicago Commerce* (January 5, 1924): 13; "Big Industrial Buildings of 1930," *Chicago Commerce* (January 11, 1930): 122; "Industrial Building Figures Record Number of Small Structures," *Chicago Commerce* (January 18, 1930): 10, 29–31.

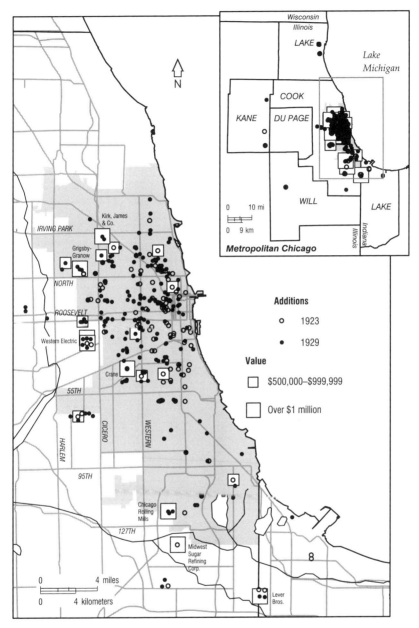

FIGURE 11. The Geography of Additions to Plant in Chicago, 1923 and 1929. Source: "Industrial Building Sets Great Record," *Chicago Commerce* (January 5, 1924): 13; "Big Industrial Buildings of 1930," *Chicago Commerce* (January 11, 1930): 122; "Industrial Building Figures Record Number of Small Structures," *Chicago Commerce* (January 18, 1930): 10, 29–31.

This geography of where manufacturers decided to invest indicates the reduced importance of some areas (the Loop and the central area in general), the reinforcement of existing factory districts (Milwaukee Avenue, Packingtown, and Garfield Park), and the development of new or less developed districts (Portage Park and Clearing). This geography was the product of a long history of investment in the industrial landscape. In turn, the decisions that manufacturers made in the 1920s as to where and how they would invest in plant facilities shaped the geography of metropolitan manufacturing for a generation.

MOVING THE FACTORY

While additions, new starts, and new plants were common forms of firm behavior, factory relocation was another. Moving the factory in the 1920s was an extremely important process shaping the metropolitan district's geography. While economic decline was not widespread in the 1920s, factory moves did have important effects on different parts of metropolitan Chicago. Manufacturers were mobile. In some ways this hardly needs stating, for the moving of plants from one location to another underpins location theory. The evidence to document the facts of this mobility, however, is hard to find. Little is known about the reasons for and the direction of industrial moves within a metropolitan area. The general consensus is that firms moved because they needed one of two things—more space or cheaper land—and that they moved away from the city center.

In an early industrial location analysis, Glenn McLaughlin noted that suburban manufacturing between 1899 and 1929 grew because of the construction of new factories, the expansion of existing suburban factories, and the relocation of firms to the suburban districts. Other writers point to a more complex picture. One of the earliest statements was given by Charles Colby, who argued that company executives faced a set of push-pull factors. Ever-increasing property values, traffic congestion, high transportation costs, the difficulty of securing special types of sites, and the desire to avoid nuisance complaints and "irksome" legal restrictions pushed manufacturers to uproot from downtown and to migrate to the urban fringe. The attractions of the fringe areas were also important. Large parcels of unoccupied and cheap land, the presence of transportation services, attractive site qualities such as level land, and control over a sizable area gave firms the freedom to lay out and operate in a manner conducive to success.[21]

McLaughlin's and Colby's point was evident in Chicago's history of companies moving from central locations to more peripheral ones. Take Chicago Railway Equipment, makers of brake beams and side bearings. Located among Packingtown's meatpackers and glue factories, it was an anomaly servicing the railroads among the stench and cries of the slaughtering and packing business. The company had not always been at this site. Starting off in the Loop in 1887, the firm made the move two years later out of the central zone, taking a new plant in an industrial district to the south. Twenty years later in 1908, Chicago Railway Equipment made its final move when it settled in Packingtown.[22]

Other firms followed a similar trajectory. The Indiana firm Sefton Manufacturing purchased an old Loop plant in 1891 to produce corrugated fiber boxes. A year later the company moved from its small premises to a much larger one. Growth was so rapid that three years later Sefton moved once again. This plant soon became too small, and over the years the factory's floor space was increased more than threefold. In 1906 the firm moved once more, this time to a five-acre site in the Central Manufacturing District, a new factory district south of Packingtown. By the 1920s the plant's 900 workers labored in seven acres of floor space. Similarly, Weller Manufacturing, first located along the river in 1886 to manufacture elevating machinery, made several moves. In 1889 Weller moved to North Avenue, where it occupied one floor of a four-story building. It eventually took over the entire building and made additions, but the space became inadequate, and in 1910 they purchased land farther north with a railroad switch. They built a conveyor department in 1911 and erected additional buildings in 1915.[23]

Moving out of the city core and settling on the urban fringe was not unique to Chicago, of course. In New York the metalworking industry was first established in the lower reaches of Manhattan. From the late 1880s, however, most of the heavy foundries moved from Manhattan to Queens, New Jersey, and Connecticut, and by 1924 only five remained in Manhattan. Similarly, New York's sugar industry experienced an out-migration of refineries and ancillary firms. By 1900 not one refinery remained in Manhattan. Two decades later there were six in Brooklyn, Yonkers, Queens, and Edgewater.[24] The geography of production is dynamic.

A sense of the geography of firm moves in the metropolitan area is from evidence about 270 firms that moved a plant from one Chicago address in 1923 and 1929. In most cases, this involved moving the entire production facility (and sometimes offices) from one factory to another.

In others, companies opened a second facility, or sometimes more than two facilities, elsewhere in the metropolitan area, sometimes by moving into an existing building, sometimes by building a new one. The move to an existing building usually involved making additions or alterations. Regardless of the type of move, the major geographic pattern is the loss of manufacturing plants by the central district plants compared to the rest of the city (table 10). The center experienced manufacturing depopulation. Of the 270 moves, almost two-thirds (177) were from an address in the same area. In contrast, these central areas received fewer than a third of the moving firms (85). Moreover, almost all of the firms (77 of the 85) moving to a center address were from the area itself. In other words, the area lost plants and there was little replenishment of the center's industrial base from firms outside the district. The center fed upon itself as firms milled around in the industrial districts close to the Loop and along the nearby banks of the Chicago River.

There were good reasons why centrally located firms stayed in place. As Colby states, firms needed access to space, cheaper rents, and existing business networks. Property firms such as Hodge, Nicolson and Porter helped central companies find a new location around the corner or down a couple of blocks (fig. 12). In 1923, for example, the electric light fixture manufacturer C. G. Everson moved from West Lake Avenue, where it had been for 20 years. The need for large quarters forced the firm to take out a 10-year lease on a five-story building across the river. Because it did not need all of the space, and to help defray the annual rent of more than $11,000, Everson sublet two of the floors to enterprises that moved from other Loop addresses. Other central manufacturers left their previous address for reasons other than the need for larger space. Some moved because of unforeseen problems. In 1923 Standard Photo and Engraving were forced to leave their Wells Street plant after 15 years because a fire destroyed the premises. Wishing to stay at a central location, Standard Photo took a five-year lease on the entire sixth floor of a South Market building.[25]

The rest of the metropolitan area received more movers than it lost. Ever willing to convert the "empty" lands into undeveloped suburban districts, land speculators and real estate developers transformed Chicago's outer fringe into a prime resource for manufacturers seeking new locations. However, fringe areas were not unproblematic. Industrialists and local growth coalitions could not necessarily create and attract the locational assets appropriate for successful production. While electricity companies

TABLE 10. The Intra-Metropolitan Geography of Moves, 1923 and 1929

| Zone | Moved from Zone | | Moved to Zone | |
	No.	%	No.	%
Center	177	65.6	85	31.5
South	29	10.7	40	14.8
West	12	4.4	44	16.3
North	52	19.3	97	35.9
Other	0	0	4	1.5
Total	270	100.0	270	100.0

SOURCE: Material compiled by the author from "Industrial Building Sets Great Record," *Chicago Commerce* (January 5, 1924): 13; "Big Industrial Buildings of 1929," *Chicago Commerce* (January 11, 1930): 122; "Industrial Building Figures Record Number of Small Structures," *Chicago Commerce* (January 18, 1930): 10, 29–31; and the "Machinery News" section of *Iron Age* for the years 1922–1924, 1928–1930.
NOTE: See table 9 for breakdown of zones.

would welcome the opportunity to expand their market, streetcar companies were not so accommodating. Nor was the mismatch between labor supply and demand easily solved. Even when land was available, it could take quite awhile for the appropriate houses, schools, and shops to be constructed. Moreover, workers were not easily pulled away from their central neighborhoods, with their well-developed social, political, and economic worlds, to the rude pleasures of the urban fringe.

Fringe development was uneven; there was great variation between the different zones. The city's southern part did not experience much movement, one way or another. Only 29 of the 270 firm moves were from southern addresses, while 40 were to southern addresses. The west end (west of Western Avenue) acquired more firms (44) than it lost (12) as industrial real estate firms such as J. H. Van Vlissingen aggressively promoted industrial property on the western fringe. Firms seeking addresses in the south and west were moving to some of the districts formed in an earlier generation such as Bridgeport and South Chicago. Others moved to the new and rapidly growing organized districts, Clearing and the Central Manufacturing District. In most cases, they sought more room as they expanded or built new premises to introduce new production facilities. Other factors were in play in other cases. Strom Ball Bearing, for example, wanted to stay in the suburbs. The company built a large factory in Cicero to replace its Oak Park one that was razed to the ground.[26] In other cases, manufacturers moved because of the opportunity to locate close to similar companies.

FIGURE 12. To Rent, 1928. The old factory building at Clark and Nineteenth recently vacated by the enamelware manufacturer Lalance and Grosjean was in a prime part of the city for light manufacturing firms or wholesalers seeking a central and "convenient location" with access to the New York Central Railroad line. Source: *Chicago Commerce* (May 26, 1928): 3.

Industrial real estate companies made areas such as the west end attractive to moving firms. In the nineteenth century, industrial developers such as Bogue and Hoyt opened up the city's fringe districts to manufacturing, warehousing, and transportation firms. In the first half of the twentieth century, this tradition was continued by enterprises such as Hodge, Nicolson and Porter and J. H. Van Vlissingen and Company, who helped firms find new locations, turned land into industrial subdivisions, and planned large-scale developments. As already noted, from its downtown office the former company controlled a range of land transactions that furthered the movement of manufacturing out of the older industrial districts. The Vlissingen company also played an important part in the shaping of the metropolitan area's industrial landscape. Not only did they sell and broker individual lots; they also laid out and developed the Kenwood and Grand Avenue planned "industrial districts" (fig. 13). One of the family members, Arend Van Vlissingen, was responsible for a 1920 plan for Lake Calumet. The plan, which became the official plan of the city of Chicago, was turned into the Lake Calumet Harbor Act by the state legislature in 1921 and enabled the city to link water and land transportation facilities; to modernize the harbor through dredging, landfills, and the building of terminal facilities; and, of course, to open up the lake region to industrial development.[27]

Companies such as Hodge, Nicolson and Porter and J. H. Van Vlissingen allowed the city's northern zone to be the recipient of the largest number of relocating firms. A busy area, accounting for almost 20 percent of the city moves, the north became home to more than a third of the moving firms. Like those moving south and west, northern firms sought new industrial land. But unlike the two former zones, which were already heavily built up, the northern zone was relatively free of manufacturing establishments. The major trend of industrial development in the city before World War I was in the districts strung along the canal, the southern part of the Chicago River, and the railroad tracks leading out of the city into Indiana. In the 1920s firms moved northward for several reasons: the existence of large tracts of available land, the activities of a new set of industrial and residential promoters, and increasing dependence on the truck at the expense of the railroad. They clustered close to Elston, Milwaukee, and Lincoln avenues, the diagonal roads leading northwest out of the central city. This northward trend continued over the next generation of industrial development so that by the end of World War II the northern part of the city and the northern suburbs such as Skokie had become major factory districts.[28]

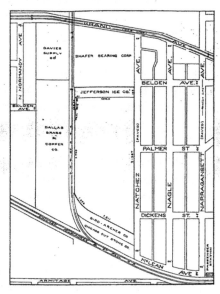

FIGURE 13. Van Vlissingen's Grand Avenue Industrial District, 1928. J. H. Van Vlissingen and Company was one of Chicago's most active industrial property promoters. Among other areas in metropolitan Chicago, the firm developed the Grand Avenue Industrial District. Source: *Chicago Commerce* (March 31, 1928).

The changing geography of moves, new starts, and additions in the 1920s illustrates the various ways that past and present capital investment in factory plant shaped the urban-building process. While there can be no doubt that there was a long-term trend toward industrial decentralization, there was no simple, unilinear process of deconcentration. Rather the formation of new factory districts on the urban fringe and the continued viability of existing ones occurred through the incremental buildup over time of an assortment of decisions made by manufacturers about where to establish a new plant and what to do about existing ones. The appearance of industrial space was not a functional response to the need for greater manufacturing and warehousing space. Manufacturers made decisions in response to the opportunities and constraints imposed by a capitalist economy, the character of local assets and institutions, and the organizational, managerial, and social networks of individual local entrepreneurs.

The result was an array of locational and investment choices. In some cases, the conditions under which an existing firm labored forced management to seek new sites in undeveloped areas where it laid down more effective production operations and more modern factory space. In other cases, new starts located downtown close to the whirlpool of external economies. In still other cases, manufacturers invested in place because the costs of moving—both in terms of the financial costs and of the fragility of knowledge and linkages—were too great to be borne. The overall effect of these structured investment decisions made through a dynamic yet incremental refiguring of existing and new factory districts was to reshape the geography of the entire metropolitan area.

Starts, additions, and moves reflect the changing geography of employment growth and decline. Expansion of manufacturing employment through the arrival of new plants or additions to existing stock had enormous repercussions for the working-class neighborhoods and community structures. Greater employment typically led to the building of more homes and an increase to a place's total income pool. These had multiplier effects through the neighborhood and the surrounding districts, and in the process added to the place's economic, social, and political endowments. New firms were responsible for the emergence of new populations, which in turn could lead to the establishment of new institutions based on ethnicity, race, and class. New suburban industries created the basis for a different set of industrial relations both inside and outside the workplace.

Locational decisions also had important implications for the employment opportunities of factory workers in the areas that experienced both absolute and relative decline. As factories were closed and companies moved production to another district, they often left behind a declining industrial base, increasing unemployment, lower wages, and longer journey-to-work travel times. As this trickle becomes a flood, the community as well as economic structures become severely compromised, leading to a spiraling and structural economic decline. As many writers have noted, this was characteristic of the postwar metropolis. However, the seeds of postwar deindustrialization of the central districts of the American metropolis were sown as early as the 1920s.

SECTION II

NETWORKING
the Industrial
Metropolis

CHAPTER 5

The Metropolitan Geography of Firm Linkages, 1872–1901

In the previous chapter, the examination of factory districts after the Civil War was framed so that firms were considered as having very little interaction. However, how does this account for the case of Beatty Machine and Manufacturing, a small metal producer located in the industrial suburb of Hammond? In a special edition of the *Calumet Record* published in 1925, it was reported that Beatty supplied local firms with punches, presses, and fabricated structural steel, despite demand from national markets. Not only did Beatty furnish car and shipbuilding firms; it also relied on local steel mills for its raw materials, residential districts for its labor force, and local institutions for legal and financial support. Beatty was just one of several local metal fabricators. The Hammond company was well embedded within the industrial fabric of the Calumet steel district and was emblematic of the importance of inter-firm linkages in the making of metropolitan manufacturing.[1]

Beatty Machine was not unique. All manufacturers—from the large, vertically integrated, multi-unit corporation to the small proprietorial workshop with one or two workers—interacted with firms, institutions, and individuals. Bewildering in nature, scale, and scope, industrial interaction ranged from the formal and routine to the unique and infrequent. Routine interactions included cases such as where company managers and owners met with landlords, obtained weekly deliveries of coal, or made bank deposits. Others were less regular but still required meetings, such as with subcontractors about the nature, specifications, deadlines, and the price of the work to be performed; with lawyers and bankers to discuss lawsuits, patents, capital flow, and debts; with wholesalers to acquire quick delivery of parts; or with machinery makers to go over the details of a new machine. In other cases, many inter-firm relations were onetime, involving little other than delivery of a good or the provision of a service.

An astonishing array of differences characterized the functional relationships among businesses. Some firms—especially those in industries centered on style such as clothing and furniture or in the manufacture of one-of-kind products such as heavy equipment—depended on interaction all along the production chain. Others operating as satellite suppliers to a central firm, such as a small wheel maker with a large carriage manufacturer, delivered most if not all of their output to that company. Similarly, some manufacturers brought in most of their inputs from a small number of suppliers. Large, vertically integrated, multi-plant petroleum refining, brewing, and meatpacking corporations tended to have a limited number of suppliers as they performed many production steps within house. In these cases, however, many ties occurred between the different plants of the corporation. On the other hand, large companies relying on a variety of raw materials or assembling many semi-processed inputs into a large complicated product were much more reliant on links outside the firm.

An examination of bankruptcy records reveals how a small metal fabricator such as Beatty Machine played an important part in the economic history of metropolitan Chicago.[2] Two elements of Chicago's manufacturing history are identified, documented, and developed: the existence of an extensive metropolitan-wide industrial complex of interlinked firms; and the manner in which local manufacturing firms were embedded in a regional circuit of business linkages. Chicago's manufacturing cluster formed an industrial metropolitan district that was city-centered, while forging specialized links with the Manufacturing Belt, the industrial center of the United States that ran from Wisconsin and Minnesota through the Great Lakes states to the Mid-Atlantic and New England.[3]

THE GEOGRAPHY OF CHICAGO'S BUSINESS LINKS

The operations of an urban-based manufacturing cluster directly tied into a wider regional world are illustrated by the voluntary bankruptcy of the Richards Iron Works, a downtown machine shop, in April 1878. The enterprise owed 71 creditors almost $14,000, including $2,400 to 35 Chicago workers, $10,400 to 28 Chicago firms and individuals, and the remainder to

8 creditors from outside Chicago. Richards Iron's largest debt was a mortgage to De Witt Curtis, a self-styled Chicago capitalist. Among the various local manufacturing and wholesaling creditors, Richards Iron was in debt to a foundry, several coal firms, and a sheet metal shop. The company owed money to various firms for advertising, telegraph, security, and financial services. Creditors from outside the city included a family member from Newark, Ohio, and the Moline Governor Works. On the opposite side of the ledger, Richards Iron had a few assets at the time the bankruptcy was filed. The largest, a mortgaged property, sold for $4,766, while tools, implements, machinery, and patterns were auctioned off for $546.[4]

In March 1900, more than 20 years after Richards Iron went under, Chicago Shoe and Slipper filed for voluntary bankruptcy. Chicago Shoe's proprietor Allen White, like the owner of Richards Iron, owed money to associational, financial, and service organizations: the National Shoe and Leather Exchange, several insurance agents, a local bank, and a towel supplier. Its largest number of creditors, however, were manufacturers and distributors who supplied an assortment of raw materials and semi-processed products: leather, tacks, shoe machinery, lasts, disinfectants, boot blacking, rubber heels, wooden boxes, paper bags, and thread.[5]

The range of businesses, institutions, and individuals that Chicago firms were part of is illustrated by the bankruptcy proceedings of Richards Iron and Chicago Shoe. Chicago bankrupts owed money for everything from club memberships to building repairs, steel billets to buttons, and advertising space to legal advice. There were important differences in the sectoral distribution of creditors (table 11). The 90 percent of creditors whose business operations could be determined were as follows: manufacturers (48 percent of firms and 28 percent of value), distribution (24 and 16 percent), service (19 and 8 percent) and financial (9 and 48 percent) firms.

Nineteenth-century Chicago enterprises were heavily in debt to institutions and individuals. The legal structure of most bankrupt firms was proprietorial, not corporate. Many had their origins in loans taken out from a small number of local banks, family members, and, in a few cases, owners of other businesses from along the production chain. Before World War I, company growth typically occurred by reinvesting retained earnings, and few firms obtained any capital from industrial securities or large loans from banks. Some manufacturers, however, did have direct financial dealings with banks and national capital markets. Firms received loans from banks such as the Corn Exchange National, the German National, and the Merchants' Savings, Loan and Trust to pay for the necessities of production—machinery, mortgages, materials, and labor.

TABLE II. Sectoral Distribution by Chicago and Non-Chicago Firms, 1872–1878, 1898–1901

Sectors	All Creditors		Chicago Creditors		Non-Chicago Creditors	
	No.	$	No.	$	No.	$
Service (%)	19.4	8.0	23.7	10.8	8.0	.8
Distribution (%)	24.0	16.2	26.5	15.8	17.3	17.1
Finance (%)	9.1	47.7	10.3	58.4	6.0	20.9
Manufacturing (%)	47.5	28.1	39.4	15.0	68.7	61.2
Total (%)	100.0	100.0	100.0	100.0	100.0	100.0
Not Available (%)	10.3	6.0	6.7	4.9	18.4	8.7
Total (No.)	2,267	1,297,053	1,577	918,580	690	378,473
Total (%)	100.0	100.0	100.0	100.0	100.0	100.0

SOURCE: Compiled from 44 bankruptcy cases for 1872–1878 and 1898–1901 taken from Record Group 21, Records of the United States District Court, Northern District of Illinois, Eastern Division at Chicago, Bankruptcy Records, Act of 1867, 1867–1878; Bankruptcy Case Files, 1872–1878; and Bankruptcy Records, Act of 1898, 1898–1972; Bankruptcy Case Files, 1898–1946.

In a few cases, firms drew upon credit from financial institutions outside Chicago. Schureman and Hand, a marble furniture maker, had accommodation papers with two Cincinnati banks. The link with the Ohio city was through Sylvester Hand, the company's president, who lived in Cincinnati. As well as from banks, bankrupt companies borrowed from other financial institutions, most notably small venture capitalists and brokers. Schureman and Hand had a large mortgage ($17,260) with a Chicago mortgage broker, and smaller loans and accommodation papers with local banks.[6]

Similarly, Henry Bennett utilized resources out of the city to finance his planing mill. In 1876 when his enterprise failed, he owed almost $30,000 to seven people, five of whom were from out of town: two from Boston, and one each from Philadelphia; Lawrence, Massachusetts; and Springfield, Kentucky. Most likely these were family members or friends. If not, they may have been individuals or companies who bought "shares" of the company in return for a reliable and fixed source of goods or a market for their products. However, even though Chicago firms secured loans from outside the metropolitan area, most financial links were local. Despite the regional orbit in which Bennett operated, the local relationships that he created were fundamental to the supply of loans.[7]

In some cases, Chicago manufacturers received loans from other businesses. One of Henry Bennett's creditors, the "merchant-capitalist" W. D.

Houghteling, lent $1,260. He was also president of one of the city's largest lumber firms, Menominee River Lumber Company, and probably invested in the box company because it provided an excellent market for his company's lumber. In another case, Warner Proprietorial Medicine owed $8,109 to Edward Lawrence, the owner of the Northwestern Distilling Company, for cash advances and notes, and for legal support through access to his lawyer. Moreover, as with Bennett and Houghteling, the Lawrence-Warner link is a case of vertical integration along the supply chain, as the distilling firm supplied Warner with spirits. Similarly, Frank Thomas, a Des Plaines hotel keeper, supplied Wheeling Brewery with money so that it could supply his hotel. In other cases, businesses may have switched capital from one sector to another in order to widen their investment portfolio, such as a Chicago boot and shoe merchant who invested in Schureman and Hand.[8]

Chicago manufacturers borrowed money from sources other than banks and companies. One was industrial associations. These institutions negotiated labor-manufacturer relations, coordinated information on markets and prices, and provided insurance and capital to its members. For example, the Independent Brewing Association lent money to Wheeling Brewery, while Independent Electrotypers was in debt to the Electrotypers Association. Family members were another source of capital, many of whom were female.[9] In one case, the Union Screw and Bolt Company was financed by 16 Chicago-based merchants, bankers, lawyers, and real estate agents who had "bonds in the firm." Accounting for $109,902 of the total debts of $123,425, the bonds debt ran as high as $24,000 to Peter Page, a Chicago capitalist, and $12,550 to Henry Waller, the company president, and as low as $500 to T. Prosser, a machinist and inventor. Two banks, the Marine Company of Chicago and the Mechanics National Bank, lent $10,000, while iron merchants, attorneys, and capitalists enjoyed varying stakes in the company. Union Screw was an enterprise bolted together by a consortium of Chicago entrepreneurs and financial institutions. Some, such as Prosser and E. K. Rogers, an iron merchant, had direct metalworking experience. In other cases, such as with Edward Waller, a real estate dealer and brother of the company president, they were simply interested in making money out of screw manufacture. While it is possible, but unlikely, that Page and the Waller had ties with the metal industry, they did make links with people like Prosser and Rogers who did.[10]

Most service and wholesaling inputs were supplied by local businesses, individuals, and institutions (table 12). This was partly to do with the ubiquitous nature of these goods. Manufacturers, however, were intimately tied to local non-manufacturing sectors because spatial proximity was essential to their interaction. All industrial centers, regardless of their size, had plumbers, financial institutions, coal dealers, and transportation services. In some cases, services were directly place-related. Membership in a local association, owning property, or the provision of electric power and telephone service was linked to the business owner's social world, rooted in a set of land and property transactions and upkeep, or wired to infrastructures that were only accessible through a material landscape bounded by the technology of the day. In other cases, most notably advertising, financial, and legal matters, the importance of face-to-face contacts was critical. Traded interdependencies such as legal support for a host of issues, from applying for a patent to filing for bankruptcy, required extensive face-to-face contact that was extremely difficult to substitute for at a non-local level.

Spatial proximity was also critical for firms requiring just-in-time delivery of raw materials, semi-processed inputs, and finished parts. For various reasons, ranging from the need to keep inventories low to unforeseen demands for their products, many manufacturers expected immediate delivery of inputs. The ability to do this depended on freight time. In the pre-truck era this was severely constrained, despite the revolutionary impact of the railroad. At the regional level, the railroad, despite frequent service and quick delivery times, did not have flexibility outside the narrow corridors of the iron rail. Moreover, entrepôt costs associated with transferring products from the train to another transportation mode were time-consuming and expensive, especially for bulky or one-off items. At the metropolitan level, the horse and cart was slow, clumsy, and expensive for small lots, while the streetcar was rarely used and suffered from being attached to the rail line. Under these conditions, a successful urban economy demanded local facilities that delivered dependable products in quick time and at low cost.[11]

Three types of suppliers controlled the delivery of inputs: the specialized retail store, other manufacturers, and jobbers. The former, while important in the city's early days, became redundant as Chicago developed a more specialized economy. Most businesses acquired their goods directly from local manufacturers and wholesalers. Between them, they accounted for 61 percent of all the links between bankrupt companies and creditors (table 12). In most cases, the creditor was used by only one failed firm. This was typically the case with specialized products such as fruit syrup for confectioners and printing presses for printers. In other cases, some businesses were used by more than one firm. Eugene Arnstein, the bicycle

TABLE 12. Chicago Creditors by Business Sector, 1872–1878, 1898–1901

Sectors	Creditors		% of Total	
	No.	$ Owed	No.	$
Manufacturing	581	130,863	36.8	14.2
Finance	129	492,267	8.2	53.6
Distribution	390	138,716	24.7	15.1
Service	371	111,338	23.5	12.1
Advertising and Publications	15	3,331	1.0	.4
Contractors	21	6,590	1.3	.7
Legal	32	18,834	2.0	2.1
Real Estate	22	17,251	1.4	1.9
Services, General	246	58,984	15.5	6.4
Transportation	7	749	.4	.1
Utilities	28	5,599	1.8	.6
Not Available	106	45,396	6.7	4.9
Total	1,577	918,580	100.0	100.0

SOURCE: Compiled from 44 bankruptcy cases for 1872–1878 and 1898–1901 taken from Record Group 21, Records of the United States District Court, Northern District of Illinois, Eastern Division at Chicago, Bankruptcy Records, Act of 1867, 1867–1878; Bankruptcy Case Files, 1872–1878; and Bankruptcy Records, Act of 1898, 1898–1972; Bankruptcy Case Files, 1898–1946.

sundries jobber, supplied four of the bicycle firms in the sample, while machine supplies wholesalers and a large hardware distributor sold goods to several metal fabricators. The same was true for manufacturers; some furnished specialized goods for a single firm, while others supplied, among other things, paper, leather belting, emery board, brass goods, tacks, glass, and oil goods to many of the bankrupt companies.

There was a metropolitan geography to the supply of goods and services (table 13). The Loop had both the largest number of creditors and the greatest concentration of services of all kinds (general services, utilities, legal, newspaper, advertising, finance, and real estate). While it had a significant wholesaling complex, it was well underrepresented by manufacturers. The rest of the central zone, in contrast, had low shares of many of the specialized services found in the Loop, and the largest single cluster of manufacturing creditors along with numerous wholesalers. In the more distant industrial neighborhoods, all sorts of creditors supplied the bankrupt firms. While manufacturers clearly dominated in terms of numbers, neighborhood newspapers placed advertisements, lawyers offered legal services, wholesalers furnished goods, and banks and family members supplied loans.

TABLE 13. Distribution of Creditors by Geographic Districts, 1872–1878, 1898–1901

Sectors	Number and Share in the Loop		Number and Share in the Center Zone[a]		Number and Share in the Rest of the City	
	No.	%	No.	%	No.	%
Utilities	26	93	0	0	2	7
Legal	25	78	3	9	4	12
Newspapers	14	78	0	0	4	22
Advertising	7	58	0	0	5	42
Wholesale	212	56	94	25	71	19
Finance	66	54	19	16	37	30
Real Estate	10	50	8	40	2	10
Contractors	9	47	5	26	5	26
General Services	75	41	55	30	53	29
Manufacturing	195	34	228	40	144	26
Total	639	42	412	27	469	31

SOURCE: Compiled from 44 bankruptcy cases for 1872–1878 and 1898–1901 taken from Record Group 21, Records of the United States District Court, Northern District of Illinois, Eastern Division at Chicago, Bankruptcy Records, Act of 1867, 1867–1878; Bankruptcy Case Files, 1872–1878; and Bankruptcy Records, Act of 1898, 1898–1972; Bankruptcy Case Files, 1898–1946.
NOTE: Taken from Chicago firms only, and not from non-Chicago firms having a Chicago office or warehouse.
[a]Center Zone = Northwest Side and West Town.

The business links of Chicago companies, however, did reach beyond metropolitan limits. Close to a third—30 percent of firms and 29 percent of money owed—of all firm transactions were with non-local creditors. But local relations differed in important ways from non-local ones. One major geographic difference had to do with creditors who supplied inputs. Firms and individuals outside of Chicago were much more likely than their local counterparts to supply manufactured goods: more than two-thirds of the former were manufacturers, while the corresponding share for the latter was less than 40 percent (table 11). In contrast, the number of non-local creditors providing services, finance, and jobbing was much lower than local ones. In other words, spatial proximity was not the only factor driving the decision about where to obtain their manufactured inputs. Chicago firms also based their purchasing decisions on close linkages that were relational (at a distance) rather than spatially proximate; they targeted specific manufacturers outside of Chicago.

Several reasons explain the decision to acquire semi-manufactured goods and finished parts from outside metropolitan Chicago. One is that local proprietors had past or ongoing personal or business connections with manufactured-goods suppliers outside of Chicago. As Gordon Winder demonstrated in his study of a Brockport reaper maker, process-based trust with its faith in the integrity of a long-term partner was one element of nineteenth-century production practice. This is impossible to verify with bankruptcy records. There can be little doubt, however, that Schureman and Hand's strong financial and manufacturing links to Cincinnati played a pivotal part in that Chicago company's business dealings. If it was possible to disentangle the biographies of the local bankrupt firms' proprietors based on correspondence or court records, there can be little doubt that a similar strong link would be found in other cases under consideration here.[12]

A second reason was that more than a third of non-local manufacturing creditors kept a Chicago office, warehouse, or agent. This allowed them to be competitive with local manufacturers. By keeping stock and an office in Chicago, the non-local enterprise effectively and quickly fulfilled local demands, even though their plant may have been more than 1,000 miles away. Take the case of the publisher Lakeside Publishing, which filed for bankruptcy in 1877. It acquired paper from Cleveland and New York City, printers' ink from Philadelphia and New York City, and type foundry goods from New York City. All of these firms maintained offices and showrooms in the Loop, a short distance from the Lakeside plant. In some cases, the firms offered Lakeside immediate delivery from stock; in others, they gave expedited service. This close geographic relationship was true for others. Atlas White Lead bought linseed oil, machine instruments, varnish, and lead from firms in Cleveland, Detroit, Newark, and Boston, all of which kept offices downtown. From their office in the Loop, the clothing manufacturer Buchanan and Reens acquired trimmings, wool, cotton, and felts from 16 Boston; Philadelphia; Plymouth, Massachusetts; and New York City manufacturers.[13]

A third reason Chicago firms reached far afield was that manufacturers had to acquire specific and locally unavailable products. The development of a specialized regional industrial system after the Civil War laid the basis for the geography of supply. Other than New York and perhaps Philadelphia, there can be little doubt that the range of products available in Chicago could not be replicated anywhere else in the country. This did not mean, however, that all inputs were available at the needed specifications or at the desired cost. Using trade journals, past personal connections, or word of mouth, and having a sense of the regional geography of their industry, manufacturers sought out specific products from a wide set of places.

The reach of Chicago's firms outside the metropolitan area was not arbitrary. Specific places in the Manufacturing Belt supplied each industry with specific products. Take Chicago Shoe and Slipper's relationship to Boston. As the nation's major shoe center, Boston maintained the greatest selection of shoe-related goods in the country. Chicago Shoe took advantage of this, depending heavily on the Massachusetts city for leather goods, shoe rivets, shoe fasteners, shoe blacking, rubber heels, shoe machinery, cotton waste, and thread. The Chicago firm was directly plugged into the shoe-making capital's emporium. Boston was not unique in this regard: New York City and Philadelphia supplied textiles, trimmings, woolen goods, and dress goods from their massive textile complexes; Pittsburgh, Cleveland, and Toledo furnished products from their steel mills, foundries, and metal shops. Chicago's relations with businesses outside the metropolitan area followed distinct channels greatly influenced by urban specialization.[14]

Chicago firms were rooted in the Manufacturing Belt. More than 92 percent of creditors were found in America's major industrial region (fig. 14). The South and the West regions were not important suppliers. While these vast areas were markets for Chicago goods and served as suppliers of raw materials, they offered few inputs for the city's production system. More than 60 percent of the firms furnishing goods to Chicago firms were from New York (26 percent), Ohio (15 percent), Massachusetts (12 percent), and Pennsylvania (9 percent). Easily the most important metropolitan area was New York with 159 firms, followed by Boston (59), Philadelphia (38), and Cleveland (32). Most other major industrial centers in the Manufacturing Belt—from Akron to Cincinnati, Indianapolis, Detroit, St. Louis, and Pittsburgh—supplied Chicago firms with an assortment of manufactured products.

LINKAGES IN CHICAGO'S BICYCLE INDUSTRY, 1898–1901

Five bicycle companies taken from the 1898–1901 sample illustrate the metropolitan and regional linkage lines of Chicago manufacturers. The five, all of which were located in the Loop or the adjacent industrial districts, had a total of 388 creditors to whom they owed just more than $50,000.[15] The Chicago bicycle industry developed in the late 1880s and

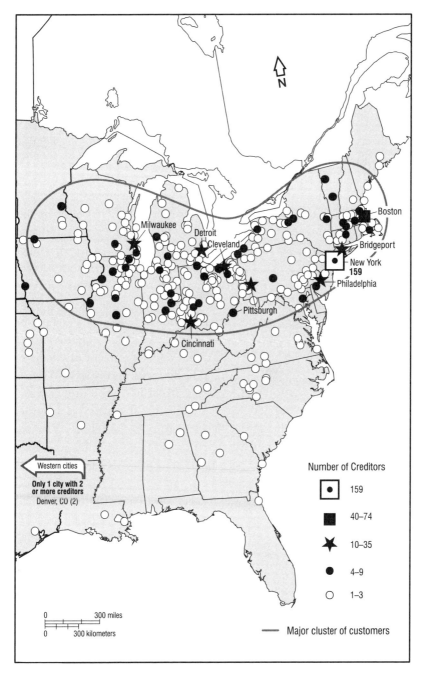

FIGURE 14. The Location of Non-Chicago Creditors, 1872–1878 and 1898–1901.
Source: Records of the United States District Court, Northern District of Illinois,
Eastern Division at Chicago, Bankruptcy Records, Act of 1867, 1867–1878, and
Act of 1898, 1898–1972.

the mid-1890s when the "bicycle craze" fueled growth of a small industry. A handful of firms across the nation at the time of the 1890 census developed into 300 producing more than a million bicycles by 1896. While a small number, such as Pope Manufacturing, were large and vertically integrated, most were small operators. The interchangeability of specialized bicycle components permitted small enterprises to be competitive with the large integrated ones. After 1897, however, the market for bicycles dissolved, resulting in a major industry shakedown. Despite consolidation at the end of the century, the bicycle market continued to tumble, and for those Chicago firms unable to switch production lines or forge a niche in the restructured industry, the only recourse was bankruptcy.[16]

The five bicycle makers' creditors were from all walks of business life. The most common and important were manufacturing (193; $24,151), distribution (89; $3,860), service (37; $1,531), and finance (14; $15,376), while a small number were found in the utility, advertising, and real estate sectors. The bike companies obtained financing from several quarters. In some cases, banks such as the National Bank of Chicago supplied loans, while in other cases, money originated from individuals (a teacher and a pawnbroker) and family members (Mrs. J. Kass of Lake Park, Iowa). Loans also came from two bicycle businesses. Frank Wenter, president of Gladiator Cycle, provided Artemis with $700, and R. Boericks, the owner of Western Wheel, was owed $1,200 by Hero Cycle. Most likely the former was a retailer ensuring the supply of bicycles and parts. The latter, on the other hand, involved one of the largest bicycle makers in the country. Western Wheel was either buying into a competitor and thus extending its market or purchasing a market for the parts made at its factory. Other than financial support, manufacturers, wholesalers, and service providers furnished the companies with the necessary production inputs. The bicycle firms were reliant on a host of suppliers, from coal to advertising, real estate and legal services, and the full gamut of intermediate inputs (bicycle hardware, shafting, steel balls, leather seats, and tires).

The Chicago bicycle businesses were also deeply imbricated in the regional manufacturing sphere. More than 120 non-local creditors accounted for 31 percent of the firms' total number of creditors and 38 percent of the money owed. Many were Manufacturing Belt suppliers of specialized goods: armatures from Syracuse and Fulton, New York; bicycle sundries from Buffalo, Milwaukee, Cleveland, and Kokomo, Indiana; handlebars from Shelby, Ohio, and Syracuse; rubber goods from Akron, Trenton, Boston, and Erie; steel from Pittsburgh, Cleveland, Boston, Toledo, and De Kalb, Illinois; and wire from Jonesboro, Indiana, and Worcester, Massachusetts. As in other industries, bicycle manufacturers obtained parts from specialized producers outside the metropolitan area. In particular, New England, the industry's main production center, was home to many creditors. Chicago Tube received drop forgings from a specialist company in Springfield, one of the major bicycle-producing centers in the United States and home to Pope Manufacturing, the country's largest manufacturer. Likewise, numerous towns in Massachusetts (Ayer, Boston, Framingham, Northampton, Upper Falls, Worcester) and Connecticut (Banberg, Bridgeport, Hartford, Salisbury, Torrington) supplied a full selection of semi-manufactured parts.

Chicago bicycle makers had a two-tier set of business relations. They forged direct links to Manufacturing Belt manufacturers and depended heavily on local companies and individuals for almost every production input. Chicago's 266 creditors accounted for 69 percent of firms and 62 percent of the money owed. Unlike non-local creditors, which almost entirely supplied manufactured goods, Chicago ones furnished goods and services along the entire production chain; they offered all of the advertising, financial, real estate, contracting, and legal services, and supplied most of the goods obtained from wholesalers. For example, 30 wholesalers supplied Chicago Tube with an assortment of products, while various service companies provided, among other things, telephone, livery, and horseshoeing. The frame maker, like the other four firms, acquired semi-manufactured products from numerous local firms. In August 1900 F. Waters, the trustee of Chicago Tube's estate, paid A. Bagley for nickel-plating 328 frames, 250 sets of chain adjusters, 161 crowns and tips, and 75 seat posts; Independent Supply for head fittings; Edwin Hartwell Lumber for 700 wooden crates; Morgan and Wright for expanders; Beckley and Ralston for guards, rims, clusters, and hubs; International Stamping for ladies' guards; Chicago Enameling for enameling 153 frames and 50 pairs of rims; H. W. Fauber for hanger fittings and brackets; Ewald Manufacturing for "eccentric" hangar fittings; and Bicycle Equipment for 125 pairs of rims and 10,000 balls. Chicago Tube was tightly entrenched in a significant local complex of manufacturers and wholesalers that serviced Chicago's bicycle industry.[17]

The five bike companies' locations in the Loop and the adjacent industrial districts ensured that they had close connections to numerous suppliers (manufacturing and wholesaling), markets for finished goods (wholesalers, retailers, and other manufacturers), financial, legal, and real estate services, and transportation facilities (rail and water). Most Chicago creditors were close to the five firms (fig. 15). A belt running west from

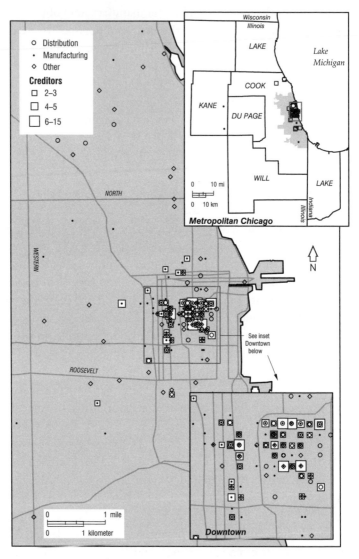

FIGURE 15. The Metropolitan Chicago Geography of Five Bicycle Firms' Creditors, 1898–1901. Source: Records of the United States District Court, Northern District of Illinois, Eastern Division at Chicago, Bankruptcy Records, Act of 1898, Act of 1898, Bankruptcy Case File 1995, *In the Matter of Northern Cycle and Supply Company*, "Petition" (November 1, 1899) and "Affidavit of Mailing Notices of Final Meeting of Creditors" (July 8, 1901); Bankruptcy Case File 2451, *In the Matter of the Artemis Plating and Manufacturing Company*, "Petition" (February 8, 1900) and "Affidavit of Mailing Notices of First Meeting of Creditors" (February 9, 1900); Bankruptcy Case File 2579, *In the Matter of the Hero Cycle Company*, "Schedule" (February 8, 1900); Bankruptcy Case File 3433, *In the Matter of Woltz Wheel Works*, "Petition" (April 17, 1900); Bankruptcy Case File 4008, *In the Matter of Chicago Tube Company*, "Schedules" (July 14, 1900).

the Loop across the South Branch and north into the West Town and Near North sections of the city accounted for 221 of the 259 firms (85 percent) for which an address could be found. Each company had their own bounded supply field that it drew upon. The Near North Side's Hero Cycle, for example, drew materials from the neighboring districts in which other competitors were not active. As well as accounting for most of the Near North Side creditors, Hero also bought goods from firms located in the district to the north, Lincoln Park.

LINKAGES OF BENNETT BOX AND J.G. HOFFMAN IN 1899 AND 1900

Two firms illustrate the local and long-distance ties of Chicago firms and the types of relational and spatial proximity that played a part in their industrial history. The first is Bennett Box Manufacturing, a wooden box maker employing 53 workers. The Bridgeport company, at the time of the voluntary bankruptcy in February 1899, was operated by two partners, James Shields and David Bleloch. Bennett Box had been around in various guises for some time. It was the latest manifestation of Henry Bennett's planing mill, which was established as early as 1874 at the same address as Bennett Box. Henry Bennett went bankrupt in 1876, but the firm was restarted by Alexander Bennett as early as 1881 and continued operating under his control until the mid-1890s, when it was taken over by Shields and Bleloch.[18]

With extensive assets and a large workforce, Bennett Box was a substantial firm with a long lineage. The company's smaller assets were $40 cash on hand, $225 of stock (lumber, oil, grease, wool waste, box racks, and "unfinished boxes"), 17 horses worth $500, and wagons (for delivery of finished goods) and trucks (for moving parts around the box mill), valued at $500. Of much greater value was the firm's machinery, valued at $3,500. Bennett Box had a full selection of woodworking machinery, including a 30-inch planer, two molding machines, two sliding saws, various saws (box, re-saw, and rip), two grooving machines, a boring machine, a box matcher, and a printing press. Finally, the firm had company stocks worth $5,000 in the Shaving Sawdust Company, a neighboring firm across the river, which gave it access to a captive and stable market for its sawdust and shavings.

When it failed, Bennett Box did not have a large client base. Only 41 individuals and businesses with debts just more than $1,500 owed money to the box manufacturer. All of the debtors were from Chicago and owed small amounts, the largest being $165 by the department store Marshall Field's. The firm supplied manufacturers, wholesalers, and retailers with boxes of various grades and sizes. Paint, confectionery, and iron wholesalers; hop merchants; shoe and tobacco manufacturers; confectioner suppliers, printers, and department stores all bought boxes from the Bridgeport mill. While it was not a large client base, Shields and Bleloch serviced many different parts of the Chicago economy and contributed one element to the production chain of many local companies.

Shields and Bleloch voluntarily filed for bankruptcy, even though their assets and debts due were valued at more than what they owed to other firms. Why this was the case is unclear. One answer could be that the difference between money owed and owing as shown in the case file did not take account of earlier debts that were assumed before the company filed for bankruptcy. This may be unlikely, however, as the 1898 Bankruptcy Act created mechanisms to stop favored payments before the businesses went bankrupt.[19] Records reveal that Bennett Box owed $3,002 to 22 creditors at the time of the bankruptcy. The largest debt was $750 for a chattel mortgage on machinery taken from David Bain, paid by reassigning the debts on open accounts of several firms due Bennett Box. The next largest debt, not surprisingly, was $656 to four lumber companies, which supplied Bennett Box with its major raw materials. Other creditors provided an assortment of goods (hay for the horses, emery wheels, metal strapping, oils) and services (railroad cartage, rented space) that supported the making of wooden boxes.

Many of Bennett Box's creditors were part of a locale effect—the geographic relations in which workers and suppliers were either located in close proximity to the planing mill or the Loop (fig. 16). Reflecting the short employment field of the nineteenth century, most of the 53 workers lived within a very close distance of the plant: 42 lived within a mile, and 10 lived between one and two miles from the plant. Only one had to commute more than two miles to get to work. All workers were within easy walking distance to the plant, thus ensuring that one input into the production process was local. The lumberyards of CL Willey, Globe Lumber, Edward Hines Lumber, and L. Williams were also in close vicinity to the mill. Spatial proximity allowed Bennett Box to quickly acquire its major raw material at a moment's notice. Equally important given lumber's size and weight was that proximity cut down on transportation costs. Given the great variability of lumber, proximity allowed mill owners to check the grade and quality before they purchased. Similarly, a neighborhood plumbing supplier, machinist, blacksmith, and printer supplied their wares

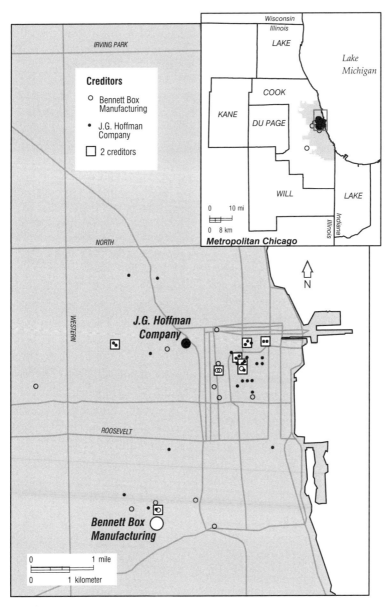

FIGURE 16. Metropolitan Chicago Geography of Bennett Box and Hoffman Creditors, 1899 and 1900. Source: Records of the United States District Court, Northern District of Illinois, Eastern Division at Chicago, Bankruptcy Records, Act of 1898, Bankruptcy Case File 490, *In the Matter of Bennett Box Manufacturing Company*, "List of Creditors and Debt Amounts" (February 8, 1899); Act of 1898, Bankruptcy Case File 2317, *In the Matter of J.G. Hoffman Company*, "Lists of Creditors and Debts Amounts" (February 1, 1900) and "Debtors Sheet" (November 27, 1900); Act of 1898, Bankruptcy Case File 2338, *In the Matter of J.G. Hoffman Company*, "Petition" (January 23, 1900).

to the box manufacturer. This locale effect—linking firm and suppliers, workers and customers—was important for many firms other than Bennett Box. Providing quick service, and in many cases built on face-to-face negotiation and trust, the locale effect linked together and strengthened local neighbor economies.

Like Bennett Box, J.G. Hoffman Company was a woodworking company, but one at a different place on the production chain.[20] The former played an intermediary role: it took a raw material, lumber, and transformed into a product, wooden boxes, which then went as an input to other manufacturing industries. As a picture frame and art novelty maker, however, Hoffman directly supplied the retailer and the final consumer. The company had lots of scattered customers. At the time of the bankruptcy in January 1900, it was owed money by 247 firms and individuals to the tune of $4,643. None of the debtors were in debt for particularly large amounts; the largest were $211, $150, and $133 by individuals from Wilkes-Barre, Pennsylvania, Chicago, and Milwaukee, respectively. The market for its goods was extremely wide; only a quarter (63) were from Chicago, while the majority were from the Manufacturing Belt, although clients were located as far away as Atlanta and Greenfield, Texas.

Hoffman's list of creditors reflects the nature of the business. Several production inputs came from a small number of non-Chicago enterprises: chemicals from Boston, varnish from South Bend, Indiana; paints, fancy brass goods, and shellac from New York City; and bronze powders from St. Louis. But most goods and services were obtained locally. An assortment of institutions, individuals, and businesses furnished wholesale goods (plate glass, paper, and lumber), financial support (First National Bank and a capitalist, George Williams), and services (teaming, telegraph, and stenography). Most local suppliers provided semi-finished goods, from paper and paints to moldings and metal goods, that went directly into Hoffman's production process. In this way, Hoffman was typical of so many other Chicago firms. It was also typical in that most of the creditors were to be found a short distance from Hoffman's workshop at Kinzie and Peoria. Of the 29 Chicago addresses that can be verified, 24 were from the Loop and the districts close to Hoffman. Like with Bennett Box, a locale effect was part of the frame maker's business dealings.

Firms were embedded in the business world at various geographic scales as their production chain pulled them into spheres of interaction that operated outside their proprietorial boundaries. Most companies at this time functioned at three spatial scales: the local, the metropolitan, and the regional. There is compelling evidence to demonstrate that relational proximity was fundamental to how firms operated. Relational proximity did not necessarily involve spatial proximity, but revolved around the ability of manufacturers to construct relationships with other businesspeople that were not dependent on geographic closeness. In many cases, these inter-firm relations were short distance; while in others—whether it was the link of Chicago Shoe and Slipper to Boston, bicycle makers to New England, or Schureman and Hand to Cincinnati—inter-firm relations extended across metropolitan districts and regions.

Spatial proximity was also important. The effective working of an enterprise depends not only on its ability to function competitively in the marketplace. Firms also had to be part of a viable local environment; they did not function in a placeless world. Without the appropriate external surroundings, a business's ability to generate competitive strategies was severely constrained, if not impossible. A firm's working environment had to contain many locational assets, from employment agencies, land developers, and ethnic churches to cheap water supply and an interventionist and highly regulated municipal government. Many of the Calumet suburbs were examples of this locational world. In other cases, such as the establishment of the Corn Products Refining Company at Argo on the metropolitan area's western edge just after 1900, the appropriate environs involved minimal trappings, a relatively hands-off local government, and the barest range of infrastructures. In other cases, large corporations, such as Gary's U.S. Steel and Pullman's Pullman Palace Car, created extremely well-planned environs for their industries and, sometimes, for their workers.[21]

BUSINESS LINKS AND LOCAL ASSOCIATIONS IN CHICAGO HEIGHTS AND HAMMOND

Each type of locational asset creation involved a different set of inter-firm relations. While a simple market-based corporation shaped the building up of the metropolitan edge, the development of joint state-private ventures was by far the most common form of new industrial settlement on Chicago's fringe. These joint ventures involved a variety of private interests and public institutions working together in a local alliance. Together these alliances formed locational assets that underpinned the emergence of industrial suburbs on the metropolitan edge. The manner in which the inter-firm networks operated in this suburban building is examined through the lens of two suburban companies that went bankrupt in 1899 and 1901: the Hill Cart and Carriage Company of Chicago Heights and the Hammond

Rolling Mill Company of Hammond.[22] The two towns, Chicago Heights and Hammond, developed as diversified industrial suburbs of the Calumet district, the great industrial complex south of the central city running from Gary, Indiana, to Blue Island, Illinois.

Both suburbs were established in the last quarter of the nineteenth century by land associations and individuals. Formed in 1891 by a Chicago syndicate that included the well-known Charles Wacker as well as Martin Kilgallen and William Donovan, the Chicago Heights Land Association built a manufacturing center on the edge of metropolitan Chicago. Wacker (1856–1929) was one of the city's leading civic citizens at the turn of the century. Born in Chicago to German immigrants, he made a fortune in brewing. By the 1890s he had switched his attention from manufacturing to investing in suburban real estate and engaging in civic activities. Along with his interest in turning sleepy Chicago Heights into a modern manufacturing suburb, he was a member of the board of directors and various committees of the 1893 Columbian Exposition, the chair of the Chicago Plan Commission (1909–1926), and the secretary of the Chicago Zoning Commission. His interests in shaping metropolitan Chicago, made manifest through his work on the Exposition and the *Chicago Plan*, were further extended through the creation of manufacturing suburbs on the metropolitan fringe.[23]

Wacker's consortium quickly transformed Chicago Heights. The Land Association took advantage of new railroad connections and mounted an aggressive sales promotion to offer 1,260 acres to manufacturing firms and to build up population to work in the factories. In light of the problems associated with single-industry towns, a major objective of the Land Association was to provide a diversified industrial base. There were several reasons for not being dependent upon any one line of manufacture, including not being controlled by one large corporation, the provision of a range of working opportunities, and cushioning the blow of economic decline in any single industry. The Land Association was successful. As one contemporary noted, "The work originally mapped out has been carried on tenaciously and aggressively, yet always conservatively, in the face of many inevitable obstacles and disappointments." Even though the initial development led to overspeculation—more than 14,000 residential lots were put on the market—several manufacturers did move to the area. This generated a multiplier effect through the new suburb. As workers moved into the district, housing demands outpaced supply and the Land Association was forced to build 500 houses to remedy the situation. By

1896 about 22 industries were established, including American Manganese Steel, American Brake Shoe, Inland Steel, and Victor Chemical.[24]

The formation of the industrial suburb of Hammond was tied to the ability of one entrepreneur to bridge his interests with those of other manufacturing and financial agents. In 1869 Marcus Towle—in association with George Hammond, a meat businessman, and several other investors—opened a slaughterhouse on the Grand Calumet River. Towle, like many Chicago entrepreneurs of his day, originally came from out east. Born in Danville, New Hampshire, he moved to Massachusetts as a teenager, where he learned the butchering business. After moving to Detroit in 1865, he worked with Hammond and Caleb Ives, a local banker, to ship dressed meat to the large urban market of the eastern seaboard. The investors traveled to Chicago, where they chose the site for the packinghouse and bought most of the land that would become Hammond. At first there was little settlement, but in 1875 Towle platted the town. To ensure the fledgling town's success, Towle sold lots to packinghouse employees and furnished the lumber and money to build the houses. Over the next quarter century, in association with other business interests, he built a dock on the Grand Calumet River and opened a lumberyard, saw and planing mill, and sash and door factory. In 1886 he organized the First National Bank and used it to finance his share in the establishment of the Chicago Steel Company. Towle's business contacts and access to financial resources allowed him to fund, along with James Young, another Hammond investor, the building of the Chicago and Calumet Terminal Railroad in 1888. The railroad connected Hammond with the Chicago trunk-line railroads. A year later Towle and Young sold their interests to Joseph Torrence, one of the founders of East Chicago.[25]

By the late nineteenth century, these initial alliances had outlived their usefulness, and a new set of associations and individuals took up the task of building up Chicago Heights and Hammond. The Industrial Club of Chicago Heights was formed in 1911 in order "to advance the knowledge pertaining to the industries of this district, broaden acquaintance, exchange experience and extend the general welfare and social courtesies between the representative business and professional men of the city." This broke into two different organizations. The Committee for Chicago Heights developed the local economy by attracting new plants. Complementing this, the Manufacturers Association of Chicago Heights worked to increase production efficiency, inform manufacturers of legislation, promote good public relations, and assist in labor training and recruitment. Similar events were occurring in Hammond. By the end of the century, a new breed of local entrepreneur, a mix of American-born and German immigrants, had taken over the running of the town. Over the next 25 years, the reins of local control shifted to the Hammond Industrial Commission, which—in conjunction with a leading real estate company, Gostlin, Meyn Realty—drove the city's industrial development.[26]

Chicago Heights's and Hammond's diversified local economies were influenced by the tightly bound set of associational relationships that developed in the two suburbs. Despite diversity, the two suburbs' industrial development coalesced around specific sectors, which firmly situated them as specialized local economies within the wider Calumet production complex. This in turn was built on firms that operated at various spatial levels. The earliest days of the suburbs were bound up with the development of local inter-firm linkages, which continued over the generations. Hammond's first plant, the slaughterhouse of George Hammond established in 1869, set up a path-dependent trajectory around animal by-product manufacture. J. M. Hirsh opened an albumin factory in 1874, which was later converted to a glue and fertilizer factory. By 1915 it employed 400 workers and produced about 5 percent of U.S. total glue production. By the 1920s Hammond had another glue factory, three animal-feed makers, a livestock remedy laboratory, and a boiler-feed chemicals plant.[27]

This industrial and promotional world reveals the geography of the two bankrupt firms' business decisions. Hill Cart and Hammond Rolling Mill generated inter-firm linkages that functioned at three geographic scales. First, they were part of the wider Manufacturing Belt regional flow of goods. As with the firms discussed earlier, there was a specialized geography within the wider regional sphere. Hill Cart received carriage parts and metal goods from Cincinnati, a major carriage-making center, and other carriage-related products from other carriage-making districts—Detroit, Milwaukee, New York City, and Moline. In contrast, Hammond Rolling Mill obtained most of its non-Chicago inputs from the Pennsylvania heavy industry and specialized metalworking districts of Pittsburgh, Erie, and Philadelphia. This was true for other suburban firms.[28]

In the second place, the firms were linked to Chicago's downtown, despite being more than 20 miles away. Here a multitude of wholesalers, banks, service providers, utilities, and manufacturers furnished the suburban firms with goods and services not available locally. As the commercial, financial, legal, and administrative center for all of metropolitan Chicago, the Loop's business and financial interests played a pivotal role for most suburban firms. Similarly, the central factory districts with their well-stocked supply of firms from most industrial sectors furnished an assortment of manufactured inputs. Hill Cart's and Hammond Rolling Mill's webs of networks were spun as far away as thousands of miles to the east and the south and 20 miles north to Chicago's central business district.

There was also a distinct local geography. The Calumet district in general and the two suburbs in particular supplied the two firms with workers and manufacturing inputs. There was a tight employment field. Seven of the carriage maker's eight workers lived in Chicago Heights, while the other lived in nearby Glenwood. This was not unusual. As other writers have noted, workers in South Chicago and other Calumet industrial districts tended to live in very close proximity to their workplace. Even as late as the 1950s, more than 78 percent of workers employed in Chicago Heights factories lived within five miles of their workplace.[29]

Local firms supplied the two factories with raw materials and semiprocessed goods. Coal was bought from the Hammond Coal Company. The carriage firm obtained, among other things, carriage axles, tacks, paint, drills, and lumber from Chicago Heights, Pullman, and South Chicago. Similarly, machinery, foundry supply, hardware, and belting firms from the two latter districts as well as Hammond furnished the rolling mill. Hill Cart and Hammond Rolling Mill were not unique in this regard. Many Chicago Heights and Hammond firms used steel produced in the Calumet district, scrap material obtained locally, and brass, aluminum, and other nonferrous metals obtained from the area. All of the metal works were fabricators and received their semi-processed steel from local primary producers, especially from the U.S. Steel Mill in Gary after 1908. The metalworking complex bridged to the local railroad equipment sector, while chemical firms used local materials entirely and were dependent upon by-products, hydrogen, and solvents, received from the Whiting refineries.[30]

The Chicago Heights and Hammond firms were part of local, metropolitan, and regional circuits of manufacturers, wholesalers, finance, and services. This was true for other firms. Before 1900 the business contours of an industrial district were discerned. Associated with the rapid growth of metropolitan Chicago's industrial economy was the development of a distinctive array of local inter-firm relations. Chicago was like Philadelphia and Detroit, whose extensive range of textile and automotive firms were deeply embedded in a distinctive array of local suppliers, manufacturers' and workers' associations, financial organizations, service providers, and institutions. The local world of large metropolitan districts such as Chicago,

Philadelphia, and Detroit, with their wide and unique sets of formal and informal structures, formed the basis for industrial district development. But industrial districts were more than areas formed around local external and transactional economies. The linkages found in industrial districts such as Chicago, Philadelphia, and Detroit were not just local. Firms did not function in an enclosed world but had manifold links to non-local firms. Manufacturers sought out capital, services, and manufactured goods from firms far afield. In this period these links were almost entirely made with firms and individuals located in the Manufacturing Belt.[31]

CHAPTER 6

Forging the
Calumet District,
1880–1940

In a June 1925 radio broadcast, Theodore Robinson, Illinois Steel's vice president, announced that Chicago had become America's largest producer of pig iron. This was a cause for celebration in the Midwestern metropolis. Not only had Chicago steel firms outproduced their archrivals in Pittsburgh, the country's largest producer for more than half a century, but the news reinforced what local boosters believed to be true, that the Calumet region, Chicago's steel-producing district, was the "workshop of the world."[1] Chicago's success did not surprise Colonel A. Sprague, Chicago's commissioner of public works. In his opinion, "The possibilities of the Calumet region are without a discernible limit—the location of factories for the manufacture of steel and iron products. By an intelligent utilization of this opportunity, a system of vertical combination can be effected among groups of seemingly independent industries with practically all the economies which could be derived from such combination, if practiced within the major industry, and with the added advantage of broadening the source of capital."[2] The Calumet district had extraordinary supplies of the major locational factors—land, labor, capital, transportation, and "intelligent" entrepreneurs.

Nor did success surprise Arend Van Vlissingen, author of the 1920 Lake Calumet Plan and a member of J. H. Van Vlissingen and Company, one of Chicago's largest industrial real estate businesses. More than 10 years earlier than Robinson's crowing, Van Vlissingen asked if President Theodore

Roosevelt's assessment that "the Calumet Region is destined to become the greatest center in America" was true. His answer, not surprisingly for one of the Calumet district's most energetic land developers, was yes. Like the *Iron Age* writer, the reasons that he gave for this assessment were the district's "unexcelled railroad facilities, deep water navigation, proximity to market, accessibility to raw materials and nearness to the unexcelled labor market of the great city of Chicago." And like Robinson, he reveled in Chicago's rapidly expanding industrial economy.[3]

But there was more to the building of Calumet as an industrial district than the ability of local business leaders to harness land, transportation, labor, and markets in a planned development. Critical to the creation of a highly integrated industrial complex after 1880 was the building of strong intra- and inter-firm ties between district firms. This was not lost on contemporaries. In Chicago, according to the writer George Plumbe, "surrounding industrial plants are, almost invariably, a cordon of industries that draw their material from the parent plant." He must have been thinking of Calumet's steel mills and fabricating factories when he stated that "adjacent to great iron industries there will grow up a collection of collateral interests, such as factories for making machinery, large foundries, and mills of various kinds using the pig iron for the manufacture of iron implements of all kinds, from wire nails to steamship plates."[4]

For Plumbe, it was not just a matter of laying out the land, dredging the river, or building workers' homes that made Calumet successful. Nor was success due to the mere clustering of firms in the district. Just as importantly, it was interaction between plants. The movement of goods, information, capital, and other assets between firms within a district not only sustained existing relations and kept profits, wages, and innovations within the district; it also provided the basis for an important set of multiplier effects throughout the region. The communities—from paternalistic Pullman to corporate Gary and the developer-led towns of Harvey and East Chicago—provided the built environment in which firms were established, developed, and interacted with other firms, both within the Calumet region and elsewhere. Each community had a distinctive history and employment structure, and was connected in quite different ways to the rest of the Calumet district, as well as metropolitan Chicago and the Manufacturing Belt. The formation and development of the Calumet steel production complex between 1880 and 1940 was linked to businesses operating at various spatial scales. The geographical embeddedness of Calumet business was central to the district's rise to industrial prominence.

The Calumet production complex was put in place in the late nineteenth century. Before the 1880s the Calumet region consisted of a smattering of small satellite districts and a few firms. In Illinois the main districts were South Chicago (South Works and Iroquois Iron), Pullman (Pullman Palace Car), and Whiting (Standard Oil), while a few Indiana industrial suburbs, most notably Hammond and East Chicago, began experiencing the first pangs of industrial development.[5] Over the next 40 years, the numbers employed in Calumet mills grew from a few thousand workers to almost 160,000 (table 14). By 1924 the district accounted for almost a quarter of the metropolitan area's manufacturing workforce.

Calumet was a specialized district. Its heavy industry structure was dominated by the metal, transportation equipment, and chemical sectors. Blast furnaces, rolling mills, and foundries accounted for more than half of the district's employment, while the fabricated metal and machinery sectors employed 18,000 workers. The most important segment of the transportation sector was the railroad equipment industry with more than 26,000 workers. The manufacture of petroleum products and industrial-based goods such as paints was the major component of the chemical sector. While Calumet was dominated by a few heavy sectors, the district also accounted for a disproportionate share of employment in a few industries. Three—primary metal, railroad equipment, and petroleum—had more than 80 percent of all employees of the metropolitan area's industry working in Calumet. This is in contrast to other important Chicago industries, of which Calumet had very few firms and workers, most notably clothing, printing, furniture, leather, and meat.

Calumet was dominated by a small number of very large firms. The district's median firm size was 70, more than three times that of the metropolitan area.[6] From the U.S. Steel mill in Gary to the Standard Oil refinery at Whiting, the American Steel Foundries plant at Indiana Harbor and Pullman's railroad car shops, Calumet was home to some of the largest firms in the United States. Despite the large overall size in Calumet, the majority was much smaller. Almost 6 out of every 10 had fewer than 100 workers, while another fifth had fewer than 250 workers. Small and medium-size businesses played an important role in some industries such as fabricated metal, food, chemicals, and machinery. In other words, the Calumet industrial sector consisted of the large, vertically integrated, multi-unit corporation with millions, even billions of dollars in capital, the

TABLE 14. The Industrial Structure of the Calumet District, 1924

Sectors and Industries	No. of Workers	No. of Firms	Median Firm Size	No. of Firms by Workers			% Share of City Workers	Location Quotient	% Female
				>999	100–999	1–99			
Metal									
Primary metal	85,553	46	250	14	18	13	81.5	3.7	1.2
Fabricated metal	13,976	56	62	2	20	29	16.8	.8	11.4
Machinery	4,534	23	33	1	4	15	12.0	.5	3.6
Transportation									
Railroad equipment	26,284	24	320	5	15	3	81.8	3.7	2.5
Automotive	5,752	16	32	2	1	9	25.3	1.2	2.5
Chemical									
Petroleum	7,086	10	194	2	2	4	80.0	3.6	.6
Chemical	4,190	28	42	1	8	15	17.6	.8	5.3
Other									
Food	5,583	31	30	1	6	22	6.4	.3	19.0
Non-metallic	1,656	22	26	0	5	14	17.3	.8	3.9
All others	5,175	85	23	0	16	59	1.6	.1	26.6
Total	159,789	341	70	28	95	183	21.9	1.0	4.0

SOURCE: *Directory of Illinois Manufacturers, 1924–1925* (Chicago: Illinois Manufacturers' Association, 1924).
NOTE: Only 306 of the 341 firms had employment figures.

small custom enterprise employing one or two workers, and the independent batch or routinized producer employing a few hundred workers in a specific market niche.

Calumet grew from an unsettled suburban area in the 1870s to one of America's leading industrial districts by the interwar years. As Sprague and Van Vlissingen noted, firms such as Brown Steel and Iron (1875), North Chicago Rolling Mill (1880), Pullman Palace Car (1881), Standard Oil (1890), and Inland Steel (1893) were attracted to South Chicago, Pullman, Whiting, and Chicago Heights to gain access to a combination of large amounts of cheap and flat land, excellent transportation facilities, and access to labor, raw materials, and markets. At the far eastern end of the Calumet district, the building of Gary in 1906 with its huge complex of steel-related plants consolidated the region's industrial might. By the 1920s, even though Calumet was still at a disadvantage with Pittsburgh in some respects, its later development gave it certain advantages. Most Calumet mills were modern compared to their Pennsylvania and Ohio counterparts. Chicago firms were much more likely to have more modern equipment, permitting the adoption of the latest practices for reducing costs of production, while eastern firms had many small, old-fashioned blast furnaces.[7]

The construction of elaborate and effective business relationships underpinned the development of the Calumet complex. These relationships involved the supply of finance, information, raw materials, and semi-processed products between different company plants (intra-firm) and between different companies (inter-firm). These interactions operated at various spatial levels; some businesses were vertically integrated and had many of their raw material, financial, and research links outside of the Calumet district, while others had few outside interactions and were locally based. The fact that most Calumet industries drew upon numerous suppliers and sold in different markets permitted co-industrialization to take place. Firms tend to be attracted to areas with firms making similar products or goods that are inputs into their production process. This generates a multiplier effect in the form of facilities and services within the area that are formally and informally organized for the industries: a specialized labor force, local agencies with contacts for particular skills and credit facilities, public utilities and infrastructures, and residential districts.

At the end of 1909, a survey of Gary reported that several companies, most of which were subsidiaries of U.S. Steel, had purchased land from the Gary Land Company "for plants which will surround the steel works, radiating like the spokes of a wheel from the source of supply for their raw material." Here was a planned industrial complex centered on a rational geometry of location. Leaving nothing to chance, the company orchestrated several links in one locale for the transformation of iron ore and coal into semi-processed steel products. U.S. Steel not only built the largest steel complex in the world at Gary, but assembled the constituent parts of a highly integrated firm in one place. In the process, it created the single most important company in Calumet's history. Combining the Illinois Steel mill (South Chicago), one of the earliest firms in the district, with an assortment of new steel mills, fabricating factories, and infrastructures, U.S. Steel constructed an elaborate Calumet-based intra-corporate flow of materials. The corporation's three main Calumet fabricating subsidiaries—American Sheet and Tin Plate, National Tube, and American Bridge—operated as downstream recipients of the ingots and merchant bars pouring out of the Gary and South Chicago blast furnaces. The Buffington Portland Cement plants were designed to produce cement from the steel mills' blast furnace slag. All of Calumet's U.S. Steel plants were directly linked by the Elgin, Joliet, and Eastern Railroad belt line.[8]

U.S. Steel was not alone in assembling different parts of the production chain in Calumet. The Grand Crossing Tack Company, which began production in 1883, made its own steel in a South Chicago mill built in 1902. The new plant's open-hearth furnaces and blooming mill sheared steel billets and shipped them off to the Grand Crossing fabrication plant. American Steel Foundries expanded after 1902 by constructing a new open-hearth mill at Indiana Harbor. Expansion also involved the acquisition of Calumet downstream clients. In 1905 it bought out Simplex Railway Appliance, a Hammond railroad car spring maker, and 14 years later, the Griffin Car Wheel Company, a car wheel manufacturer with a plant in Kensington. In another case, Inland Steel built an open-hearth furnace in 1907 at Indiana Harbor to ensure that its Chicago Heights rail mill received better and more uniform quality pig iron than was available on the open market, and the company bought Red Top Steel Post (Chicago Heights) as a market for its steel. Not only would Inland supply the "best steel for the purpose," but it informed potential customers that it would "serve you quickly" (fig. 17). From U.S. Steel to Inland Steel, the largest firms in Calumet were part of a deeply embedded network of local intra-corporate interactions.[9]

Calumet's machine shops, foundries, and factories functioned as both satellites to and customers of the large integrated mills. According to Phyllis Bate, "The mill locations were in close connection with the . . . consuming fabricating industries which soon built their factories in the areas

FIGURE 17. Bring Your Steel Problems to Inland, 1921. Not only did
Chicago's large steelmakers offer a large amount of steel in various
shapes and sizes; they also promised that they would provide good
service. Source: *Chicago Commerce* (April 23, 1921): 21.

surrounding the mills." Most specialized fabricating firms "were users of
raw iron or steel, and, in turn, essential constituents of the large fabricators
of steel products." In other cases, satellite firms serviced the demands of
the larger firms. They performed many tasks, from regular maintenance of
machinery to taking contracts when demand was too great for the bigger firms. This involved direct interactions between the firms, including
negotiations about machinery repairs and the specifications, costs, and
delivery dates of outsourced work. In both cases, an intimate relationship

existed between Calumet's large steel producers and the smaller, specialized foundries, machine shops, and factories making railroad equipment, agricultural implements, and heavy producer equipment.[10] Calumet's big-firm-led industrial growth depended on the development of strong intra-firm relations between producers of all sizes.

An intricate local geography of production developed that linked metal-related firms from various segments of the production chain. As early as the 1880s, firms moved to the Calumet district to provide products to local firms. In 1888 William Graver built a factory at East Chicago to make steel tanks for the new Standard Oil refinery at Whiting. The vice president of Harvey's Whiting Corporation, makers of electric traveling cranes, told the Federal Trade Commission's hearings in 1922 that "all the steel used by his firm is purchased in the Chicago district." Another Harvey firm, Bliss and Laughlin, a large producer of cold finished bar sheet and shafting, acquired hot-rolled bars and rods from local steel producers. The company distributed its products to many industries, including automotive, farm implement, electric motor, and household appliance. More generally, the Calumet suburbs had a symbiotic relationship to the steel industry. Specializing in metal consumer goods, Hammond's firms and workers had strong links to Calumet's steel and related metal producers, especially the large primary steel firms in Gary. Likewise, Calumet dominated the supply of raw materials for Blue Island metal-machine firms.[11]

Several specialized satellite businesses serviced Calumet's production complex. The Hubbard Steel Foundry supplemented its regular production with one-off (or jobbing) work. The major markets for the foundry's machinery castings, large steel gears, and steel locomotive parts were the local steel mills and fabricating plants. In 1925 they constructed a machine shop with an extensive array of equipment. According to Roger Fiske, a writer for the trade journal *Iron Age*, the equipment was "large enough to handle the heaviest class of work required in the steel mills of the Chicago district." Along with the regular work, the foundry performed a jobbing business, including machine work for firms without the equipment to tackle certain tasks.[12] Other companies did likewise. American Manganese supplied several standard lines (steel tracks, centrifugal pumps, steam shovel dippers), undertaking a good deal of work in its gray iron foundry for the local railroad repair shops. The Riverside Iron Works specialized in the manufacture of white-iron castings for steel mills and blast furnaces. The South Chicago Pattern Works made patterns of all kinds for steel mills, as well as marine machinery, engine, and turbine makers. Valley Mould designed and built molds for all types and grades of steel.[13]

Other companies catered to the specific needs of the Calumet steel industry. The economies of scale generated by the industry's large size provided scope for individual firms to satisfy the demands that large steel mills did not house internally. One niche was the construction, repair, and maintenance of steel mills. Hibben and Company built and installed most of the region's coke ovens and blast furnaces. According to one observer, "The rapid, accurate and reliable construction of many local plants is due entirely to the wide experience, high character and determination of this firm, who, by reason of their location in the Calumet region, can give prompt and efficient service." Hibben did more than build and install, however. They also repaired and provided quick service. To this end, the company maintained a "'flying squadron' ready for service at a moment's notice."[14]

The cumulative effects of these relations and their drawing power for fabricating firms were reinforced by the dropping of the Pittsburgh-plus pricing system in 1923. The ending of the freight rate differential on steel that worked in favor of the Pennsylvania mills not only restricted the Pittsburgh steel mills' market to their local territory, but also enabled Chicago to be a "far greater center both of basic steel production and of steel fabrication." One feature of this was that Pennsylvania- and Ohio-based steel producers—like Jones and Laughlin, and Youngstown Sheet and Tube—considered investing in Chicago. While Jones and Laughlin never did, even though it acquired land in Hammond, Youngstown bought two Calumet plants in 1923 and made important additions over the next few years. Freed from the freight cost restrictions imposed by the Pittsburgh-plus, several other firms reopened their plants or expanded production.[15]

The federal government's expenditures on harbor improvements opened the Calumet region to capital investment. Beginning in the late 1820s, the federal government underwrote the creation of a harbor at the mouth of the Chicago River. This remained the region's main port until the end of the century when port facilities in Calumet took over this role. The first large-scale appropriations occurred in 1869 when Congress approved funds for building the harbor on the Calumet River. More money flowed in the following years, leading to, among other things, the dredging and straightening of the Calumet River and the building of the Indiana Harbor Canal (1906) and the Calumet-Sag Channel (1922). The importance of the Calumet Harbor to Chicago was understood by the promoters of the 1909 *Plan of Chicago*. In their opinion, and this was written into the plan, the downtown should be improved for leisure. Building up the Calumet region, on the other hand, would not only reduce inner-city congestion and create noncommercial spaces downtown; it would also allow the harbor to directly service the heavy industries springing up at the south end of the metropolis. Both small local firms and large non-local corporations would be able to use the modernized harbor facilities to their own profitable ends.[16]

The importance of the Calumet fabricators for eastern producers was clear. As one commentator noted, eastern steel mills did not "consider locating in Chicago unless there was a probability of increased fabrication in this district which would insure a constant consumption of their production." Boosted by investment from eastern and local interests and the federal government, Calumet's growth was so great that the already brash tone of local boosters became more abrasive. As one promoter crowed in the summer of 1925, "Steel producers in the Calumet district are perfecting arrangements for 'diversification' which will place Chicago in a position of absolute independence in the way of industrial steel requirements. . . . The Calumet district then will be a complete storehouse for everything needed in steel, and Pittsburgh will be further removed as a competitor in any producing line."[17]

THE CALUMET, PULLMAN, AND PATCHWORK INDUSTRIAL COMMUNITIES

In his 1937 history of Chicago's southern portions, local historian Theodore Longabaugh called the Calumet district a "patchwork of nearly a score of separate communities."[18] In so doing, he recognized that the area was not a homogeneous region dominated by a few industries functioning as autonomous entities. Rather, the region's industrial success was rooted in a patchwork that connected goods, raw materials, and people at various geographic scales. Using the well-known case of Pullman as an illustration, this section looks at how Calumet's industrial locales were rooted in material places operating at different spatial levels. Since the opening of the Pullman Palace Car shops in 1880, the town of Pullman was considered a one-industry town. Writing just before World War I, Arend Van Vlissingen believed this was wrong, as each Calumet district was different: "The Pullman-Roseland District . . . is confused in the minds of many with the South Chicago District." But, he argued, Pullman "is nearly a separate city; quite different in the character of its labor and the nature of its manufacturers, and situated miles away." In his opinion, the economy had shifted "from the one car shop to the many varied industries now scattered throughout the district."[19]

While it is hard to quarrel with Van Vlissingen's point about Pullman's uniqueness, it is more difficult to accept that it was not dominated by the railroad equipment industry. Certainly, with a substantial share of the district's workers and a narrow range of industries, the Pullman car factory was the industrial anchor of the Pullman area's industrial landscape. But he had his reasons for emphasizing Pullman's multi-sectoral character. First, his vision distanced the town from dirty South Chicago, where steel mills and blast furnaces loomed literally and metaphorically over the poverty-stricken inhabitants, noisy railroads cut through the neighborhood's residential fabric, and polluted air was ever-present. Second, Van Vlissingen's story presented Pullman as a balanced, rapidly growing district with its own character, one that was not only free from its own infamous past, but that stood as a contrast to its poor, smelly, noisy, and dingy neighbor to the east. As a real estate developer with major holdings in the area, Vlissingen was at pains to ensure that Pullman was viewed favorably by industrialists, merchants, politicians, and workers, even if it meant he had to pare away at the character of Pullman's industrial structure.[20]

But there was an element of truth to Van Vlissingen's statement. Pullman was home to other industries. To a significant extent, this variety was made up of firms linked to railroad car manufacturing. One way that industrial areas developed was through the links between a magnet firm and its satellites. Pullman was an area in which a magnet firm (Pullman) connected with dependent satellites: 4 other railroad equipment firms manufactured an assortment of products, including chilled iron car wheels, brake shoes, and car specialties; and 12 fabricated metal companies, 5 machinery makers, 4 steel makers, 5 chemical firms, 5 lumberyards, and 18 miscellaneous firms made automotive parts, electrical switches and signs, radio units and cement blocks.[21] But even with this degree of industrial variety, Pullman remained first and foremost a railroad center.

From the very beginning, Pullman's role as a railroad center was tied to its links with other local businesses as place-based relations developed around magnet-satellite interactions. Many plants were created because of their material and product links to Pullman Palace Car. In the 1880s the Bouton Foundry was linked to the Pullman car shops, which consumed most of the foundry's castings, while Allott Paper Car Wheel made wheels for Pullman's sleeping cars. The attraction of Pullman for railroad-equipment-related firms continued into the next century. The Ostermann Manufacturing Company was established in West Pullman in 1906 to build and repair freight cars for both Pullman and other local railroad companies. In the same year, the Detroit-based firm Griffin Wheel bought

27 acres in the Kensington district and built a plant to manufacture car wheels for Pullman. Close by was Chicago Malleable Casting, one of the principal sources for railroad car castings.[22] New sets of manufacturing facilities were put into place, in some cases replacing old linkages with new ones, and in others ensuring that older networks were consolidated.

By the 1920s the Pullman shops were tied to a complex of metropolitan manufacturing and transportation firms. According to a 1923 survey, Chicago's 559 railroad equipment producers made it the largest passenger and freight car manufacturing district in the United States. Along with the sprawling Pullman plant, the Calumet district was home to Western Steel Car (South Chicago), Chicago Steel Car (Harvey), General American Tank Car (East Chicago), North American Car (Blue Island), and Standard Steel Car (Hammond). The city manufactured goods along the industry's entire production chain. As a writer for *Chicago Commerce* noted in 1925, the city's manufacturers produced "all the way from the apparently insignificant, the all-important spike and washer to the glistening steel rail and palatial Pullman coach." Not only did Chicago railroad equipment firms make these products; they, as well as the railroad companies, also heavily relied on the supply of semi-processed products from local factories and warehouses for their production needs. As the survey reported, "A large proportion of the rails, spikes, bolts, angle bars, fillers, axles and wheels for the American railroads are manufactured in the Chicago district." The Calumet district was the nerve center for these intra-metropolitan linkages. Chicago's primary steel mills supplied steel while plate mills furnished "millions of tons of steel plates annually for the railroad cars manufactured by the forty-six car building concerns, large and small, in metropolitan Chicago."[23]

Calumet was also home to many railroad equipment manufacturers and dealers. Harvey's Buda Company made railroad supplies and motors. In response to growing demand for freight and passenger car axles and the shift from wrought iron to steel axles, a car axle mill was one of the first plants built in the massive U.S. Steel complex at Gary. The Indiana location was the best possible one given that "more than half of the axles required in the United States are used on railroads which center at Gary and Chicago." From Morden Frog and Crossing (track equipment) to Simplex Railway Appliance (car springs) and Northwestern Malleable Iron (car appliances), a constellation of railroad firms clustered around Pullman. Together, these firms created a massive two-way market for railroad equipment. Pullman, being the largest railroad firm in the metropolitan area, drew a substantial part of this production into its shops and was the hinge of a network of inter-firm linkages stretching from the local neighborhood to the entire metropolis.[24]

Pullman's railroad equipment sector had direct and indirect spin-offs to other sectors. Paint manufacture emerged in the district because of links to the railroads. In 1879 the Cleveland firm Sherwin-Williams established a warehouse in Chicago. Nine years later, partly in response to growing demand for railroad car paint, it opened its first plant outside of Cleveland. Seeking a site with large local demand and quick access to the burgeoning western market, it chose Chicago and Pullman. The acquisition of the Calumet Paint company gave the Cleveland firm a strategic location to serve railroad car and carriage manufacturers. From its initial focus on a small set of product lines, it expanded the plant and added new lines such as lead corroding in 1914 and lacquers in 1922. The links were felt beyond the paint factory. As A. Schneider related in 1929, "There is a very definite tie-up betwen [sic] the chemical industries locally and our great industries, such as iron and steel, packing, machinery, etc."[25]

Sherwin-Williams bridged to another cluster of firms. It not only serviced the railroad industry but supplied industrial paint to the district's farm implement industry. Several agricultural implement producers, such as Whitman and Barnes Manufacturing, had plants in the vicinity. The most important, International Harvester, had several Chicago plants, two of which were in Calumet. Its Wisconsin Steel plant in South Chicago sent pig iron to the company's Plano Works at Pullman to be made into gray iron for use in the manufacture of manure spreaders and wagons. Even though the steel mill was enlarged, it continued to send most of its output to the fabricating firm. The distribution of materials was facilitated by the 36-mile Chicago, West Pullman, and Southern Railroad tracks that connected the two International Harvester firms with each other and with adjacent railroad lines.[26]

The heavy industry and metalworking culture associated with the Pullman district attracted other firms. Typical of the metal-making firm that settled in Pullman was Niagara Radiator and Boiler. In 1921 it purchased 10 acres and built a factory to manufacture boilers, radiators, and furnaces. Like other firms, Niagara Radiator was attracted to Pullman by its access to "Chicago's wide-spread local market." As well as the desire to rapidly build up its local and national markets, National Radiator sought large swaths of cheap industrial land such as could be found in Pullman. As one commentator noted in 1924, "Plans for the future are now given serious consideration in the construction of an industrial plant, for the time has passed when a manufacturer will submit to the difficulties incurred in attempting to add to the factory built without regard to possible future extensions."[27]

The ties of Pullman's firms stretched outside the neighborhood and the metropolis. In some cases, there were interlocking directorships, such as between Pullman Palace, International Harvester, and the Buffalo firm Lackawanna Steel. This had direct impacts on markets and prices for the products flowing between the Pullman and Buffalo firms, and allowed for the development of new organizational capabilities between the firms. In other cases, corporate offices and financial institutions were located elsewhere. The banks servicing Sherwin-Williams's capital stocks operated out of Cleveland (Cleveland Trust) and New York (Central Union Trust). When Pullman was created as a holding company to oversee its expanding empire in 1927, it was organized by a New York venture banking house, which worked with financial institutions in New York (National Bank of the City of New York) and Chicago (Illinois Merchants Trust). Similarly, Highland Iron and Steel was part of an intricate network of firms located in Connecticut, Ohio, Pennsylvania, Ontario, and New Jersey operating out of offices in Bridgeport, Connecticut, and serviced by New York financial institutions (Chemical National Bank and Central Union Trust).[28]

As the Highland Iron case suggests, Pullman firms were part of an intra-firm production chain operating at national and international scales. Sherwin-Williams's Chicago plant was one element of a national and international sequence of production. It acquired lead and zinc from its mines at Magdalena, Mexico. Zinc was sent to the company-owned smelter at Coffeyville, Kansas. Linseed oil was secured from its Cleveland plant. Its products were sold by company branches and stores throughout the United States and Canada. Similarly, Carter White Lead was part of an intra-firm network linking raw materials (tin mines and smelters in Bolivia and Argentina) and manufactured markets (expansion bolts, ammunition, and paint) with several intermediate manufactured input producers (lead pipe and white lead) in England and Canada as well as the United States.[29]

The Pullman district was representative of Calumet's patchwork communities. With industrial manufacture centered on railroad-equipment-related products, the area formed a specialized role in the larger Calumet production complex. Internally, the area had a distinct core-periphery relationship based on the Pullman Company, which was directly and indirectly linked to many satellite firms. The railroad company was deeply embedded in a set of interlaced relationships that tied local firms to each other and with other Calumet producers. Calumet's ability to develop into such a successful production complex rested on specialized districts. Other areas within Calumet also developed a distinct array of production activities that set them off from each other. Each place, however, had in common direct yet varying links to each other and to the rest of Chicago and the United States. The creation of patchwork places made the Calumet district.

THE CALUMET DISTRICT'S EXTERNAL LINKS

The Calumet steel complex was also linked to other factory districts in metropolitan Chicago. The Calumet district was Chicago's major supplier of raw steel. More than 5 million tons of iron ore melted in the Calumet mills in 1911 were "made into steel rails, structural and every other variety of mercantile iron, which has found a ready and active market almost at the doors of the blast furnaces." Nine years later, 91 percent of all U.S. Steel sales to Illinois and Indiana came from its Chicago mills, while 58 percent of Illinois Steel production went to the metropolitan area. Similarly, close to half of Inland Steel's finished steel was delivered to the Chicago Switching District, while another quarter went to industrial districts in nearby Indiana and Wisconsin. Local steel mills, as Phyllis Bate noted, "undoubtedly found their greatest market in the needs of local industries." A strong symbiotic relationship linked Calumet steel and the metropolitan area's metalworking, electrical, automotive, and construction industries.[30]

Chicago metal fabricators depended on Calumet mill products. The steel drum maker Wilson and Bennett was a longtime customer of Inland Steel. Chicago Roller Skate Company informed the 1922 Federal Trade Commission that all of its steel originated from local mills. Suburban Cicero firms bought large amounts of steel from the Calumet mills. The city's railroad equipment makers, foundries, and machine shops were large users of steel, while the city's electrical and automotive industries demanded sheets, cold rolled steel, bars, and wire products. If they did not get steel directly from the mills, local firms turned elsewhere. Chicago's wholesalers were close to the "mills from which their stocks may be replenished daily if required. They have in Chicago and its metropolitan area thousands of machine shops and miscellaneous industries using steel products as raw materials."[31]

The flow of inputs went both ways. Lasker Iron obtained steel from the Calumet mills and also sent specially ordered goods—oil refinery equipment, stacks, and tanks—to the petroleum and steel mills of Whiting, South Chicago, and Gary. The Hodgson Foundry made copper and bronze pieces for an assortment of industries, including U.S. Steel and

Youngstown Sheet (fig. 18). Cicero firms supplied producers' goods to the Calumet district. Hundreds of railroad equipment firms scattered across Chicago supplied Calumet's railroad car shops with signal boxes, brass engines, machine castings, electric headlights, heating apparatus, railway jacks, car curtains and fixtures, and ventilating floors, while plate mills annually supplied them with "millions of tons of steel plates."[32]

Demand by the construction industry spurred production of structural steel, a Calumet staple. This was evident from the late nineteenth century with the increase in the number of steel-framed industrial and commercial buildings. In his review of the 1906 steel trade, A. Backert, the *Iron Age*'s Chicago correspondent, reported that "the number of office buildings and warehouses erected contributed largely to the heavy consumption of the

---

HODGSON FOUNDRY CO.

2010-12-14-16 West 13th Street　　　**Telephone Canal 1359**

——————————— FOUNDED 1902 ———————————

| JOHN W. HODGSON, President | JOHN HODGSON, Vice-Pres. |
| M. M. HODGSON, Secretary | S. G. HODGSON, Treasurer |

Producing pure copper tuyeres, bosh plates, coolers, monkeys, monkey coolers and valve seats for:

St. Louis Gas and Coke Corp.	Carnegie Steel Co.
U. S. Steel Corp.	Otis Steel Co.
Columbia Steel Corp.	Jones & Laughlin
Bethlehem Steel Co.	Youngstown Sheet & Tube Co.
American Steel & Wire Co.	

The company also made all of the copper for the first furnace in Europe to use a copper lining. The Freyn Engineering Company, who are now developing the Iron Industries in Russia, are among our customers.

We are also making Bronze Trunnion Bearings and All-Bronze Castings for bridges in Chicago and elsewhere and producing Bronze Gears for

Foote Bros. Gear & Mach. Co.
Foote Gear Works
W. A. Jones Foundry & Mach. Co.

and pump castings, bronze impellers up to one ton each for

Fairbanks, Morse & Co.
Yeoman Bros.

General jobbing of brass, bronze and aluminum castings of all kinds ranging from one ounce to 3,000 pounds each.

FIGURE 18. Hodgson Foundry Advertisement, 1929. As the 1929 advertisement makes quite clear, the Hodgson Foundry made goods "of all kinds" for a variety of local and regional manufacturing and engineering firms. Source: *Chicago Commerce* (December 7, 1929): 365.

structural steel." He listed 22 buildings requiring more than 45,000 tons of steel in their construction, including banks, government buildings, office towers, warehouses, factories, and, of course, Calumet mills—the Gary steel works, the South Works of Illinois Steel, and Griffin Wheel. Two years later, some of the largest consumers of structural steel were the new suburban plant of Corn Products Refining Company and the downtown La Salle Hotel, City Hall, and People's Gas Light and Coke building.[33]

The Calumet mills were also linked financially to Chicago. Not surprisingly, Calumet's leading steel firms were active in local banks. Eugene Buffington, president of the Indiana and Illinois Steel companies for more than 30 years, sat on the board of three Chicago and Calumet banks. Inland Steel's Leopold and Joseph Block were directors of the First National Bank of Chicago, a leading underwriter of new securities. As Inland Steel's expansion before World War II occurred by investing retained earnings rather than mergers and takeovers, it not only relied on First National but also used Chicago's First Trust and Savings to float bonds based on a first lien on all of its properties. Similarly, the Buda Company, a Harvey equipment maker, used Chicago's Northern Trust and Illinois Merchants Trust to service their capital stock and liberty bonds.[34]

While metropolitan Chicago was important for the sustenance of the Calumet industrial complex, the Calumet mills were intimately involved in the emergence of an extensive national and international system of financial services. Industrial securities were rare before the early 1890s. After that, however, a growing number of industrialists sought new sources of capital to enlarge or improve their business. Several types of financial institutions—brokerage houses, investment banks, and commercial banks—orchestrated the formation of industrial combinations and provided industrial capital. Originating in the eastern seaboard, these enterprises drew Calumet steel mills and fabricating factories into the orbit of national and international capital flows and financial institutions. In the process, this underwrote a massive program of factory construction from the 1890s and helped establish Calumet as a leading steel producer by the 1920s.[35]

Perhaps the single most important financial institution underpinning the rise of Calumet's production complex was the House of Morgan. Under J. P. Morgan, New York and Philadelphia became important United States financial nodes connected to European capital. In the last decades of the nineteenth century, Morgan, along with William Vanderbilt and Edward Harriman, channeled capital into railroad expansion and in the process spurred Calumet's production of iron and steel. As the railroad industry was the largest market for the nation's iron and steel producers at this time, Morgan became active in the machinations of the turn-of-the-century steel industry. In 1898 he financed the formation of Federal Steel, the merging of Illinois Steel and other midwestern producers. Three years later, working with Andrew Carnegie and John Rockefeller, Morgan consolidated several manufacturing firms, iron mines, and transportation companies into U.S. Steel, America's first billion-dollar corporation. In addition, Morgan formed mergers in tinplate and steel sheets with the Moore steel syndicate and the financier John Gates. All of these financial dealings reinforced the corporate dynamics driving Calumet's rise to steel preeminence and forged a district on non-local capital.[36]

Local and non-local capitalists worked together to finance the Calumet steel industry. Inland Steel relied on retained earnings and gold bonds underwritten by Chicago banks to finance the expansion of its Chicago area mills while investing new capital. The construction of the open-hearth steel mill at Indiana Harbor in 1901, for example, was partly financed by R. Beatty's acquisition of a large share of the company's new stock. Beatty accomplished this because he sold his Muncie firm, Midland Steel, switched the capital from that sale to Inland stock, and joined the Chicago firm as director and general manager. Twenty years earlier, Cleveland and Chicago capital was behind the move of a Cleveland machine firm to the Chicago Drop Forging Company's new plant in Kensington. This amalgamated firm quickly became an important supplier of railroad car couplers, transportation hardware, and agricultural implement parts to other Calumet producers.[37]

The rise of the Calumet district also depended on the transfer of managerial knowledge and production skills from other locales. Before coming to Illinois Steel as president in 1899, Eugene Buffington learned his trade in the American Wire and Screw Nail Company. George Danforth, the company's general superintendent, was trained in various steel mills in Birmingham, Alabama, before moving to Chicago in 1903. He was responsible for numerous changes to the company's open-hearth practices in the following 33 years. At Inland Steel, Joseph and Leopold Block brought skills developed in Cincinnati and Pittsburgh iron firms. Managers were recruited from hardware and farm implement companies, and superintendents, foremen, and skilled workers moved with Beatty to Chicago from Muncie. Finally, the making of American Steel Foundries involved the merger of a Chicago firm with most of the nation's leading makers of railroad cast products. The company's new 1904 foundry at Indiana Harbor was plugged into a nationwide research complex. To ensure that it remained competitive, the company opened research centers at Alliance

(1905), Granite City (1910), and Indiana Harbor (1926) and transferred expertise from firms that it acquired, such as Simplex Railway Appliance and Griffin Car Wheel, to the parent company. The improvements to both foundry practice and product lines developed at these centers found their way to the Indiana Harbor plant.[38]

The district's elaborate transportation system linked Calumet to its suppliers and markets. Most importantly perhaps was the link between the district's mills and non-local raw material suppliers. Calumet was the locus of the most intense concentration of railroad tracks in the United States. These intricate networks brought in huge amounts of raw materials, a considerable portion originating directly from corporate-controlled raw material deposits. U.S. Steel obtained its primary raw materials from several iron ore and coal regions, while Sherwin-Williams secured lead and zinc from its Mexican mines. International Harvester drew upon its Hibbing, Keewatin, and Chisholm iron mines and Kentucky coal fields. The Calumet district was linked to these raw material deposits through corporate-controlled water-based lines. U.S. Steel, Youngstown Steel, and Inland Steel all operated their own Great Lakes shipping company. In some cases, firms such as U.S. Steel and International Harvester had their own ore docks and railroad lines (figs. 19 and 20).[39]

The making of Calumet as one of America's premier industrial districts depended on the ability of manufacturers and financiers to forge business links across different spatial scales. Calumet's patchwork places were part of an intricate production chain covering other Calumet areas as well as other parts of metropolitan Chicago, the Manufacturing Belt, and the rest of the United States as well as overseas. Calumet's success rested on a dynamic set of inter- and intra-district business linkages. Behind the development of the district's industrial suburbs, however, were local alliances of investors, local entrepreneurs, and land developers. Despite their interest in local, regional, and national manufacturing linkages, they were just as much concerned with building a social and economic environment that facilitated the development of place-based industrial relations.

TREAT MANUFACTURING VERSUS STANDARD STEEL AND IRON

In the spring of 1888, Cornelius Treat—a successful car wheel manufacturer with a casting factory and machine shop at Hannibal, Missouri—traveled to Chicago by train. During the trip he struck up a conversation with the Chicago real estate broker Valentine Surghnor, an ex-Hannibal

No.383.Dec.19.07. SLIP & HARBOR.

FIGURE 19. The Slip and Harbor of U.S. Steel under Construction, 1907. The harbor, which was one of the first parts of the Gary U.S. Steel plant to be built, linked the steel mills with customers and suppliers. Source: U.S. Steel Gary Works, Photograph Collection, 1906–1971, Calumet Regional Archives, CRA-42-102-064.

resident who told Treat about Joseph Torrence, an industrial promoter of the industrial suburb of East Chicago. This chance meeting set in train a series of events that ended in Treat bringing a lawsuit against Torrence for fraud. At the heart of the case was the clash between two key elements of American business and how they figured in the making of industrial places. On the one hand, there was the question of trust between business partners and, on the other, the issue of no-holds-barred competition. Treat believed that the verbal agreement he made with Torrence to relocate his factory to East Chicago in exchange for land and low freight rates was not only valid but inviolate. Torrence denied that he ever promised such things. Unfortunately for Treat, the contract that he signed with Torrence in 1888 did not specify the exact details of the freight rates. Verbal commitments without a signed contract do not go very far in American law, and the judge was obliged to find Torrence not guilty. Regardless of the verdict, the case reveals a great deal about the nature of business relations and the making of an industrial district in Calumet.

FIGURE 20. The Coal Yard of U.S. Steel's Coke Plant, 1919. The railroads connected U.S. Steel at Gary to, among other places, the Pennsylvania coal fields. Source: U.S. Steel Gary Works, Photograph Collection, 1906–1971, Calumet Regional Archives, CRA-42-III-032.

Intrigued by his train conversation with Surghnor, Treat traveled to Chicago to meet with Torrence. The July 1888 meeting at Torrence's downtown office was obviously successful, for the next day the two took the train to East Chicago to visit the site running along the railroad freight of way. Treat was impressed. Torrence offered the Hannibal manufacturer 10 acres of prime industrial land that, in Treat's words, "would be the best kind of location."[40] More importantly, and central to the ensuing drama, was Torrence's alleged promise that he would supply Treat with freight rates at $3 per car, plus a 50-cent switching fee. According to Treat's testimony, Torrence told the car wheel manufacturer that he controlled the rail line running from Whiting to Chicago and had a contract with the Chicago and Atlantic Railroad to carry freight between East Chicago and Chicago for $3. Torrence purportedly also stated that he had the contract for 20 years and that all firms in East Chicago had the benefit of it. The problem for Treat was that even though he did receive good industrial property, he did not get the promised freight rate, and that without the lower rate, he could not undertake profitable business. As a result, Treat sued Torrence for fraud for $100,000 in 1890.

Joseph Thatcher Torrence was an operator, a man with a finger in many different pies. Born in 1843 in Pennsylvania, Torrence worked in the blast furnaces and steel mills of Pennsylvania (Sharpsburg and New Castle) and Ohio (Brier Hill Furnaces) at an early age. Like many other easterners, Torrence moved to Chicago seeking fame and fortune. In 1868 he began work in the furnaces of the Chicago Iron Company, where he remained for four years. Between 1872 and 1874, he was involved in various projects, including superintending furnace construction at Joliet. While an officer in the Illinois National Guard between 1874 and 1881, he was involved with iron firms in Chicago and Ohio, and became a consulting engineer for Calumet's Brown Iron and Steel Company. In the 1880s he was associated with the construction of the South Chicago and Western Indiana Railroad and an Evansville rolling mill.[41]

In 1887 Torrence and three Chicago investors—Marcus Towle, James Young, and Jacob Forsythe—formed Standard Steel and Iron. Despite the name, the company was never anything but a real estate firm. The enterprise began operations with 1,000 acres, and six months later Caroline Forsythe purchased 637 acres from the Calumet Canal and Improvement Company for $290,000. Quickly turned over to Standard Steel, the land became the core from which East Chicago developed. The boosters organized the Chicago and Calumet Terminal Railway Company, which laid the basis for the belt lines that encircled the city. As well as this, Torrence worked with others to construct essential infrastructures, including a ship canal from Lake Michigan to the Grand Calumet River, a pier, and the Chicago Elevated Terminal Railway Company to connect the suburb to the city's main lines.[42]

All of this industrial and building promotion was part of a history of land speculation and place-making that went back a decade. In 1881 Jacob Forsythe sold 8,000 acres to William Green of New Jersey for $1 million, who, on the same day, transferred the land to the East Chicago Improvement Company, a business incorporated in New Jersey and owned by a London banking house. Even though the firm was intent on speculation in industrial sites, no sites were sold. Six years later in 1887, East Chicago Improvement transferred the tract for $1 to Joseph Torrence and his wife, Libbie. They then turned it over to the Calumet Canal and Improvement Company for $1. This latter company was organized in the same year for $2 million, and its principal stockholders were Towle, Torrence, the Forsythes, and English interests operating through Robert Tod of New York.[43]

The speculative history of what became East Chicago was well advanced by the time that Treat became interested. In 1887 the Torrence companies and new land syndicates such as the East Chicago Company began clearing land and laying out subdivisions. In 1888 the Chicago and Calumet Terminal Railroad tracks were extended from the Indiana-Illinois line to the town. In the same year, work began on the ship canal but was quickly discontinued. Torrence and his colleagues tried to convey land to the federal government so that it would take over the construction of the harbor and ship canal. They were successful. Meanwhile, Standard Steel organized special trains to bring prospective land buyers from Chicago. In May 1889 the town was incorporated and the basic forms of institutional order were put in place: the land company gave land for a town hall and a fire department, promoters attached the numerical system of streets used in Chicago to East Chicago streets, churches were built, and a school was opened. In 1893 it became a city.[44]

Alongside the institutional forms of place-making developed the beginnings of an important industrial base. In 1888 William Graver opened his steel tank works. Graver had moved his factory from Allegheny, Pennsylvania, to Chicago in 1884. A year later he shifted his plant to Lima, where he constructed tanks for the Standard Oil Company. Three years later Graver built an East Chicago plant to service the newly constructed Standard Oil refinery at Whiting. In the following year, Famous Manufacturing (hay balers), Chicago Horseshoe Foundry, and National Forge and Iron were established. Treat was attracted by this busy hive of industrial and land development. East Chicago offered manufacturers the traditional advantages of a suburban greenfield site—accessibility to water, excellent railroad transportation, and industrial sites for reasonable prices. Just as importantly, Treat gained access to a rapidly growing steel supply and railroad equipment complex. At a new site in East Chicago, Treat would be in close proximity to Illinois Steel, Pullman Palace Car, and the city's railroad companies. Few if any other places offered Treat such a medley of locational advantages.

Treat signed a contract on September 11, 1888, at which time the East Chicago entrepreneur allegedly made the same promises about land and freight rates. The contract, however, did not live up to his promises. It contained three key points: Treat was to build a factory capable of making 200 car wheels a day, and Torrence was to supply a 300-by-1,140-feet site (instead of the 10-acre one he had promised) fronting the railroad, and he was to furnish freight rates. Unfortunately for Treat, the specific cost of the latter was left vague. Despite the less than desirable contract,

Treat built the factory, which was soon ready for operation. He transferred enough machinery from Hannibal to the new factory to build 100 car wheels a day, a quarter of the plant's capacity. The firm began production in September 1889 but closed down in April 1890 after making fewer than 6,000 car wheels. According to Treat, the failure was due to high freight rates. Rather than the $3 promised by Torrence in the summer of 1888, the company was forced to pay $7.

It is not difficult to understand why Treat signed a contract that did not stipulate exactly what he believed he had verbally agreed upon. He put his trust in Torrence's words and a vague contract because, as he told the court, "I did not suppose that General Torrence or any other man could afford to tell me as big a falsehood as he did in regard to it, occupying the position that he did, that of president of those economies down there. I understood those companies to be responsible companies, composed of responsible men." This was not left to chance; Treat looked into Torrence's character. According to his testimony, he heard from his son, James Treat, who was in New York at the time, that Standard Steel had "responsible men behind it." In August 1888, a month before the signing of the contract, James had visited the New York office of the law firm Alexander and Green and met with the company's general counsel, Colonel McCook, who convinced him that Standard Steel and Torrence were respected business operators.[45]

Treat's actions were rooted in a strong sense of what he considered to be the appropriate business boundaries; his actions were not a sign of weakness. It was quite clear to him that the successful businessperson, as he believed Torrence to be, had to be trustworthy. He was not unique in making the decision that he did, based on the evidence that he had gathered. As Gordon Winder concluded for a Bridgeport, New York, firm, nineteenth-century manufacturers combined calculated trust, mutual dependencies, contractual obligations, and tacit knowledge to shape their decisions. While Treat was willing to accept that negotiation, evasion, and even small lies were part of everyday business affairs, he was unwilling to believe that responsible people propagated big falsehoods. In Treat's view, Torrence was worthy of trust because he had a long history in the steel industry, was part of a reputable firm, was dependent on Treat and his like for the success of the industrial suburb, and was supported by other businesspeople.[46]

The trial offered little reason to question that Torrence indeed told Treat that he would receive a $3 rate if he built a plant in East Chicago. The witnesses on Torrence's behalf were, with the exception of William

Graver, all employees of Standard Steel. The best that they could say in his defense was that they never heard Torrence guarantee $3. In contrast, Treat's witnesses were numerous, and their evidence was overwhelming in its condemnation of Torrence. According to Surghnor, Torrence spoke to Treat of a $3 rate several times. Milo Hascall, an East Chicago lumber and real estate dealer, testified that Torrence made the same inducements to him as he did to Treat. On the strengths of this representation, Hascall bought 40 acres and built a plant. His lumber firm did secure a $3 rate, but not through Torrence. He made his own arrangements with the railroad company. A similar story was related by Frank Felt, vice president of National Forge. The railroad axle maker never received the promised rate and pursued a similar action against Standard Steel in the Indiana courts. The testimony of executives and superintendents of several railroads confirmed the $7 rate and that Torrence did not have a special arrangement with them. Finally, Oliver Forsythe, the son of Torrence's own partners, testified that Torrence promised the $3 rate.

Irrespective of what happened to Treat, East Chicago developed from these dubious events to be an important part of Calumet's production complex. Like many other industrial places, it was built on a strong entrepreneurial presence that promoted the industrial, residential, and infrastructural elements of a previously undeveloped area. Torrence and his colleagues did what most other promoters did: using the networks they had constructed over time, they built transportation networks, subdivided land for residential and industrial functions, enticed manufacturers with free land and low freight rates, and boosted East Chicago as a model of suburban life. They constructed this picture on a mélange of truths, half-truths, and lies. This was not unfamiliar to experienced and successful manufacturers such as Treat and Hascall. Accordingly, they decided to invest in East Chicago only after serious and careful consideration about the locational advantages of East Chicago and the trustworthiness of the promoters. While the locational assets of the rapidly growing Calumet district could not be gainsaid, the idea that the building of successful places and businesses could take place outside the network of "responsible men" and "responsible companies" was incomprehensible to them.

The same was true for Calumet's other patchwork areas. From Pullman to the South Chicago–Gary steel corridor, Chicago's business and social interconnections were critical to the building of the region. Calumet, as Homer Hoyt noted in 1942, was "the meeting place between the iron ore of the Mesabi range, brought down the lake on huge ore vessels, and southern Illinois coal conveyed on an almost endless stream of freight cars."[47]

More than this, the Calumet district was an elaborate production complex forged by combining the endless stream of raw materials with an intricate network of intra-firm linkages operating at various spatial scales. Calumet was one of America's greatest factory districts. As Henry Lee, editor of the *Calumet Record*, crowed in 1908, Calumet's destiny was to be the "great crossroads of America." Four years later I. Hardin, a participant in one of the industrial inspection tours, exclaimed as the train crossed over the state line near Hegewisch that "out here in this territory you have a river which with its branches and its basin furnishes facilities that cannot be surpassed, and it would seem that here there is bound to come, as has in the past, immense development."[48] Growing from "empty" prairie, the Calumet district developed as both a hub linking Chicago to the wider industrial world of the United States and elsewhere, and a manufacturing center in which locally based inter-firm flows fed local needs and propelled the district's economic growth.

CHAPTER 7

Chicago's Planned Industrial Districts

CLEARING AND THE CENTRAL MANUFACTURING DISTRICT

In 1905 the U.S. Leather Company moved into a factory on South Morgan Street. This was a historic moment since the factory was the first one built in the United States with financing from a planned industrial district. The guests of the industrial inspection tour taking a group of Chicago's industrial, railroad, property, and political elites to the city's southwestern fringe may have seen the factory from the window as their train traveled though what was known as the Central Manufacturing District (CMD). This district, American's first operating planned industrial district—a full-service, industrial real estate development—was sandwiched between the Union Stock Yard and Bridgeport and was founded by railroad interests to increase their freight volume and to take advantage of their land hold-ings. The CMD grew quickly and by World War I the district's 300 acres of industrial property were home to 200 concerns and 15,000 workers. By the early 1920s, as Charles Bostrom, chair of Chicago's zoning commis-sion, noted, there were "many great manufacturing plants located in what is known as the Central Manufacturing District. . . . This district is given up entirely to industrial pursuits, and the centralization of these great in-dustries proves that it is not a wild dream to segregate residence, business and manufacturing."[1]

The CMD was not the only point of interest for the industrial tour's guests. Here among the many other suburban industrial attractions were the Clearing Yards. Still under construction, the yards eventually covered

FIGURE 21. The Clearing Yards, July 1914. Located on the western edge of the city, the Clearing Yards were the largest switch and classification yards in the world when they were built in the early twentieth century. Source: *Chicago Commerce* (July 24, 1914): 187.

an area of almost five square miles. With 150 miles of track and a capacity of 10,000 cars daily, they were the largest gravity switch and classification yards in the world (fig. 21). The tour guests, however, were interested in an aspect of the district other than its railroad possibilities. As Dr. W. Evans, a member of the Chicago Association of Commerce's Civic Industrial Committee, told the guests at lunch, "When the necessary kinks are ironed out and the great Clearing yard is built, it will be well for the entire southwest district because it is certain there will group around those yards a lot of industries that can be well developed in that vicinity" (fig. 22).[2]

Clearing and the CMD are among the earliest examples of planned industrial districts, or, as they are otherwise known, organized districts, industrial parks, and science parks.[3] What they all had in common was joint private and public involvement in the creation of space that was solely given over to industrial functions. In this way they differed from the Calumet suburbs, which combined industrial, residential, social, and commercial functions. The construction of industrial landscapes in Calumet and elsewhere centered on the building of working-class housing,

FIGURE 22. Local Industries and Civic Industrial Committees Have Luncheon at the Club of the Central Manufacturing District, 1931. Chicago's bourgeoisie gathered to exchange ideas about a wide range of social, economic, and political topics. Source: *Chicago Commerce* (June 13, 1931): 27.

social institutions, local retailing establishments, and an industrial base by a number of local alliances and a much large number of individual actors. Despite informal and formal zoning, each separate district featured a multitude of land uses. The planned industrial district differed from these other types of industrial areas in three critical ways: first, its functions were exclusively industrial—residential and other non-industrial uses were not permitted; second, it was more heavily planned and involved fewer coalitions and individual actors; and third, it had a more narrow industrial structure.

PLANNED INDUSTRIAL DISTRICTS: PLANNING, FIRMS, AND BOUNDARIES

Although Clearing and the CMD were two of the earliest industrial districts to be planned, controlled, and shaped by corporate design, they were not the first. This honor probably belongs to the New England industrial towns of Waltham and Lowell, which from the 1810s were planned by

Boston-based corporations and their resident agents. Following in their footsteps, other districts emerged, such as other New England mill towns (Manchester, New Hampshire), company suburbs (Pullman), and satellite cities (Gary). These were founded to maintain strict hierarchical control over capital investments, labor, factory space, and urban land-use. The districts were laid out according to pragmatic principles, typically comprising the separation of industrial and residential spaces, the laying down of the street network, and the building of operatives' housing. While different in some important ways, these nineteenth-century towns and suburbs were precursors to twentieth-century planned industrial districts.[4]

Twentieth-century planned industrial districts like Clearing and the CMD, however, differed in two important ways from their nineteenth-century forerunners. In the first place, unlike the nineteenth-century variants with their attempts to maintain control over both home and work, the primary focus of the early twentieth-century district was on industrial space. Unlike earlier attempts to police labor supply outside the factory walls, corporate promoters of the planned industrial district did not construct working-class housing. Secondly, firm owners and managers entering the planned industrial district relinquished some power over their work space. Combining legal obligations such as restrictive covenants with comprehensive control over financing and factory design, district promoters forced managers and owners to cede some of the control that they normally exercised over the workings of their firms to the overall planning requirements of the planned industrial district.

Planned industrial districts also differed from other twentieth-century factory districts. They furnished financial support as well as architectural and construction assistance for firms seeking industrial property and plant. Planned industrial districts offered an extremely well-coordinated and comprehensively planned community, thus providing firms with a regulated work environment. District promoters, who typically operated out of railroad-related companies, developed an extensive set of locational techniques, linking the district with wider metropolitan, regional, and national spheres. Planned industrial districts also supplied packages of collective benefits that were unavailable in most areas of the metropolis.[5] In combination, these factors formed a unique example of metropolitan infrastructure, institutions, and firm behavior.

District promoters drew on some of the important streams of urban planning thought before World War II.[6] The rationale for the districts was rooted in the notion that urban areas could be designed around community services and functions. In their early, rudimentary forms, planned industrial districts predated Clarence Stein's neighborhood unit, but nevertheless built on the same planning and reform currents that underpinned his work. The central tenet was that industrial districts or neighborhoods had to be internally coherent if they were to be effective work spaces within the rapidly expanding industrial city. Despite parallels to planned residential areas, planned industrial districts were not derived from the cultural ideals that drove the building of new neighborhood units. Unlike the statements made by Stein and his predecessors about the role of neighborhoods within the larger urban world, there was no attempt, other than booster rhetoric, to address the relationship between industrial district development and overall metropolitan growth. Unlike Stein and other planning and urban reformers with their search for an ordered social solution to the contradictions of the capitalist city, the operating principle of planned industrial districts was the search for profit, which was to be obtained from increases in railroad traffic volume.

Planned industrial district promoters believed in the necessity of land-use segregation.[7] For many, a major problem of the rapidly expanding city was the intrusion of manufacturing and related functions, like railroads and warehouses, into residential and retail districts. The commingling of dissimilar land uses undermined the sound and necessary relationship between factory districts and the wider urban community. Zoning, with its separation of land uses, became the main mechanism to deal with this problem. The principle of separation between competing land uses underpinned planned industrial districts. From as early as 1905, when the first factory was built in the CMD, district promoters strove to separate their firms from the rest of the surrounding metropolitan area by imposing a strict set of regulations excluding all functions other than factories, warehouses, and transportation facilities Planned industrial districts were early examples of the zoned metropolis.

Separating factory from residential areas enabled district promoters to plan specialized industrial landscapes. To achieve this, promoters created locational assets.[8] Although these varied from district to district, promoters built up a core set of place-bound infrastructures (water, power, lighting, streets, and freight tunnels), institutions (architectural office, banks, and clubhouses), and practices (land-use covenants, traffic planning, factory design, and financial services). To ensure that a district's locational assets were implemented in an effective manner and to enable firms to reduce the hazards of locating in a large urban complex, a set of codified rules that operated within the boundaries of the district was established by the district corporation. Once in place, these assets generated and sustained industrial growth over a generation or more, and offered manufacturers and wholesalers access to a rich and proximate milieu of workers, firms, institutions, and infrastructures.

Although planned districts were not part of a metropolitan-wide comprehensive plan controlled by formal planning institutions, they were shaped by "deliberate and systematic decisions" made by political and economic agents after 1900. In the development of these organized areas, local alliances consisting of railroad, manufacturing, commercial, property, financial, and political elites designed and assembled a new form of industrial space, and in the process reshuffled the relations between firms and industrial territory. Typically, one agent took the lead role. For example, railroad companies such as the Atchison, Topeka and Santa Fe Railway created the Central Manufacturing District in Los Angeles and the Airlawn Industrial District in Dallas, respectively. Industrial real estate developers such as Industrial Properties Corporation and Cabot, Cabot and Forbes were responsible for the Trinity Industrial District and the New England Industrial Center. In some cases, local government took the lead role, such as when the city of Chico's built the industrial district at the Chico Municipal Airport in California.[9]

Alliances of land developers, railroad companies, and local and regional governments established the rules and regulations under which district institutions and firms operated. This in turn modified firm-district boundaries. As members of a geographically bounded form, promoters applied proprietorial control over district firms in ways that set them off from other firms and provided them with specific modes of operating. Through the institutional structures they created, promoters built upon preexisting notions of power and property, and established rules about firm behavior.[10] As planned industrial districts were legal corporations or common law trusts with large financial backing and a high degree of control over property and firms' actions, promoters had the ability to plan urban space in a way that had not been previously possible.

The heightened control that promoters had over district firms, together with the techniques they developed to attract firms, set planned industrial districts apart as an institution involved in plant location. Acting as an intermediary, they brought together industrialists seeking plant sites and landowners seeking to sell their holdings. The leasing of land allowed firms to invest freed-up capital elsewhere, gave flexibility as to the amount of land and buildings required over time, and placed the risks and obsolescence of the property on the district rather than the firm. The planned character of the district enabled firms to benefit in ways that they could not otherwise. Being plugged into an ordered industrial environment with a set of regulations controlled by economic actors sympathetic to the needs of industry and geared to industrial production made possible a set of inter-firm relations that could not be replicated elsewhere. The district's regulated environment gave firms access to financial aid, construction services, banking and club facilities, and social and cooperative activities, all under the control of promoters who knew that the success of the planned industrial district depended on the success of individual firms (fig. 23).[11]

The inter-firm relations offered by planned districts were very attractive to small and medium-size companies. Forming the bulk of the planned district's client base, most smaller firms relied on a range of external economies and business networks that typically located in the city center. The ability of the planned district to replicate many of these was extremely advantageous to the smaller enterprise looking for a location outside the city's business core. The array of services the district provided, from financial aid for building construction to access to transportation facilities, contributed to a firm's ability to reduce the time and expense of transaction costs by linking customers and suppliers closely in space. Particularly important for businesses dependent on monitoring local price, market, and labor conditions, the services established by district promoters permitted manufacturers to overview competition through close proximity of firms operating in the same industry. In the process, they allowed firms to take advantage of all forms of knowledge spillover, especially through spin-offs, collaboration (both intentional and unintentional), and mimicry.

In exchange for these advantages, executives relinquished some of their managerial power and influence. In so doing, they changed the nature of what they did as an independent entity. This was especially the case with control over land-use and financial issues. Entry into a planned industrial district came through leasing or buying of industrial property. For most firms this was achieved by taking one of the several types of financial assistance plans offered by the district. Companies looking to settle in Clearing and the CMD, for example, could buy land outright and improve it themselves or have the district make the improvements. Before World War II, promoters sold land, with or without improvements, on a deferred payment plan, a quarter to a third of the purchase price down, with the rest paid over 4 to 10 years with interest at 6 percent. Firms could also gain access to land by short- or long-term leases, with the improvements undertaken by the district.

FIGURE 23. The Central Manufacturing District Bank and Club House, 1915. The business and social activities of the Central Manufacturing District's managers and owners were centered at the district's bank and club. Source: *The Central Manufacturing District* (Chicago: Central Manufacturing District, 1915), 8.

The relationship between manufacturers and the operators of the planned industrial district centered on the latter establishing territorial control and imposing rules that all firms had to observe. Regardless of how property was acquired, firms had to follow a specific set of rules. Property was developed according to a comprehensive plan policed by the district's managers. Buildings were built to specific physical layouts, plans of operation, and financial arrangements. Zoning was implemented by excluding certain industries and regulating building appearance, type, and location. District promoters constructed and maintained railroad spurs, streets, and utilities. Many factories and warehouses were built by the districts' own architectural and construction departments. Promoters strongly influenced the actions of district firms in the very process of administrating property development, manipulating land use, and monitoring property financing. In this way, manufacturers, by allowing districts this authority, allocated control over their firms' workplace and boundaries to another entity to a degree unknown before. The very practice of locating in a district established rules based on a specific set of power relations between companies and the proprietorial group.

Clearing and the CMD were two of America's most significant early twentieth-century planned industrial districts. They were planned factory areas in which district promoters offered various forms of direct assistance (financial, design, and construction) and locational assets (from water to street layout) that could not be found elsewhere in Chicago at that time. With almost 30,000 manufacturing employees working in 160 factories, plus those laboring in warehouses, the two districts were important planned nodes on the metropolitan landscape in the interwar period (table 15).

Clearing and the CMD emerged out of an urban political economy in which skills, knowledge, and capital were funneled into an expanding range of manufacturing and non-manufacturing uses. Chicago was a hive of innovation. Not only did it capture inventions from other parts of the United States; innovation was a critical part of the practices of local firms and entrepreneurs. Large firms such as International Harvester, Western Electric, and the meat packers provided a stream of product and process innovations from as early as the 1840s. Innovation, however, was not left to large producers. In other important Chicago industries, such as furniture and radio, innovations in style, organization, and mechanics were part of their everyday culture.[12] To be successful in this changing world, companies had to make frequent adjustments to and to be constantly on

TABLE 15. Manufacturing at Clearing and the Central Manufacturing District, 1924 and 1940

Industrial District	Initial Development	1924		1940	
		No.	Workers	No.	Workers
Clearing: Total		25	2,409	71	13,928
Original district	1909	25	2,409	62	12,804
Other districts	1939+	0	0	9	1,124
CMD: Total		84	16,682	89	15,903
Original & Pershing	1905–16	73	15,584	54	10,818
Kedzie Avenue	1920	11	1,098	20	2,320
43rd Street	1926	0	0	9	1,706
Crawford Avenue	1931	0	0	6	1,059

SOURCE: *Directory of Illinois Manufacturers, 1924–1925* (Chicago: Illinois Manufacturers' Association, 1924); *Manufacturers in Chicago and Metropolitan Area* (Chicago: Chicago Association of Commerce, 1940).

top of shifting industrial processes. Planned industrial districts offered a place where firms could find order within an established factory area and implement these changes.

An active set of institutions promoted Chicago's development. Intermediary organizations—ranging from Chicago's large selection of financial institutions to its business associations, industrial marts, and wholesalers—channeled commodities through the industrial system and lubricated dense competitive and cooperative inter-firm relations. Chicago's industrial fabric was also tied together by an array of formal institutions and regulatory supports. Institutions—ranging from the Board of Trade to the Chicago Harbor Commission, the Chicago Switching District, the Chicago Plan Commission, and the Sanitary District—established a supportive framework for the rise of dense and sticky social and business relations.[13] For firms unable to tap into these institutional networks, especially those that were new to the city or outside the established framework of organizational life, it was important to seek out other avenues of firm interaction. Planned industrial districts with their strong set of intra-district linkages and planning offered business a good location.

The construction of metropolitan Chicago as a place and people's identification with that place provided a boundary around which industrial activity, promotion, and interests were contained. The geographic extent of this boundary was not predetermined. Each new cycle of growth reframed what was metropolitan Chicago, usually by incorporating greenfield residential and industrial suburbs, in the process creating new boundaries and reshuffling the character of Chicago's internal spaces. For example, the upper-class South Side residential area became a light industry district by World War I, the residential and warehouse area along Michigan Avenue north of the river was converted into a high-end retail and financial district after the war, and industrial suburbs such as Bridgeport were turned into inner-city slums. Chicago's planned industrial districts were part of this process of changing land use. They were attractive areas for firms seeking particular set of locational assets.[14]

Established in 1902, the original CMD tract was boxed in by Packingtown and Bridgeport (fig. 24). The impetus behind its establishment was the need of the Chicago Junction Railway and the Union Stock Yard to develop their real estate interests and to increase the volume of freight traffic. With the establishment of the Union Stock Yard in the 1860s, the railroad company perfected facilities for handling livestock, secured direct connection with every Chicago railroad, and constructed enormous freight yards. Despite the volume of trade generated by the stockyard, the

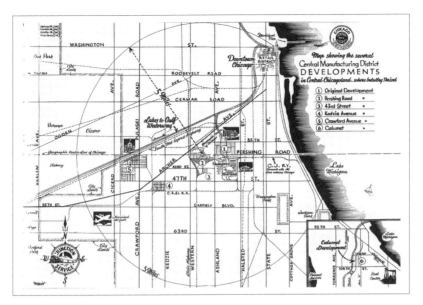

FIGURE 24. The Central Manufacturing District Developments. The several developments established by the Central Manufacturing District between 1909 and 1960 are shown on this map. The Clearing district is at the bottom left hand side of the map south of the Municipal Airport. Source: Robert Boley, *Industrial Districts: Principles in Practice* (Washington, DC: Land Institute, 1962), 31.

railroad's capacity for handling business was far in excess of the traffic offered. This was partially solved as some land was occupied by lumberyards. By the end of the century, however, the lumberyards had shut their gates or moved elsewhere. At the same time, the threat of the packinghouses, the railroads' major client, to leave the stockyard further accentuated problems, forcing the railroad to find another way to ensure a more reliable source of business. The solution was the development of America's first operating planned industrial district. As one promoter put it, the intent was to turn the "barren waste" into a "great industrial center" by "developing this section with tonnage-producing industries."[15] The result was the CMD.

The railroad financed the CMD in part through returns from the sale or lease of its properties and in part by a long-term bond issue. Once established, the development of the Original tract was rapid. By 1910 there were 25 factories, and growth continued.[16] As the Original tract filled up, a new one, Pershing Road, opened, and by 1918 the two tracts' 200 factories and warehouses employed 15,000 workers. Producing more than 100,000 carloads of freight a year, these firms, according to Francis

Stetson Harman—the CMD's assistant industrial agent—were "drawn to this spot in the center of Chicago, because they [were] able to obtain efficient freight service and the other advantages that spell economy and contentment to them."[17] By the mid-1920s, the Original, Pershing, and the Kedzie tracts were home to an estimated 250 firms, with the Kedzie tract opening in 1920.[18] According to the 1924 directory, 84 manufacturers employed close to 17,000 workers (see table 15). Growth continued, and by the early 1930s three additional tracts on the edge of the city (Forty-third Street, Crawford, and Calumet) opened. Despite a slowdown in the 1930s, growth continued and 111 manufacturing firms accounted for more than 23,000 workers in 1950 (fig. 25).

Even though it would not be developed until the early twentieth century, Clearing originated in 1890 when several promoters including the Chicago Great Western Railroad formed the idea of a "clearing yard" to handle the growing problem of freight interchange among the various railways serving Chicago. To cope with Chicago's massive rail volume, several belt lines were built connecting the railroads without the necessity of bringing the traffic into the city proper. The first was the Belt Railway Company of Chicago in 1880, which formed a semicircle with a radius of about seven miles from the city center. This company was the pioneer in providing an intermediate connection avoiding the city terminals, thus reducing central-city congestion. In order to link up with the Belt Railway, promoters acquired a large tract of land on the edge of Chicago extending from Fifty-fifth to Seventy-ninth streets. Times, however, were not propitious for freight and industrial development. The project was suspended in 1893. Railroad yard construction and the conversion of the land into industrial properties would have to wait a decade.[19]

A planned district emerged at the turn of the century. In 1898 the Clearing Industrial District was formed as a common law trust to develop the land for manufacturing, and the Chicago Transfer and Clearing Railway Company was incorporated to furnish capital for property promotion. Four years later the first extensive yards at Clearing were constructed. Initial funds were obtained through sale of stock and augmented in 1912 when portions of the original tract was sold to the Belt Railway for use as classification yards. Development was slow, however, and by 1912 only a few firms had taken advantage of the industrial sites lying on the city's built-up edge. In that year the Belt Railway was reorganized when 12 railroads took out a 50-year lease. This was the culmination of the search for a way to eliminate the use of local terminals for handling through cars and providing some means of interchange between the inbound and

FIGURE 25. Aerial View of the Central Manufacturing District, c. 1960. The separate character of the wall of factories running along Thirty-fifth Street and the residential districts to the south are clearly evident from this aerial photograph of the original Central Manufacturing District taken about 1960. Source: Robert Boley, *Industrial Districts: Principles in Practice* (Washington, DC: Land Institute, 1962), 37.

outbound carrier that would give prompt service without interfering with local traffic. It also laid the foundation of the industrial district.[20]

Following the 1912 reorganization, Clearing became one of Chicago's fastest-growing districts as it opened up industrial property for industrial and storage purposes. In the next few years, manufacturing and wholesale firms appeared. The potential of Clearing for industrial development on the cusp of World War I is evident from the remarks made during a 1914 industrial inspection tour. Frank Spink, the Chicago and Western Indiana Railroad's traffic manager, pointed to the "improvements" being made at Clearing, while others referred to how the yards would facilitate inbound and outbound railroad traffic, thus attracting industries that desired a Chicago location but served regional and national markets. This was certainly what happened. By 1922 the tract was home to 34 companies, and over the next six years the pace of growth accelerated, before tapering off with the Depression (fig. 26). It rebounded in the late 1930s and during the war.[21] Before World War II, most manufacturing activity occurred in the original tract, while growth accelerated in the new tracts during and after the war (table 15 and fig. 27).

The tremendous growth of the two planned industrial districts in the first decades of the twentieth century was spurred by the fact that they were not Packingtown, Bridgeport, or the other older industrial areas. In contrast to the uncontrolled environment of other industrial areas,

A Modern Industrial Center Planned Specially for Warehouse and Factory

Aerial Photographic Service

LOCATED ON THE BELT RAILWAY OF CHICAGO

OFFERS the best in Freight Service for Carload or Less Than Carload Traffic ∽∽Is within the Chicago Switching District taking Chicago rates∽∽Building construction financed, conserving working capital∽∽Has all facilities and conditions essential to successful plant operation.

LARGE local working population within walking distance of plants∽∽Now has Seventy-five Manufacturing Plants and has added Twenty-two New Industries during the past Sixteen Months∽∽Is sold at Moderate Ground Values of from twenty-five cents to fifty cents per square foot.

Responsible Manufacturers are Invited to Investigate

THE CLEARING INDUSTRIAL DISTRICT

Telephone Randolph 0136

1007 First National Bank Building

CHICAGO

FIGURE 26. A Modern Industrial Center, Clearing, 1927. The "scientific" character of the Clearing Industrial District is clearly shown in this advertisement. Source: *Chicago Commerce* (September 17, 1927).

Clearing and the CMD offered a new relationship between industrial development and planning. By providing coordinated close proximity, the districts presented firms with a dense network of facilities, including storage, transportation, infrastructures, and labor that yielded inter-firm dependencies. The districts offered specialized spaces that accommodated and furthered inter-firm specialization and helped create networks of

FIGURE 27. Aerial View of the Clearing Industrial District, c. 1960. The strong relationship between the railroad tracks, the clearing yards, and the factories are clearly visible in this aerial view of the Clearing Industrial District taken about 1960. Source: Robert Boley, *Industrial Districts: Principles in Practice* (Washington, DC: Land Institute, 1962), 37.

reciprocity. The planned coordination that went into district construction facilitated the development of interaction outside the district. The orderly, well-regulated character of districts allowed firms to pursue a range of business networks, from the local to the international, while being situated within a controlled milieu. Unlike other metropolitan factory districts, they provided a planned, ordered, and regulated space with the benefits of geographic proximity and shared institutions.

INCUBATORS, CLUBHOUSES, AND DAYLIGHT FACTORIES

As consciously planned industrial spaces, Clearing and the CMD were part of the growing trend in urban America toward greater institutional control over metropolitan space. In the process of doing this, promoters built a centralized management structure that coordinated a program of standardized and uniform factory construction; oversaw the building of district-operated transportation, water, and power facilities; laid down rudimentary landscaping; and established features that mimicked those of central Chicago. By bringing plants together in a select location and maintaining a high degree of corporate planning over firm selection, building type, and district layout, planned industrial district promoters built a particular set of business networks and boundaries, as illustrated by the districts' incubator buildings, clubhouses, and factory design.

A defining feature of the central-city economy was its role as an incubator or seedbed for small and new enterprises.[22] Typically, because of their small scale and vertically integrated structure, new firms sought out the external economies of the city's central area. Planned district promoters replicated these central-city advantages outside the central city. To this end, Clearing and the CMD constructed unique locational assets and provided specialized work spaces for small and new businesses. Promoters consciously sought to make the planned district an industrial incubator in a non-central location. To do this, they built specialized incubator buildings in order to replicate the resources present downtown. These buildings were composed of several units of small and different sizes and were typically occupied by new firms seeking specialized industrial space.

Several incubators were built in the CMD during the interwar period. The first, a one-story "bungalow" constructed during the war for government storage, was converted into a number of small working spaces for small companies after the war. The bungalow's success convinced some manufacturers that a non-centrally located incubator could thrive. Soon after, a large munitions factory was altered to accommodate several fledgling firms, many of which, once they had been established, moved out to larger premises. For example, when it first set up in Chicago in 1926, the Lapham-Hickey Company, distributor of high-grade steel, leased space in the building. This allowed the firm to gain a foothold in the Chicago market. As Lapham-Hickey grew, however, the space was no longer adequate, and it was forced to build a one-story warehouse on a large tract of land in another part of the district. When it moved out, there were many small firms queuing to move into the incubator factory. In 1929 there were no vacancies for the bungalow's 23 "apartment' spaces."[23]

While firms such as Lapham-Hickey moved into the CMD incubator buildings, others moved into three Clearing incubator buildings built between 1924 and 1929. The first was a one-story factory with 35,000 square feet of floor space designed to attract existing and start-up firms seeking small amounts of space. A year later a one-story incubator factory with 60,000 square feet of floor space was built so that space could be leased in 10 units of 6,000 square feet each. The third incubator was erected four years later and was occupied by, among others, a sash and door warehouse, heating contractors, electrical contractors, sewing machine makers, and an electric dishwasher firm.[24]

Incubators, as an anonymous commentator stated in 1928, were "constructed for the purpose of providing small units of one story on short term leases to companies who are starting new business in an experimental way and to companies coming into the Chicago district who desire to try out a Chicago district operation before definitely determining the extent of their operations here."[25] The point here is that two major types of enterprises would benefit from a district location. The first, the small local firm, could establish itself before seeking a more permanent site. The second, the non-local firm, could focus on finding its bearings in Chicago while having all of its basic needs met at a well-serviced site.

Incubators offered firms close proximity to other businesses, and the process tied them into a larger industrial complex. Once established in an incubator building, new plants had access to an array of services, suppliers, clients, and collaborators, a few of whom were in the same building, while the rest were located within a very short distance. Moreover, the excellent transportation links that district promoters created enabled incubator firms to make direct and rapid links to metropolitan, regional, and national locational assets. Once firms had forged links with other metropolitan firms and had outgrown the incubator space, they could gain access to industrial property elsewhere in the district. Clearing (and this was also the case for the CMD) "provides in the leases arrangements for constructing permanent buildings on a large scale for industries locating in these buildings at such time as their plans have developed to the point of determining their permanent requirements."[26]

District clubhouses had a similar but yet different role to play in the building up of local assets. The incubator building allowed small enterprises to be embedded in the industrial fabric of both the district and the wider metropolis itself. The proprietary-controlled clubhouse allowed firm executives to monitor local business conditions, exchange information, and forge social as well as business ties. In the process, the clubhouse offered an institutional center around which managers organized their economic affairs and could keep in touch with events outside of their own immediate networks. Entry into the club ensured inclusion into a select group. Discussions about labor conditions, firm innovations, seasonal styles, or marketing strategies were facilitated by immediate access to a pool of people, thus allowing individual managers the opportunity to raise important long-term issues and to air everyday concerns.

Clearing's first district club was built just before World War I. Despite being small, the club accommodated the community's business in the early years. As the number of firms increased in the 1920s, however, the building could not hold the executives who wished to use it nor provide enough space for the growing range of facilities offered by the club. Newer and bigger premises were necessary. The Clearing Industrial Association Club, representing local manufacturing, transportation, and utilities interests—notably the First National Bank, the Belt Railway, the Continental Can Company, the Chicago Transfer and Clearing railway, and the Public Service Company of Northern Illinois—was opened in April 1928.[27]

Standing at the district's entrance, the clubhouse both guarded the district and stamped its position within Chicago. It was the headquarters of the Clearing Industrial Organization, the local business venture that coordinated district activities. Each company had two or more executives who were members of the Industrial Organization and used the institution to socialize, discuss business, and conduct personal affairs. As one commentator noted, "It is expected with the added facilities the new club provides, attendance will largely increase and benefits flow from better acquaintance and mutual exchange of experiences."[28] The Chicago Transfer and Clearing Company and other transportation firms had their business offices there. All members had access to the club's services, including a dining room able to seat 350, a lounge room, pool and billiard rooms, a card room, bowling alleys, and sleeping rooms.

The club not only offered a setting for the transmission of ideas; it also built a sense of place centered on the district itself. While the club was used by members for having lunch and cocktails, it also generated local pride and a sense of belonging. The club reinforced the idea that the district was different from the rest of the metropolis, while at the same time replicating the Loop's business clubs. As one commentator noted, "All of the facilities of the large downtown club are provided next door to the work shop of industrial executives of this growing community."[29] The club was a local place that merged the interests of proprietorial control with those of the resident firms and mimicked the central district in the range of advantages that it offered. As the central node of common business affairs, the club also allowed managers to handle affairs within the area's boundaries and to link with concerns and issues outside the district.

One of the subjects of conversation that would have come up at the clubhouse's lunch tables was factory design. During the interwar period, the question of how to shape the productive and aesthetic character of the factory became a topic of interest among business executives.[30] Manufacturers locating in Clearing and the CMD were cognizant of the modernist River Rouge plant built for Ford by Albert Kahn as well as the hundreds of modernist single-story structures constructed throughout America's industrial districts. Through trade journals, newspapers, and word-of-mouth, manufacturers recognized that the factory as a form was experiencing tremendous change. Indeed, their need for a new type of factory space helped redefine factory design and replace the old multi-storied mill structure with the sleek single-story factory (fig. 28).

The single-story factory and the resulting new industrial landscape highlighted the networks into which district promoters were plugged and how they gained access to ideas from outside the district itself. All of the elements of new factory design, even those designed by the district's architectural office, were taken from outside the district. In their use of professional architects and university-trained mechanical engineers, rather than the traditional millwright and self-trained engineers, the districts constructed a daylight factory that was a break with past factory design.

One of the CMD's first architects was Alfred Alschuler. Born in Chicago in 1876, he trained in some of the city's leading architectural firms before opening his own company in 1907. Cognizant of changing designs and construction methods, Alschuler was conversant in all forms of architectural styles. He brought these skills to the CMD, where he produced several mill-type factories. Another was S. Scott Joy, who became the district's architect in 1913. Taking a degree in architecture from the University of Illinois in 1900, he moved to Birmingham, Alabama, where he built among other things warehouses and cotton mills. As the CMD's chief architect, he was responsible for overseeing an office of 20 assistants and for designing and overseeing the construction of a large number of buildings, including the gigantic Midland warehouse, which at a cost of $1 million and at 600 square feet was the largest building of it type at the time.[31]

By the 1920s the building of daylight factories by Clearing's chief architect, Fred Foltz, became commonplace. The daylight factory was part of a wider information network about what to build and how to build. By taking ideas from national industrial journals such as *Factory and Management*, distilled in more local business magazines such as *Chicago Commerce*, and formulated in managerial and professional word-of-mouth, local promoters were agents transmitting ideas about factory design circulating at regional, national, and international levels to the local level.

The types of factories and thus the character of the industrial landscape constructed in Clearing and the CMD differed despite their common

DAYLIGHT FACTORY
for
MAY FIRST OCCUPANCY

New daylight sprinklered one story factory building of 156,000 sq. ft. including two story office building of 16,000 sq. ft. Two depressed truck driveways. Two depressed switchtracks of 12 car capacity. Belt Railway Co. of Chicago, Universal L. C. L. Service. Oil fired heating plant. Constructed in 1930.

For Lease (Short or Long Term) at Reasonable Rental
Can Be Divided

THE
CLEARING INDUSTRIAL DISTRICT
of CHICAGO

Telephone Randolph 0136   38 South Dearborn Street
CITY FACILITIES at COUNTRY PRICES

FIGURE 28. A Daylight Factory in Clearing, 1932. Clearing promoted itself as a modernist industrial district. Essential to this was the most modern of factory forms, the "daylight" factory. Source: *Chicago Commerce* (February 1932): back page.

origins and similar functions. By the 1920s, when it was beginning its major expansion, Clearing's architects were involved in a heavy construction boom, building new factories and warehouses and making additions to existing ones. In contrast, the focus of activity in the CMD's Original and Pershing tracts was moving new firms into existing premises. While 42 of the 48 Clearing firms built new structures, either as entirely new buildings or as additions to existing ones, only 16 of the 53 CMD firms did. In Clearing the single-story building was, almost without exception, the only factory type that was built. As Perry Phelps, the district's vice president, noted in his 1936 overview of the district, "Single-story buildings predominate." In Clearing, as elsewhere, the factory-design trend was to a modernist building, whose one-story frame was a sleek, streamlined production machine. As Foltz stated in 1937, firms were moving "down to the first floor. Factory buildings are definitely on the way down." This was not just a matter of aesthetics. According to Foltz and others, under the right conditions the one-story building generated operating costs that were significantly lower than for a similar-size multi-story building. A better laid-out floor plan resulted in decreased cost of supervision, simplified handling of goods, and increased storage capacity.[32]

Some examples illustrate these points. The cosmetic firm Lady Esther moved to Clearing from Evanston in the mid-1920s. The new single-story factory was an up-to-date design in which production facilities were more effectively laid out. A couple of years later, Buick Motor closed a 20-year lease to purchase industrial property just around the corner from Lady Esther, where it constructed a one-story warehouse with 240,000 square feet. Freed from the congestion, older structures, and deteriorating infrastructures close to its older building in the central core, Buick obtained a switch-track arrangement allowing trains of up to 50 freight cars to be received from Flint and unloaded at one time on platforms sufficient in size to accommodate the entire shipment, all within the envelope of a one-story structure.[33]

In contrast, the earlier building history of the CMD's Original and Pershing Road tracts ensured that the district contained mostly multistory structures. This was evident in the district architect Scott Joy's 1921 description of the CMD's factories. While he makes constant reference to "mill" structures (multi-story warehouse and manufacturing buildings), there is not one reference to the single-story factory. We know that there were a few; the bungalow incubator built during the war and some buildings constructed during the 1920s were single story. But they are rare in the CMD's older sections. This was not the case in Kedzie and the Forty-third Street tracts, where the single-story factory was the norm.[34]

The promoters of the two planned industrial districts were innovative yet cautious factory builders. From the prewar period, when the CMD architect Alschuler built modern three-story mill structures, to the large and expansive single-story daylight buildings constructed in Clearing and Kedzie from the 1920s by architects such as Foltz, planned district managers were alert to the financial repercussions of the work spaces that they planned and built. As evident from the various design changes and factory innovations, district promoters went out of their way to ensure that the factory space was effective, and that the buildings themselves were well planned in the broader context of the district.[35]

Despite the forms of control applied by the proprietorial group, the relation between firms and planned industrial districts was built on the relatively open set of relationships, allowing manufacturers to establish links with other firms and interests outside the district. In other words, there was a focused set of relations upon which the planned industrial district itself and the firms within it were dependent for their functioning, and a more diffuse set of relations that allowed firms to negotiate their economic performance through networks with other firms, both inside and outside the district. Firms operating in a planned industrial district functioned within another power structure from the typical one in which there is a spatially diffuse hierarchy of relations with suppliers, collaborators, and clients. While nearly all of them were plugged into this spatially diffuse hierarchy, they also functioned in a relatively autonomous one centered on the rules established and the facilities offered by the district itself. Planned industrial districts offered firms the ability to create a functional relationship with other firms, while operating within a planned, shared, and codified regulatory and institutional framework.

CHAPTER 8

Networked Space

THE CONNECTED
METROPOLIS IN THE 1920S

In 1920 Campbell Soup bought a five-acre site complete with an old two-story brick factory in the Calumet suburb of Hammond. The company, however, was unwilling to develop the site, and in 1925, tired of waiting for Campbell to act, a local syndicate bought the plant, subdivided it into smaller units, and leased space to several firms. Campbell, however, continued to show interest in the Chicago area. Less than a month after selling off the Hammond property, it closed negotiations for a $575,000 vacant site in the Dickinson Industrial District. Once again though, Campbell was not ready to move and did little to the property. Finally, beginning in November 1927, they announced that they would construct a plant with floor space of more than 1 million square feet. The first unit—a six-story concrete structure flanked by three-story buildings on either side, with 800,000 square feet of floor space—was finished a year later. Serving the Midwest market, operated by a labor force drawn from the neighboring working-class districts, and fed by Illinois and Indiana tomatoes, it began producing 3.6 million cans of soup a day.[1]

An unusual feature of the new factory was its relationship to a supplier. Cans, the company's major packaging material, were supplied by a specially constructed factory next door to the soup plant. Continental Can's four-story building would "be engaged exclusively in producing cans for the Campbell Soup company." Millions of cans daily were "sent by conveyor direct into the Campbell soup plant." Completed in early 1929, the plant allowed for "the direct transfer of cans into the new Campbell soup factory which is only about fifty feet from this new structure. . . . [A]utomatic machinery will transfer these cans from one plant to the other."[2] Canning manufacturers such as Continental were also tied to other industries. As one commentator in 1929 noted, "One finds the tin plate companies interested in producing a tin better suited for the resistance of acids, the tin can manufacturers are interested in producing a can which will facilitate the packing of food, and the can machinery manufacturers attempt to devise machinery which handles the cans with greater rapidity. The co-operation that is seen in this whole industry is not alone stimulating to an individual, but it should be an example for the entire business world."[3]

The sagas of Campbell-Continental and food processing–canning illustrate the geography of inter-firm relations. The two companies were firmly embedded in a multiple set of business relationships, which stretched across metropolitan, regional, and international industrial landscapes.[4] Success involved taking advantage of both local and long-distance flows of goods, ideas, knowledge, and capital. Similarly, Chicago's industrial economy was sustained by an elaborate set of intra- and inter-firm relations operating at different spatial scales. The evidence shows that while regional and national interactions were important to local firms and industries, a well-embedded local production complex was critical to Chicago's success as an industrial economy.

REACHING OUT

In 1928 the American Manganese Steel Company remodeled the Chicago Heights plant of the recently acquired American Brake Shoe and Foundry to manufacture manganese steel and gray iron. In the refurbished plant, American Manganese continued the tradition of producer goods production; its 800 workers made railroad track castings for local and national companies. The switch from brake shoes to railroad castings did not involve a dramatic realignment of the plant. But it did require new equipment: sand-handling machinery, an annealing furnace, core ovens, and an electric furnace.[5] The acquisition of this equipment from Chicago, Niagara Falls, Cleveland, and Pittsburgh firms is emblematic of Chicago's business world. It shows how local manufacturers were firmly embedded in both local and regional circuits and networks.

What was true for American Manganese was also true for other Chicago manufacturers. The machinery and equipment purchases of 20 metropolitan Chicago metal firms between 1921 and 1928 ranged across much of industrial America. One important supply center was metropolitan Chicago, which accounted for 41 (19 percent) of all transactions. Chicago firms looking to buy new machinery and equipment sought out local firms. When the Whiting Corporation built a new four-story pattern shop to store 18,000 live and used patterns, it bought equipment—sewer pumps, steel fire doors, card systems, and steel storage racks—from local firms. In turn, Whiting was one of the most important equipment suppliers to other Chicago firms. George Limbert's new fitting and flange plant in East Chicago was furnished with a Whiting cupola and traveling crane, while Wisconsin Steel bought job and gantry cranes from the company in 1924 and 1925.[6]

The majority of transactions, however, were made with firms outside of metropolitan Chicago. Most machinery producers supplying Chicago's metalworking firms came from the Manufacturing Belt, America's workshop (fig. 29). Pittsburgh and Cleveland, with 51 (24 percent) transactions, took the largest number, while Milwaukee, New York City, Buffalo, Philadelphia, Alliance, St. Louis, and Worcester together accounted for 28 percent of total transactions. There were another 45 places with three or fewer transactions. The supply district was bounded by Fairfield, Iowa; Petosky, Michigan; Boston; and Louisville. Concentrated in major industrial centers, firms also sought machinery and equipment from producers in small cities and towns of Ohio, Pennsylvania, New York, and New Jersey.

The dominant role of Pittsburgh and Cleveland firms points to the importance of the Manufacturing Belt's metalworking worlds. Even though Chicago by the mid-1920s was an important steel, foundry, and machinery center, the initial advantage and large-scale expansion of Pittsburgh's steel complex before 1900 ensured that the Pennsylvania city remained a major supplier of machinery and equipment to Chicago firms. Pittsburgh companies such as Abramsen Engineering, Pittsburgh Electric Furnace, Tate-Jones, Bacharach Industrial Instrument, and Mesta Machine furnished local manufacturers with rotary straighteners, electric and annealing furnaces, flow meters, and shears. Similarly, Cleveland firms, with their heavy metalworking tradition and links to the steel-working district of Youngstown, supplied products and services that Chicago firms found attractive. Chicago's pool of new machinery and equipment stretched from the local to the regional. Metalworking companies looked home and afar for their production equipment.[7]

When the New York firm Niagara Radiator settled in Pullman in 1921, it was seeking the large local market and supply base found in Chicago. But the company was also wired to networks reaching far outside the

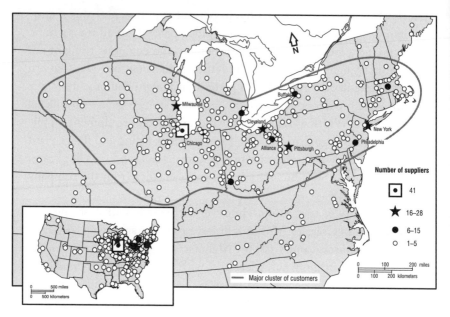

FIGURE 29. The Geography of Equipment Buying of 20 Chicago Metalworking Firms, 1921–1928. Source: *Iron Age*, 1921–1928.

metropolitan area's ambit. Most obviously, its productive strategy, financial control, and executive decision-making were decided at its New York head office, even though the Chicago manager oversaw local details. It was also linked to a wider world through machinery acquisition. Of the 14 businesses involved in equipping a 1923 addition, only one was from Chicago; Whiting furnished a monorail, overhead cranes, tumbling barrels, and a cupola. The rest were from Cleveland, Beloit, Buffalo, Detroit, Toledo, Cincinnati, Seneca Falls, New York, Pittsburgh, and South Norwalk.[8] An excellent cross section of the industrial worlds of the Manufacturing Belt, these equipment suppliers established the boundaries of Niagara Radiator's working universe and, in the process, incorporated the small Chicago neighborhood of Pullman into machine and equipment networks crisscrossing the belt.

The role of the machinery and heavy hardware suppliers did not end with the placing of the order. Large, heavy, and lumpy, machinery involved an extensive set of supplier-client relations. After the order was placed, the piece frequently having been made according to client specifications, the supplier sent in engineers to install the machines. Even after installation, suppliers were often responsible for teething problems, regular servicing, and emergency repairs. The buying and selling of heavy machinery and equipment was not a onetime deal. It involved expensive transactions, long-term relationships, and sustained interaction. Sellers had to convince clients that they could produce a piece of machinery at a competitive price and guarantee its ongoing effectiveness by their long-term commitment to the client. The very act of buying these large chunks of operating capital ensured that inter-firm networks were in place by the time the specifications were formulated and the equipment installed. The taking of a supplier contract also meant creating a stream of contacts over time as cranes, monorail systems, and electric furnaces needed frequent upkeep and service in the years following installation.[9]

A second glimpse into the operations of Chicago firms outside the local economy can be gleaned from the attendance of Chicago exhibitors at two trade meetings: the Eighth Annual Convention of the American Society for Steel Treating held in Chicago in September 1926 and the Thirtieth Annual Meeting of the American Foundrymen's Association held in Detroit a week later.[10] One important aspect of trade meetings was the ability of buyers (foundry and steel mill managers), industry observers (trade journals such as *Iron Age*), competitors (foundry supply and factory equipment firms), inventors, engineers, and others to intermingle and catch up with friends, business associates, and competitors. The conference sites—Detroit's State Fair Grounds and Chicago's Drake Hotel—were the centers of activity. Here company representatives could get an idea of what their competitors' products were, and buyers could discuss prices, specifications, delivery dates, and so forth with sellers.

Chicago firms came into contact with a wide range of firms and individuals at the two conferences. The importance is evident from the fact that Chicago had the largest single city contingent at the foundrymen and steel treaters' conferences. Obviously, the companies believed that exhibiting their goods paid off, both in direct sales and in the contacts that representatives made with other conference-goers. For Chicago executives, the meetings allowed them to highlight their product lines to competitors and prospective customers alike. Even though the non-Chicago exhibitors were obviously competition, the exhibition gave the Chicago representatives the chance to discuss with their competitors some of the important aspects of their common everyday business, such as changing trends of the industry, alterations to product design and innovation, and which customers were difficult or untrustworthy.

Just as importantly, the exhibitors paraded their wares in front of a large buying crowd. While Chicago firms may have learned a great deal about their business over the table of a neighboring exhibitor, they also had the opportunity to have face-to-face contact with thousands of potential customers. To interested clients, they demonstrated the advantages of their products and, if successful, arranged time to meet to discuss the details required for buying, delivery, and setup of their products. Even if they failed to secure the patronage of a firm during the meeting, the contacts made at the time could bear fruit later. Without the concentration of actors involved in their specific product lines at the trade meetings, foundry, steel, and machine suppliers would have been unable to conduct their business as effectively as they could.

A third way to appreciate the business geography of the Chicago heavy equipment and metalworking firms was revealed in the activities of the Whiting Corporation. The company was metropolitan Chicago's largest foundry equipment maker and was established in the Calumet town of Harvey in 1893. Harvey was an instant town created by a real estate company, the Harvey Land Association, a creature of Turlington Harvey, a Chicago lumber merchant, and a group of Chicago investors mainly composed of bankers and lawyers. It was located at the intersection of the Illinois Central and Grand Trunk railroads. Harvey's promoters used two strategies to construct an industrial suburb on Chicago's southern border. First, they sought to attract industries by providing transportation facilities, public works, and improvements paid for by the property taxes taken from the local working class. Second, they merged property and industrial interests under the guise of local betterment. When it opened in 1894, Whiting focused on making cupola furnaces but later expanded to include other types of foundry and industrial equipment. By the 1920s it was one of the country's leading heavy machinery manufacturers and, with 750 workers and $2 million of invested capital, was one of metropolitan Chicago's larger firms. Along with combustion engine maker Buda Company, it was Harvey's preeminent employer.[11]

The combination of Whiting's focus on heavy industrial equipment and the industry's specialized character ensured the firm an extensive market, one not limited to the local or even regional markets (fig. 30). Whiting was one of a handful of heavy equipment firms operating in the United States. In 1905 there were only 24 cupola manufacturers, most of which were located in the steel-making districts of Pennsylvania (Pittsburgh, Lebanon, and Philadelphia), Ohio (Cleveland and Youngstown), and Alabama (Birmingham). Whiting was the only one west of Ohio. The number did not change very much in the following years, but the industry's center of gravity did. Ten years later, 8 of the 26 cupola manufacturers were located

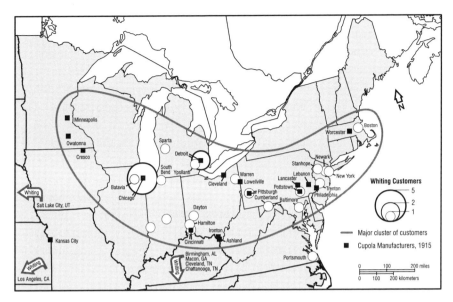

FIGURE 30. The Geography of Firms Supplied by Whiting Corp. in 1923. Source: *Iron Age*, January–April 1924; *Thomas' Register of American Manufacturers and First Hands in All Lines, 1915* (New York: Thomas Publishing, 1915).

west and south of Ohio. As the steel-making and metalworking industries moved west, so did Whiting's competition.[12]

The firms to whom Whiting sold its cupolas, elevators, tumbling barrels, and core ovens provide another glimpse into the geography of Chicago's business dealings.[13] Whiting sold foundry equipment to 37 customers in 31 locations. The compilation of these weekly sales provides a window on Whiting's territorial service area and clues to one aspect of Chicago's business world. The orders booked by Whiting covered most areas of the United States. With five different customers, Chicago had the largest single number. Whiting supplied cupolas to the Board of Education and a foundry, and tumbling barrels to three firms. Elsewhere, a handful of customers in the southern and western perimeters of the mainland United States bought cupolas and tumbling barrels. As expected, Whiting's clients from outside the Manufacturing Belt were found in the steel area of Birmingham and industrializing Los Angeles. Customers were also scattered: Utah, Tennessee, Georgia, and Virginia.

The majority of sales, however, occurred in the Manufacturing Belt. Including Chicago, 30 of the 37 customers had plants in the area stretching from Boston to St. Louis. Not surprisingly, heavy industry centers—such as Detroit, St. Louis, Newark, Toledo, and Pittsburgh, as well as the regional metalworking districts—were heavily represented. Some of the country's leading firms placed orders. For example, Ford and Singer had four No. 8 cupolas and four charging machines sent to their Michigan and New Jersey factories, respectively. Other customers were specialized heavy equipment makers from smaller urban places such as Hamilton (Ohio), Ypsilanti (Michigan), and New Brighton (Pennsylvania). Regardless of the type of order, the locus of activity was the major industrial core of the United States and demand for Whiting's products reflects the geographic cross section of manufacturing in the 1920s.

The selling of cupolas and tumbling barrels to firms in all corners of the United States was not a onetime event. Even though the small number of firms limited competition, Whiting competed with firms that were physically embedded in the industry's major market, Pennsylvania's and Ohio's steel-making districts. To be competitive, they, like other heavy machinery and equipment firms, needed to have a good reputation for their products. This had to be built on an extensive, recognized, and well-thought-of national chain, ranging from the initial selling of the idea (in terms of quality, specifications, cost, delivery date) through the manufacture of the product to the after-sales service.

Unlike its eastern competitors, Whiting did not have a large number of customers in close proximity. Whiting's early establishment, and its location before World War I, helped the company capture a significant part of the western market, while maintaining a strong showing in the national market. The fact that Whiting was the only maker of these specialized products west of Ohio at the turn of the century guaranteed a relatively open field in terms of gaining markets outside the established steel-making districts. But it did not have such a large captive market as the Pittsburgh, Cleveland, and Youngstown manufacturers, thus reinforcing the need of the company to be reliant on the manufacture of goodwill, trust, and reliability among potential customers across the United States.

Bankruptcy records provide excellent evidence for exploring both the metropolitan circuits of capital, information, and semi-finished products in various industries, and the different scales at which firms interacted (national, regional, metropolitan, local). An examination of 53 such records with 4,929 creditor links from 1920 and 1928 provides a final glimpse into the non-local world of Chicago's interests. The firms came from a range of sectors, including metal, furniture, automotive, and electrical goods. There was good geographic coverage of the bankrupt firm locations, ranging from the area close to the Loop to 10 suburban locations. At more than 90 percent, the linking of creditors to address and product was high, thus ensuring excellent coverage.[14] About three-quarters of all creditors were from Chicago, with the remainder mostly located in the Manufacturing Belt.

The bankruptcy records also reveal a large sectoral difference between Chicago and non-Chicago creditors (table 16). Only half of the Chicago creditors that the bankrupt firms owed money to were manufacturing firms (51 percent), while more than a third (42 percent) of firms and individuals supplied services (advertising, utilities, etc.), financial support, and wholesale goods. In contrast, more than three-quarters (77 percent) of the non-Chicago creditors were manufacturers, while the corresponding share of service, financial, and wholesale activity was much smaller (12 percent). Chicago firms in the interwar period received most of their services, as they did in the nineteenth century, from local firms, while material inputs came disproportionally from elsewhere in the Manufacturing Belt. Chicago entrepreneurs selectively chose the types of firms that they interacted with at different spatial scales.

Unsurprisingly, only a few of the non-Chicago creditors were firms and individuals located outside the Manufacturing Belt (fig. 31): only 157 (12 percent) had an address in the South or the West.[15] Although they supplied a range of products and services, most of the money owed was for light consumer products or services rendered as agents for the Chicago bankrupt firms. In contrast, the vast majority (88 percent) of creditors were from the Manufacturing Belt. Similar to their southern and western counterparts, these 1,107 firms and individuals furnished an assortment of products and services, although the emphasis was on manufactured goods. Reflecting the types of products that Chicago firms

TABLE 16. Sectoral Distribution of Chicago and Non-Chicago Firms, 1920 and 1928

Sectors	All Firms	Chicago Firms	Non-Chicago Firms
Service (%)	15.8	19.6	5.3
Distribution (%)	15.5	18.9	6.0
Finance (%)	3.6	4.7	0.3
Manufacturing (%)	57.7	50.9	76.7
Not Available (%)	7.4	5.9	11.7
Total (%)	100.0	100.0	100.0
Total (No.)	4,929	3,628	1,293

SOURCE: Compiled from 53 1920 and 1928 bankruptcy cases taken from Record Group 21, Records of the United States District Court, Northern District of Illinois, Eastern Division at Chicago, Bankruptcy Records, Act of 1898, 1898–1972; Bankruptcy Case Files, 1898–1946.

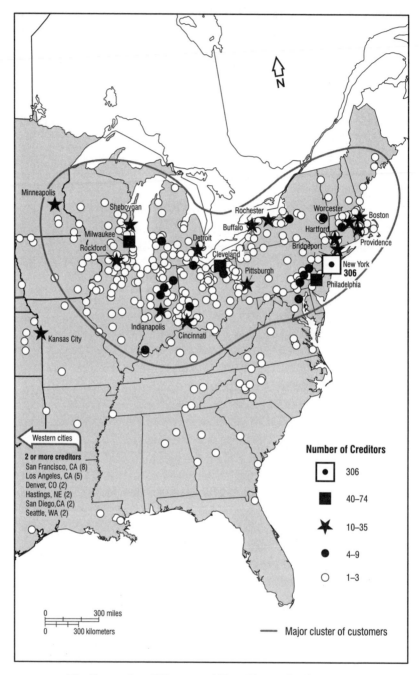

Number of Creditors

- [●] 306
- [■] 40–74
- [★] 10–35
- [●] 4–9
- [○] 1–3

Western cities

2 or more creditors
San Francisco, CA (8)
Los Angeles, CA (5)
Denver, CO (2)
Hastings, NE (2)
San Diego, CA (2)
Seattle, WA (2)

0   300 miles
0   300 kilometers

—— Major cluster of customers

Minneapolis · Sheboygan · Milwaukee · Rockford · Detroit · Cleveland · Rochester · Buffalo · Worcester · Boston · Hartford · Bridgeport · Providence · New York 306 · Philadelphia · Pittsburgh · Indianapolis · Cincinnati · Kansas City

FIGURE 31. The Geography of Chicago and Non-Chicago Bankrupt-Creditor Relations, 1920 and 1928. Source: Selected files from the Records of the United States District Court, Northern District of Illinois, Eastern Division at Chicago, Bankruptcy Records, Act of 1898, 1898–1972.

required, most creditors were from the major industrial states: New York (313), Ohio (155), Pennsylvania (128), Indiana (82), Wisconsin (78), and Massachusetts (76).

A handful of metropolitan areas accounted for a large share of the creditors. With 306 creditors, New York City easily outstripped the next important metropolitan area, Cleveland, with 72. These were followed by some of the country's largest industrial areas: Philadelphia (60), Milwaukee (45), Boston (39), St. Louis (34), Cincinnati (31), Bridgeport (28), Pittsburgh (28), and Detroit (24). Most other urban centers of any consequence in the Manufacturing Belt were connected to Chicago through a credit relationship. While a distinct hierarchy existed, both between the Manufacturing Belt and the rest of the country, and within the Manufacturing Belt, there were very few places that did not have companies and individuals supplying some service or product to a Chicago firm.

The geography of Chicago's creditors not only reflected city size, but also depended on the bankrupt companies' demand for the specific products and services they drew upon. The city's automotive industries bought tires and accessories from Detroit and Akron. Electrical instruments and machinery were acquired from Boston and Cincinnati. Chicago's automotive and furniture factories secured brass goods and machine tools from the Connecticut brass-making and machinery complex of Hartford-Waterbury, Torrington, and Bridgeport. The heavy industry centers of Pittsburgh and Cleveland supplied steel, grinding wheels, and heavy equipment. Rockford's unlikely presence in this list stems from its provision of furniture to several bankrupt firms. As well as furnishing a cornucopia of products and services, New York City supplied Chicago's advertising and publishing sectors. Scientific instruments, cameras, and optical goods came from Rochester, America's photographic center. Chicago companies were linked to other places through metropolitan industrial specializations.[16]

These different spheres of business interactions between Chicago firms and urban centers were paralleled regionally. The importance of light consumer goods supplied by southern and western creditors is illustrated by the large number of furniture manufacturers, especially those from the southern furniture belt. Chicago furniture manufacturers bought partly finished products, while dealers and retailers acquired finished goods, from companies in North Carolina (Mount Airy and Thomasville) and Virginia (Rocky Mount and Richmond). Nearly all of these furniture firms had either Chicago offices or display rooms at the American Furniture Mart, the home to year-round agents who exhibited the firm's products to prospective clients and kept up-to-date with the trends, styles, and prices in America's largest furniture-producing center.

The same was true for the Manufacturing Belt's furniture centers close to Lake Michigan. Chicago enterprises bought a great deal of furniture from specialized furniture-producing towns. The highly specialized set of suppliers from Grand Rapids furnished goods to Chicago's retailers. One of these, J. Nartzik, equipped Lemoyne Furniture with furniture veneers that it kept in stock at its Chicago warehouse.[17] Other towns and small cities in Illinois (Rockford and Freeport), Indiana (Batesville and Evansville), Michigan (Holland and Traverse City), and Wisconsin (Sheboygan and Stevensville) supplied local furniture retailers and manufacturers with a variety of goods. Moreover, many of them displayed their products in downtown Chicago salesrooms or, like Nartzik, kept stores in Chicago warehouses. In other cases, southern and western creditors worked as sales representatives and agents on commission. St. Charles Fixture Manufacturing owed sales commission to six agents, four of whom lived in Cedar Rapids, Birmingham, Kansas City, and Minneapolis, while the Cederborg Company owed two San Francisco agents money for the work they performed selling company chinaware on the West Coast. This and the other three glimpses into Chicago's interwar business world illustrate that local firms were deeply embedded in a regional and national universe of industrial linkages.

BUSINESS LINKAGES AND THE INDUSTRIAL LANDSCAPE

While Chicago enterprise was firmly linked to the larger regional and national worlds, it was even more engaged with local companies and individuals. As the bankruptcy records show, almost three-quarters (74 percent) of creditors were from metropolitan Chicago. These firms were owed more than three-quarters (78 percent) of the total money. The records also reveal that Chicago's business was firmly embedded in the local business world and covered all aspects of the production chain (table 17). Manufacturers, in terms of the number of creditors and the value of the money owed, were the most numerous. The evidence also shows that downtown (or center) accounted for almost half of all debtor-creditor transactions. Chicago's downtown with its dizzying array of business interests remained central to Chicago's manufacturing world.

Chicago's bankrupt firms were indebted to manufacturers and wholesalers that supplied everything from steel sheets to furniture hardware. They were underwritten by a small number of financial agents who sup-

TABLE 17. Chicago Creditors by Business and Location, 1920 and 1928

	Creditors		% of Total	
	No.	$ Owed	No.	$
BUSINESS STRUCTURE				
Manufacturing	1,848	569,025	50.9	38.6
Finance	171	519,534	4.7	35.2
Service	711	170,877	19.6	11.6
Advertising and Publications	133	21,129	3.7	1.4
Associations	16	1,019	0.4	0.1
Contractors	53	12,877	1.5	0.9
Legal	35	18,578	1.0	1.3
Real Estate	36	47,029	1.0	3.2
Services, General	248	53,962	6.8	3.7
Transportation	83	4,991	2.3	0.3
Utilities	107	11,292	2.9	0.8
Distribution	686	159,817	18.9	10.8
Not Available	212	54,881	5.8	3.7
Total	3,628	1,474,134	100.0	100.0
GEOGRAPHIC STRUCTURE				
Center	1,657	546,372	45.7	37.1
North Side	759	383,519	20.9	26.0
South Side	758	277,789	20.9	18.8
Suburbs	207	128,138	5.7	8.7
Calumet	22	12,480	0.6	0.8
North	39	26,436	2.6	1.8
West	146	89,222	4.0	6.1
West Side	202	118,311	5.6	8.0
Not available	45	20,005	1.2	1.4
Total	3,628	1,474,134	100.0	100.0

SOURCE: Compiled from 53 1920 and 1928 bankruptcy cases taken from Record Group 21, Records of the United States District Court, Northern District of Illinois, Eastern Division at Chicago, Bankruptcy Records, Act of 1898, 1898–1972; Bankruptcy Case Files, 1898–1946.

plied a large amount of capital—more than a third of the money owed to Chicago firms. Most manufacturers received capital from more than one source and in different ways. Representative of the range was the lamp manufacturer Metal Crafts Inc., which received money from two related individuals (almost $30,000), a self-styled "capitalist" ($2,000), and a bank (a $10,000 loan for machinery from the Lawrence Avenue National Bank), while two neighborhood firms provided insurance coverage and two downtown firms supplied auditing services.[18] Finally, an extremely wide array of service-based creditors offered Chicago manufacturers everything from advertising and membership in associations to legal services and utilities.

Three major features emerge from an examination of the geographic distribution of creditors. First, they came from all parts of metropolitan Chicago, with the greatest concentration in the central area (Loop, Northwest Side, and West Town) (see table 17). The more than 1,600 businesses here accounted for close to half of the creditors and more than a third of the value of the money owed. In the prewar years, the downtown area was the glue holding the metropolitan manufacturing world together. This stands in contrast to the present day when most American metropolitan areas feature a hollowed-out downtown area and important and rapidly growing suburban business districts.

There were two other major clusters. The South Side ran through the Near South Side before branching out southwest along the Chicago River and directly south through Douglas and Englewood. It accounted for a fifth of the firms and almost 19 percent of the value. The other, the Near North Side, accounted for a fifth of the firms and more than a quarter of the value. The remaining companies were from the West Side, mainly the Garfield Park area, and the suburbs, especially the western ones such as Cicero and Aurora. Chicago business ties stretched across the metropolis.

The second is that bankrupt firms tended to have interaction with enterprises located in their neighborhood and adjacent districts. This local effect ensured that most businesses were greatly associated with firms that were part of the same geographic zone of the metropolis. The association ranged from direct interaction, in the form of supplier-customer relations, collaboration on process or product changes, or the sharing of labor and market conditions, to the indirect advantages gained from commonly shared neighborhood facilities.[19] Businesses had a larger than expected share of links to firms in the zone in which they were located (table 18). The locale effect ran from as low as 1.26 and 1.33 for the central and north zones to 2.58 and 5.12 for the west and suburbs zones. In contrast, the locale effect for creditors outside the zone where the bankrupt company was located was, with the exception of the north-zone firms and central creditors, below 1.0. In other words, a marked spatial proximity existed between firms and the firms with whom they interacted.

Two examples illustrate this point. A third (34 percent) of all creditors owed money by American Pneumatic Chuck of Pullman were located in the south zone, compared to 19 percent of all creditor firms (fig. 32). The company's creditors varied by type. Some were financial: the Pullman Trust

TABLE 18. Locale Effect, 1920 and 1928

Bankrupt Firm Zone	No. of Bankrupt Firms	Locale Effect		
		Same Zone Firms	Central Zone Firms	Other Zone Firms
South	12	1.59	0.92	0.73
North	13	1.33	1.08	0.66
West	4	2.58	0.87	0.93
Suburbs	10	5.12	0.62	0.87
Central	11	1.26	1.26	0.78

SOURCE: Compiled from 53 1920 and 1928 bankruptcy cases taken from Record Group 21, Records of the United States District Court, Northern District of Illinois, Eastern Division at Chicago, Bankruptcy Records, Act of 1898, 1898–1972; Bankruptcy Case Files, 1898–1946.
NOTE: The locale effect is obtained by taking the number of creditors in the zone as a percentage of the total number of creditors in all five zones divided by the number of creditors of all bankrupt firms in the zone as a percentage share of all creditors. A number greater than 1.0 signifies more creditors in the zone than would be expected given the total distribution of all creditors, while a number less than 1.0 signifies the reverse.

and Savings Bank was owed $350 for a mortgage on the factory premises, while the company's owner, Herman Teninga, owed $5,000 to the Calumet National Bank. Several neighborhood enterprises provided inputs—such as steel, steel castings, chuck parts, oil products, and air compressors—and services, from towel supply to automobile and express service. In the city's north end, a third of the creditors owed money by the motorcycle maker Illinois Motor were located in the north zone (compared to 19 percent of all creditor firms). Illinois Motor relied on a local bank for credit ($9,000 to the Citizens State Bank of Chicago). Along with local printing, coal, and cartage services, the firm also drew upon neighborhood businesses for engines, metal goods, attemperators, gray iron castings, japanning, dies, and brass goods.[20]

Both Illinois Motor and American Pneumatic had wide-ranging links to spatially proximate firms. This locale effect was partly determined by the industry's geography. As the bankrupt businesses tended to be located in the Chicago clusters of the sector of which they were a part, their creditors were frequently found in that sectoral cluster. But this was not the only contributing factor, for a few bankrupts lay outside the major sectoral cluster of which they were a part. Automatic Body, an automotive body manufacturer located in a secondary automotive cluster running along the North Branch of the Chicago River, not only obtained glass, auto supplies,

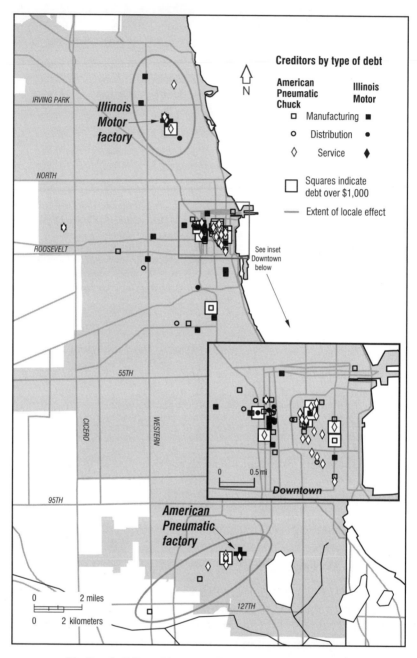

FIGURE 32. The Locale Effect of American Pneumatic Chuck and Illinois Motor, 1920. Source: Records of the United States District Court, Northern District of Illinois, Eastern Division at Chicago, Bankruptcy Records, Act of 1898, 1898–1972; Bankruptcy Case 28736, *In the Matter of American Pneumatic Chuck Company*, "Schedules A(2) and (A)3" (November 9, 1920); Bankruptcy Case 28819, *In the Matter of Illinois Motor Company*, "Schedule A(3)" (December 30, 1920).

and metal stampings from the north zone, but also secured hardware, electric motors, and windshields from manufacturers in the city's major automotive district in the Near South Side. Likewise, Illinois Motor was also situated in the northern secondary cluster. As well as receiving supplies from neighborhood firms, it also obtained japanned goods, tires, gas regulators, carburetors, auto supplies, and asbestos from Near South Side companies.

The third point about Chicago's industrial geography is that all businesses were firmly tied to metropolitan Chicago's core zone. Bankrupts, regardless of their local address, had between a third and two-thirds of their creditors located in the core zone. In other words, there was a two-way geographic tug for firms. Firms sought out the central city for publishing, utilities, wholesalers, financial services, lawyers, and advertising agencies (table 19). What they did not offer were extensive manufacturing and contracting links. The core zone was internally differentiated. While law offices, accountants, banks, newspaper houses, and utilities

TABLE 19. Concentration of Creditors by Sectors in the Core Zone, 1920

| Sectors | No. of Links in the Core | Core's Share of Chicago Total (%) | Share (%) of All Chicago Firms Located in the | |
			Loop	Rest[a]
Legal	22	100	100	0
Utilities	52	90	88	2
Newspapers	26	84	81	3
Wholesale	239	70	28	42
Advertising	26	67	64	3
Finance	46	63	66	7
Service	65	59	52	7
Associations	5	56	56	0
Real Estate	7	50	36	14
Manufacturing	417	45	19	27
Contractors	12	40	30	10
Transportation	13	36	25	11
Total	930	55	31	24

SOURCE: Compiled from 53 1920 and 1928 bankruptcy cases taken from Record Group 21, Records of the United States District Court, Northern District of Illinois, Eastern Division at Chicago, Bankruptcy Records, Act of 1898, 1898–1972; Bankruptcy Case Files, 1898–1946.
NOTE: Taken from Chicago firms only, and not from non-Chicago firms having a Chicago office or warehouse.
[a] Rest = Northwest Side and West Town.

were tightly clustered in the Loop, wholesalers and manufacturing shops were crammed along the Chicago River in the Northwest Side and West Town. Reflecting Chicago's broader geography, firms sought out different inputs from different parts of the central core. There was no simple center-periphery geography to intra-metropolitan business networks.

CHICAGO FERROTYPE, JEWELL ELECTRICAL, AND THE INSTRUMENT DISTRICT

Two substantial camera and instrument manufacturers that went bankrupt in the 1920s illustrate the dynamics of local production complexes. Chicago Ferrotype was a camera and phonograph manufacturer located in the Near West Side. With its Congress Avenue factory valued at $65,000, the firm employed 175 workers and had $500,000 of capital in the manufacture of phonograph and camera hardware and supplies in 1920. The company also had a storage facility in Forest Park valued at $15,000. The other firm, Jewell Electrical, made electric measuring instruments at Walnut and Ashland. In 1924 it employed 153 workers and had an investment of $175,000.[21]

These two firms were firmly anchored in Chicago's electrical and instrument cluster. From the early days of the industry, electric and instrument firms was concentrated in the Loop and an adjacent crescent that started in the Near West Side and continued north through West Town to the Near North Side and Lincoln Park. More than 200 instrument and electrical firms with almost 15,000 workers were located in the district. Chicago Ferrotype and Jewell Electrical lay within walking distance of phonograph and parts makers, specialty metal and wood firms, and the city's only other camera maker. Both firms were specialized manufacturers dependent on raw and semi-finished material suppliers, collaborators for innovation, clients for finished products, other businesses for a range of services, and residential neighborhoods for all kinds of workers. While there is no specific evidence from the bankruptcy records that either firm had direct interaction with competitors, it is extremely likely that there were dealings between Chicago Ferrotype and Jewell Electrical and the other neighborhood camera, phonograph, and instrument makers. While neither Chicago Ferrotype nor Jewell Electrical received merchandise from their local competitors, they did function with neighborhood firms. The instrument district, a durable goods area composed of small, vertically disintegrated firms, produced a variety of products and contained a significant share of Chicago's electrical machinery and fabricated metal makers.

The importance of this cluster and inter-firm interactions was clearly understood by contemporaries. As one planning report a few years earlier stated:

> The distribution of employment among the major metal and metal working industries might suggest that the district functions as a fairly closely integrated manufacturing unit. With the primary and fabricated metals industries serving as basic steel processing, there is an assured supply of the most essential raw materials and parts for the machinery and instruments industries. At the same time, the district contains a sufficient number of lumber and wood, as well as stone, clay and glass establishments all of which produce some parts required by the machinery and instruments industries.[22]

In other words, Chicago Ferrotype and Jewell drew upon a distinct neighborhood cluster and centrally located set of businesses. The district provided external economies that supported firms supplying semi-processed goods and services. But the two instrument makers also drew upon manufacturers, wholesalers, and services from outside the local neighborhood. As the bankruptcy records show, firms from other parts of Chicago and the Manufacturing Belt were important contributors. Chicago Ferrotype, Jewell Electrical, and Chicago's instrument district were embedded within a larger metropolitan and regional flow of materials, capital, and information.

Chicago Ferrotype provides an example of how this was played out. It owed creditors more than $35,000 for loans received. All of this capital came from Chicago sources. The largest sums were from Foreman Bros. Banking ($13,500) and Empire Wholesale Groceries ($10,000), while Equipment Investments advanced the firm $7,350 for equipment. It had legal debts—one to a patent lawyer and another to general counsel—and owed $4,800 for insurance. With their offices in the Loop, all these firms were close by.[23] Chicago Ferrotype depended on wholesalers. Forty-two local and five Manufacturing Belt creditors supplied it with factory supplies, specialized steel products from two of Chicago's largest steel wholesalers, machines parts, office supplies, and plating supplies. Wholesalers acted as intermediaries between firms, bringing products to one firm from suppliers close by and from those thousands of miles away. Local wholesalers received much of their stock from local steel manufacturers such as Illinois Steel, while non-local firms such as Niagara Fall's Carborundum Company and Cleveland's American Multigraph maintained warehouses in Chicago's main wholesaling district close to the Chicago River.

The most important group of creditors was other manufacturers. Chicago Ferrotype owed money to 165 manufacturers, covering the range of goods needed by a camera and phonograph manufacturer and from along the entire production chain. Primary inputs such as castings and cold-finished steel came from Alemite Die Casting; semi-manufactured products such as brass goods, diaphragms, and optical goods were obtained from Waterbury Brass and Vulcanized Rubber; tools and machine parts for the production process itself were acquired from a wide range of companies; while other firms provided packaging products such as paper boxes, wooden boxes, tags and cases, and office supplies. Manufactured goods came from 60 businesses in non-local centers: Bridgeport (hardware), New York City (paper, rubber, and metal goods), Cleveland (machine drills and screws), Rochester (camera and optical parts), and Worcester (grinding wheels and machines). Some of these kept an office in the warehouse district, while others sent material directly to Chicago Ferrotype's factory. Those with local factories and warehouses supplied parts and materials extremely quickly. For firms sending material directly from the factory, Chicago's excellent railroad system and downtown Chicago's set of railroad facilities offered quick and reliable service. Just in case suppliers were not quick or reliable, Chicago Ferrotype depended on more than one. For example, it acquired brass products from seven companies, machine tools from six, screws from six, paper boxes from five, and dies and tools from three. The camera maker drew on a wide radius extending over the Manufacturing Belt for production inputs.

The local instrument district was crucial to Chicago Ferrotype and Jewell Electrical as most inputs came from Chicago businesses. The district was critical to the firms that located there. Local producers drew upon a common pool of wholesalers and manufacturers, as can be gleaned from an examination of the creditors that Jewell Electrical and Chicago Ferrotype had in common. The 47 common creditors had three important features. First, they provided products for the two bankrupts. Most of the creditors supplied semi-processed products that went directly into the production of Chicago Ferrotype's cameras and Jewell Electrical's measuring instruments. Large amounts of die castings, brass, hardware, wire, machines, and etched metal products were supplied by manufacturers and wholesalers. They also had in common firms that supplied leather cases, photo supplies, tags, and envelopes. Second, most common creditor firms were Chicago-based, either as local firms or as non-Chicago firms with a city office, warehouse, or factory. Jewell Electrical and Chicago Ferrotype appear to have only taken goods from one non-local common firm: Blake and Johnson of Waterbury, Connecticut.

Finally, Jewell Electrical and Chicago Ferrotype also selected most of their production needs from firms in the instrument district. More than three-quarters of the common creditors (37 of 47) were located in the district (fig. 33). Almost all Jewell Electrical and Chicago Ferrotype inputs furnished by Chicago-based firms came from suppliers in or adjacent to the instrument district. Answers to questions about the character and the nested nature of the relationship between firms are extremely difficult to determine from the bankruptcy records. While it is impossible to ascertain the degree of interaction between firms other than a simple supplier-client relationship, there can be little doubt that the large number of firms lying inside the boundaries of Chicago's instrument district supplied most of the needs of firms such as Jewell Electrical and Chicago Ferrotype.

So important was the instrument district that the two firms remained there after post-bankruptcy restructuring. In 1920 Chicago Ferrotype was incorporated with the Mandel Manufacturing Company, which operated out of Benton Harbor, Michigan. In its new form, the company was "engaged in the business of manufacturing cameras, photographic supplies and all of the hardware materials entering into the phonographs of the Mandel Manufacturing Company, such as the motor, tone arm, reproducer, turn table, automatic stops and other hardware specialties." It had a few unstable years. By 1923 the factory was located just to the north of the original plant. A year later the firm moved into a small space farther south. As the economy boomed, it experienced dramatic growth, and by 1924 the company was located at another site in the district, where it made cameras with close to 200 workers. It did not, however, survive the 1920s. Despite the bankruptcy proceedings of 1920, Jewell Electrical was still located on Walnut in 1924 and 1928. In each year respectively, Jewell Electrical employed 235 and 300 workers. It continued to make electric instruments, and in 1928 had invested capital to the tune of $150,000.[24]

WILLIAM GAERTNER, THE RADIO INDUSTRY, AND THE INSTRUMENT DISTRICT

The inter-firm relations and locational assets of the instrument district were critical to the production strategy for firms other than Jewell Electric and Chicago Ferrotype. William Gaertner and Company, one of the country's leading scientific instrument makers, moved to the instrument

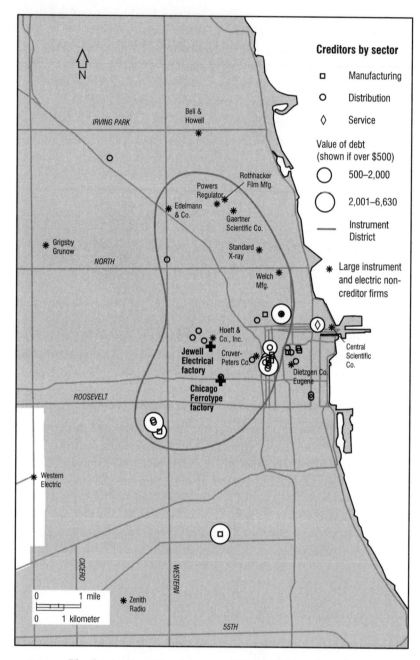

FIGURE 33. The Geography of the Creditors of Chicago Ferrotype and Jewell Electrical, 1920. Source: Records of the United States District Court, Northern District of Illinois, Eastern Division at Chicago, Bankruptcy Records, Act of 1898, 1898–1972; Bankruptcy Case 28505, *In the Matter of Chicago Ferrotype Company,* "Schedule A(3)" (July 7, 1920); Bankruptcy Case 28795, *In the Matter of Jewell Electrical Instrument Company,* "Schedule A(3)" (December 22, 1920).

district in 1923 for these reasons. For the first seven years after arriving in the United States in 1889, William Gaertner worked at the U.S. Coast and Geodetic Survey, the Smithsonian Institute, and the Yerkes Observatory of the University of Chicago. He acquired extensive experience in instrument manufacture at these institutions and gained many contacts within the wider world of science and technology. These contacts were extremely important for establishing the firm in the first place. Moreover, his connections with the University of Chicago were a major reason that he located his workshop on Chicago's South Side in 1896.[25]

Over the next 27 years, Gaertner built up a successful business making microscopes, spectrometers, cathetometers, and chronographs for private, educational (Caltech, MIT, and the University of Chicago), and government (United States military) clients. By the early 1920s, however, changing business conditions forced Gaertner to seek new premises where he could undertake "complete reorganization of the business on a more extensive scale." The need for expansion, new lines, and more modern machinery made Gaertner realize that the South Side location did not provide the types of contacts with other instrument makers, semi-made part suppliers, and a highly skilled labor force necessary for the times. The rational location was the instrument district. He moved to a new $150,000 factory on Wrightwood and Racine in 1923.[26]

At the very time that Gaertner moved from the South Side to the instrument district, the geography of the industry's locational gravity was shifting from the center to the city's northwest. While about 6 out of 10 of all instrument and electrical firms were found in the old instrument district in 1924, this had fallen to two-fifths by 1940 and to a third by 1950. New manufacturing districts appeared in Lincoln Square and Portage Park. The reasons for the shifting geography lie in the industry's changing character. Although Western Electric's move from its downtown plant to a sprawling one in Cicero between 1905–13 was a pivotal event in Chicago's industrial history, it did not redefine the industry's locational history. The shift to Cicero of all Chicago's and New York City's facilities signaled Western Electric's attempt to fully internalize the production of telephones within one gigantic workplace. A vertically integrated plant employing more than 20,000 between 1913 and 1929, it relied little on the firms clustering along the Chicago River. Likewise, these firms were unable to count on Western Electric for any sustained, long-term contracts. In effect, two nodes developed. The one centered on the Cicero plant was almost entirely independent, while the other stretching through the Loop and the central factory district was made up of a large number of closely linked firms.[27]

Several other reasons account for the shift a generation later. The growing scale and changing organizational structure of the electrical industrial in the 1920s forced firms such as Grigsby Grunow, manufacturers of Majestic radio sets, to rethink their location. In 1924 the company employed 40 workers on one floor. Five years later it employed 6,500 workers on half a million square feet of working space. To accommodate this growth, the firm reconfigured its spatial structure by making major additions to its Armitage Avenue plant, leased a portion of a plant to the manufacture of radio tubes, and broke ground for new buildings adjacent to the rest of the plant. These additions not only allowed production to be increased and widened the firm's product lines; they also facilitated the development of a large, vertically integrated company. As one commentator noted, radio manufacture by Grigsby "typifies the efficiency of American mass production."[28]

By the end of the 1920s, the company was working at such volumes and with such an ordered system of inputs, production, and distribution that the advantages of a central location were no longer essential to its locational calculus. According to one estimate, 30 carloads of raw material entered daily and 3,200 completed radios left the Grigsby plant each night. Working on this scale and in a rapidly changing and innovative industry, firms such as these were subject to constant change. In order to facilitate effective change and large-scale production, many firms questioned the wisdom of remaining locked in what they considered to be the negative externalities of a central location.

These types of significant changes to product, production, and demand placed tremendous pressure on a firm to rethink its locational choice. Unlike many of the older industries in which firms had committed a large amount of fixed capital in a centrally located site that had been built upon for a generation or more, Grigsby and other firms in the rapidly changing radio, electrical, and instrument industries were free to relocate out of older factory districts. Moreover, Grigsby's standardized array of inputs allowed it to draw on non-centrally located suppliers. The company operated five lumber mills to supply wood for its cabinet department and consumed 84 tons of steel, 16,000 pounds of tin foil, 12,000 pounds of paper for condensers, 20 tons of wax for impregnating condensers and chokes, and 5,000 pounds of aluminum daily. If the bankruptcy records of other firms are any indication, a significant number of these products must have come from local firms: steel from the mills of South Chicago and the Calumet district, condensers from local parts' makers, and aluminum from brass foundries.[29]

A similar story is true for Zenith Radio. Established as the Chicago Radio Laboratories in 1915, the firm received a large capital infusion from Eugene McDonald, a local capitalist, six years later. Renamed Zenith Radio, the firm grew dramatically at its Kedzie Avenue and Forty-eighth Street plant and used McDonald's ownership of one of the first broadcast stations in Chicago to promote Zenith. According to a *Chicago Commerce* writer, by the mid-1930s company managers found that its old "beehive of surrounding buildings into which production has splashed over" was no longer viable as "the sets have to be carted around here to there endlessly before they become a finished product. The new plant will be set up for line production with much superior working conditions both for the standpoint of economies and of employe [*sic*] efficiency. The company also will be able to manufacture some of the parts such as stampings which it now has to parcel out on the outside."[30] In other words, the firm sought better accommodation and detached itself from the district. Ironically, the new site was the former Majestic plant at Armitage Avenue and Forty-fifth Street, for which the company paid $410,000. The company refashioned the existing workplace to accommodate its own needs and by 1940 employed more than 5,400 workers at the plant.[31]

Grigsby and Zenith left the instrument district and moved to Chicago's western fringe. This was partly to do with their ability to reduce their reliance on external suppliers. But even after moving to the metropolitan fringe, they continued to depend on local inputs. Most other firms continued to have strong inter-firm networks centered on the traditional district. This is evident from testimony provided by Chicago's radio and instrument interests in a 1928 report on the radio industry. According to a city wholesaler, Peter Sampson, president of both the Radio Wholesalers Association and the Sampson Electric Co., Chicago had "a natural demand for receiving instruments . . . to an extent unparalled [*sic*] in the United States." Chicago's wholesalers were "well specialized, efficiently organized and financially able, to distribute their lines over an unusually broad territory." Three-quarters of the city's 369 electrical wholesalers were located in the city's main wholesaling district, the Loop, and the area immediately west of the Chicago River. From this district, which overlapped with the instrument district, local firms could find all of the necessary electrical, radio, and instrument parts for production, and agents willing and able to sell their products in local, regional, and national markets.[32]

Other commentators pointed to Chicago's supplier networks. Paul Klugh, Zenith Radio's vice president, wrote that local radio manufacturers had an advantage over other regions because the "raw material which goes in the making of the complete unit, is close by. The factor eliminates, almost entirely, transportation charges and enables the manufacturer to be in close contact with the raw material supplier and to maintain a rigid and fast schedule of production." The wholesaler Harry Alter stated that "Chicago has the distinction of having probably more manufacturers of parts and accessories for radio sets than any other large city of the country." W. Jacoby, Kellogg Switchboard and Supply's president, stated that "there are scores of smaller factories making parts and various types of radio apparatus." A year later B. Grigsby reiterated this when he noted that "there are countless concerns producing the various parts which go into radio manufacture and numerous other concerns in Chicago making a specialty of producing raw materials which cater to the needs of the actual producers of radios in the Chicago district."[33]

Grigsby and Zenith were not unique. Many other electrical, radio, and instrument firms grew rapidly, catered to national markets, employed thousands of workers, and relocated to large plants. In the process, they reconfigured the economic landscape of those industries. Freed from small-scale, local subcontractors and suppliers and forced to find new work space in which new innovative forms and ever-increasingly larger production processes could be planted, they moved farther out from the core's cramped quarters. The process of decentralization had been taking place in Chicago since the 1850s. This was true for most other industries as well. Indeed, the very construction of metropolitan Chicago was driven in part by the desire to leave the dis-economies of the central districts. In some cases, firms sought greenfield sites on the urban fringe. In others, they moved to existing factory districts in other parts of the built-up metropolis. All of this was dependent upon the construction of an extensive set of inter-firm linkages spanning local, regional, national, and, in some cases, international worlds.

CHAPTER 9

Manufacturing Production Chains and Wholesaling

In January 1911 R. Ardrey, *Iron Age*'s correspondent in Chicago, reported that Chicago's factories were experiencing difficulty securing crude steel from local producers. This was unusual, for there was a local substantial chain of supply from mill to workshop. Chicago's importance as a distribution center was apparent four years earlier when 100 companies, many of whom were customers of local steel mills, exhibited their wares at the Chicago Hardware Show. In the cavernous Coliseum on South Wabash Avenue, wholesalers and manufacturers from across the country arranged their products for all to see. A roaring success, despite a fire that destroyed the building housing the convention's business sessions, it brought together in one site many of the country's major hardware interests. The Chicago Hardware Show was only one of many annual trade shows held in American cities. Three months later machinery interests met at Cincinnati to view each other's wares and to discuss issues of mutual concern. After the appropriate salutary words from the president, Samuel Moyer, and reports from subcommittees of the American Supply and Machinery Manufacturers' Association, guest speakers spoke on a range of topics including resale prices, cash discounts, and dealer-manufacturer relations.[1]

The Chicago and Cincinnati shows were places where the trades came together to discuss the steel market, to view the latest innovations, to catch up with industry news, and to meet colleagues. The shows were places where different parts of the industry—manufacturers, wholesalers, and retailers—found a central mart. The conventions were an integral part of the production chain linking the raw material supplier to the final consumer. Only one aspect of trade, an effective distribution system, was critical to the ability of industrialists to reap the profits of production. Three case studies illustrate distribution's role in the functioning of the production chain: furniture manufacture provides a portrayal of an industry's web of networks and intricate set of institutional arrangements; the business mart in the 1920s highlight its role in linking the local, regional, and national scales; and the metalworking industries depict the place of Chicago's wholesaling complex in the development of the metropolitan production chain.

CHICAGO'S FURNITURE PRODUCTION CHAIN

Chicago's furniture industry had several defining features. Manufacturers built a great deal of furniture, ranging from cheap and rough machine-made chairs for the working poor to the most elaborate and ornate cabinets for America's wealthy. The industry grew tremendously, and by the late nineteenth century, Chicago was the one of the nation's largest furniture producers (table 20). The industry had an elaborate division of labor, from the small-scale firm deploying "stylistic diversity and product specialization" to the large one operating a high-volume strategy. While staple goods producers created a degree of vertical integration, most firms were small and the industry remained unintegrated. According to a 1921 survey, only 6 of the city's 294 factory makers employed more than 300 people, while 163 enterprises hired less than 40. This complex of small and medium-size businesses made numerous products under different conditions. The survey noted 12 different classes of furniture, ranging from Chicago's most important line, upholstered furniture, to the small specialized furniture novelty maker, store, and restaurant manufacturers.[2]

Piano manufacture illustrates many elements of this growth and variety. Pianos were first built in Chicago in 1884, and by 1906 Chicago producers manufactured about 75,000, or about 25 percent of the U.S. total, while another 30,000 were made within a 100-mile radius of the city. Chicago manufacturers had become major rivals of eastern piano producers. After the war, it consolidated this position; in 1919 Chicago producers built a quarter of the nation's pianos and organs, about the same amount as New York City (fig. 34). Several factors made this growth possible: the ability of Chicago piano makers to draw upon a large and rapidly growing regional market; a pool of woodworking expertise; an assortment of ancil-

TABLE 20. City of Chicago's Furniture and Lumber Industries, 1879–1939

Industry	1879 Firms	1879 Workers	1909 Firms	1909 Workers	1929 Firms	1929 Workers
Furniture	174	5,039	202	11,097	286	16,854
House Furnishings	24	392	121	2,903	142	3,475
Musical Instruments	14	226	37	5,792	16	743
Planed Lumber	42	2,741	195	11,680	147	4,923
Wood Turning	14	137	0	0	34	764
Total	268	8,535	555	31,472	625	26,759

SOURCE: U.S. Census, *Compendium of the Tenth Census (June 1, 1880)*, vol. 2 (Washington, DC: Government Printing Office, 1883); U.S. Census, *Thirteenth Census of the United States Taken in the Year 1910: Manufactures, 1909*, vol. 9 (Washington, DC: Government Printing Office, 1895); Department of Commerce, Bureau of the Census, *Fifteenth Census of the United States, 1930: Manufactures*, vol. 1 (Washington, DC: Government Printing Office, 1933).

lary trades; increased machine-based production and capital investment; excellent transportation facilities; and the switch from specialization in a poor-quality product (the "thump box") to a variety of grades. Other segments of the furniture industry paralleled the piano trade's experience. So considerable was the industry's growth that by the mid-1920s, according to a *Chicago Commerce* writer, "more furniture is made, and more sold to dealers in Chicago than anywhere else." By the end of the 1920s, some 400 firms employed 21,000 workers, and 7,000 furniture wholesale employees can be added to this number (fig. 35). Chicago was an enormous producer of all sorts of furniture.[3]

While Grand Rapids and New York, the two other leading furniture centers in the United States, may have argued with the *Chicago Commerce* writer's claim, Chicago was the largest producer of furniture for a good part of the period. For instance, on the eve of the Great Depression, the Chicago industrial area's furniture factories employed almost 20,500 workers in the manufacture of $105 million of house, store, and office furniture. This accounted for more than 10 percent of all furniture production in the United States. The New York industrial area and Grand Rapids were in second and third place with 16,000 and 13,400 workers, respectively. While other industrial centers—such as Los Angeles, Boston, and Worcester, Massachusetts—had thriving furniture sectors, no other place in the country could match the Windy City. But as the 1929 figures suggest, Chicago was more than the city. A decade earlier in 1918, Josiah Currey wrote in his great compendium of Chicago industry that to understand the city one had to understand how the city was linked to its surrounding areas. As he

FIGURE 34. The P.A. Starck Piano Company Factory, 1915. Located at Ashland and Thirty-ninth in the new Central Manufacturing District, Starck was a leading early twentieth-century Chicago piano manufacturer. Source: *The Central Manufacturing District* (Chicago: Central Manufacturing District, 1915), 70.

stated, "There is a large number of widely scattered establishments situated but a short distance beyond the city's borders, but which are in a greater or less degree tributary to the metropolis." The local furniture industrial zone, which was the largest in the country, encompassed an area running from Waukegan in the north to Aurora in the west and the Calumet region to the south.[4]

The scale and specialized structure of the furniture industry both contributed to and built upon an elaborate production chain centered on a set of specialist firms ranging from raw material supply through production and the distribution of the finished goods to consumer markets. As an industrial complex, metropolitan Chicago captured most stages along the furniture production chain. Metropolitan Chicago had a formidable array of businesses producing the furniture industry's essential inputs. The most important was lumber. Since the mid-1800s, Chicago had been the major tributary of a massive lumber region stretching across the Midwest. Despite its waning dominance after 1900, the city remained the major distribution center for the Red River district, the largest production center in the Midwest, and the largest market for its lumberyards. In 1911, according to the *American Lumberman*, the total receipts of lumber in Chicago

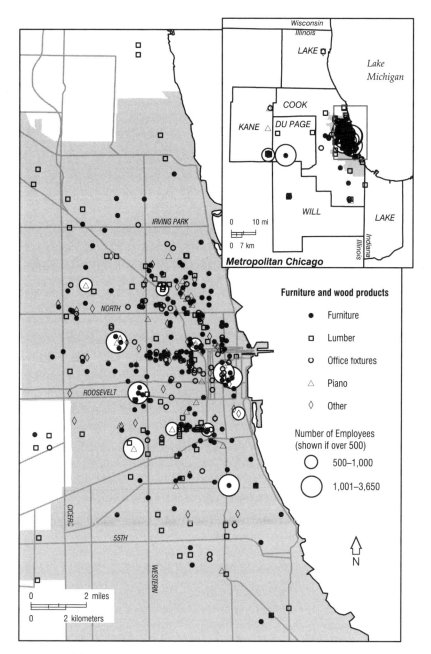

FIGURE 35. The Geography of the Furniture and Lumber Industries, 1924. Source: *Directory of Illinois Manufacturers, 1924–1925* (Chicago: Illinois Manufacturers' Association, 1924).

were more than 2.1 billion feet. Of this, 800 million were shipped out of the city, most of it to points east; the difference, 1.3 billion feet, went to local consumption.[5]

For George Plumbe, the lumber-furniture link "explains why Chicago is the largest furniture market in the country, while for pianos, organs, carriages, and other industries demanding both fine and rough, or common, lumber, it is not equaled on the continent." A generation later, Edward Hines, president of one of Chicago's largest lumber companies, noted in more prosaic but no less emphatic terms that "the city . . . has always been a large consumer of lumber." According to a 1925 survey, the city's 300 furniture factories, which bought more than $82 million (47 percent) of the city's receipts, were the local lumber industry's most important clients. When one adds the output sold by the city's planing mills (27 percent of receipts), then the amount of lumber that the furniture industry ultimately consumed was enormous. Aided by Chicago's large number of lumber wholesaling firms and commission merchants, local furniture manufacturers had access to the largest and most diverse lumber supply in the country.[6]

Furniture firms bought intermediate products from ancillary trades. Along with woodworking firms such as the planing mills, the industry looked to metal fabricators for machinery, tools, screws, hardware, springs, and steel; tanneries for leather; textile mills for cotton, curled hair, and upholstery products; and chemical makers for glue, paint, varnish, and lacquers. Several substantial industries formed out of the by-products of the meatpacking industry, notably glue, gut strings, and leather. This specialization intensified over time. In 1924 manufacturers of all sizes produced furniture hardware. Some metalworking firms were even more specialized. The American Industrial Company and Elgin Metal Novelty built piano hardware, while six firms made gold leaf, and another three focused on furniture hinges. Others concentrated on woodworking tools, such as furniture knives. As frequently stated, a major reason for "Chicago's supremacy" as a furniture-producing center was its "nearness to raw materials, wood metal, gut strings, glue, varnish, leather, ivory and celluloid."[7]

Different industry segments interacted with these ancillary industries. Chicago's largest furniture line in the 1920s was upholstered couches and chairs. These firms depended on metal firms for springs, wire, and other similar products, and textile makers for stuffing and upholstery. Similarly, some manufacturers—there were 21 operating in Chicago in 1925—made only metal furniture, including folding beds, hospital beds, chairs and tables, and kitchen cabinets. Unlike most other furniture makers whose major raw material and semi-processed product was lumber, these metal furniture makers were directly hooked into the city's vast metalworking world.[8]

The 1928 creditors of Metal Crafts, a lamp manufacturer, illustrate this furniture-metal relationship. The company acquired brass goods, stampings, and bronze powders from New York City, Providence, and Reading (table 21). Two specialized firms from another important furniture center, Rockford, supplied the company with plates, patterns, and other metal products. But it was Chicago firms that furnished most of Metal Craft's intermediate products: foundries provided castings and patterns, wholesalers supplied steel sheeting and machinist supplies, while other local businesses furnished machinery, nails, tools, soldering fluids, and general lamp hardware. In other words, metropolitan Chicago was an important repository of firms that supplied Metal Crafts with the intermediate goods essential for lamp manufacture.

The industry thrived upon a local network of inter-industry contracting relationships. Chicago firms specialized in discrete parts of the furniture-making process, from lumber preparation, fine carving, and die cutting, to the manufacture of veneers and frames. According to a 1925 Chicago Association of Commerce survey, Chicago had 13 firms that made "furniture frames and [sold] them in the white, principally to upholstered furniture manufacturers in whose factories the manufacturing process [was] completed." Another four companies built parlor frames as a sideline to their regular business and sold them to other manufacturers. Specialization and contracting out of furniture-related work varied between different trade segments. While some firms worked on staple lines, others worked on made-to-order ones. Store and restaurant furniture makers typically did "special contract work," making special furniture, and were extremely dependent on subcontracting to woodworking and metalworking firms. Malcolm Hartmann employed 50 workers in 1924 to do fine carving; his business was just 1 of 19 in 1928 that serviced the trade. More standardized machine carving was performed by other firms, while standard and made-to-order embossed moldings, halftone wood cuts, and turned wood specialties were done by a variety of small firms.[9]

Local furniture makers were embedded in a highly innovative milieu. After the Civil War, custom-made styles incorporating design elements brought in from Europe and the eastern seaboard as well as from local interior decorators, cabinetmakers, architects, and woodcarvers contributed to Chicago's dynamic furniture complex. Custom-made, high-end styles—from the Gothic through Mission and Prairie—were introduced

TABLE 21. Metalworking Creditors of the Metal Crafts Inc., 1928

Creditor Firm	Location	Owing ($)	Merchandise
Metropolitan Chicago			
Meskan Foundry	Austin	481.71	castings
North Austin Machine	Austin	43.05	machines
Obermayer Co.	Garfield Park	77.68	foundry supplies
Raymond Lead Works	Garfield Park	15.55	lead pipes
Ryerson, Joseph T.	Garfield Park	25.25	steel products
Western Plumbing Co.	Garfield Park	330.32	plumbing supplies
Keystone Mfg. Co.	Lincoln Park	295.53	screw machine work
Chicago Jewelers	Loop	na	tools
McCarthy Foundry	Lower West Side	757.79	castings
Great Western Foundry	Milwaukee	732.61	plating
Illinois Nail Co.	Milwaukee	5.60	nails
Milwaukee Grey Iron	Milwaukee	268.85	gray iron
Braun, J. G.	Northwest Side	na	ornamental iron
Chicago Metallic Sash	Northwest Side	na	metal bars
Federal Machinery Sales	Northwest Side	13.73	machinery
F.J. Phillips & Son	Northwest Side	89.33	brass foundry
Friedley Voshardt Co.	Northwest Side	165.18	sheet metal
Johnson Mfg. Co.	Northwest Side	14.50	soldering fluids
Northwestern Foundry	Northwest Side	85.88	castings
Apex Pattern & Foundry	Packingtown	358.00	foundry
Calumet Pattern Works	South Chicago	na	patterns
Advance Spring & Wire	West Side	33.31	coiled wire springs
American Machinery Exchange	West Side	168.38	machinery
Dole Valve Co.	West Side	152.82	valves
Mazel Rest & Company	West Side	37.87	hardware
Samuel Harris & Co.	West Side	74.04	machinist supplies
Sueske Brass & Copper	West Side	143.91	brass goods
Woodruff & Edwards Co.	Elgin, IL	1,110.73	castings
Keen Foundry	Griffith, IL	30.20	foundry goods
Non-Chicago			
B.F. Drakenfeld & Co.	New York, NY	39.13	bronze powders
Booker Metal Specialties	New York, NY	19.80	metal goods
T. W. Lind Co.	Providence, RI	7.50	sheet metal stampings
Crescent Brass Mfg. Co.	Reading, PA	451.74	brass goods
Liberty Foundries Co.	Rockford, IL	545.76	foundry goods
Rockford Standard Pattern	Rockford, IL	1,000.00	plates, patterns

SOURCE: Records of the United States District Court for the Northern District of Illinois at Chicago, Bankruptcy Records, Act of 1898, 1898–1972; Bankruptcy Case Files, 1898–1941; Bankruptcy Case 39124, *In the Matter of Metal Crafts Incorporated*, "Schedule A3," March 24, 1928.

to satisfy the changing tastes of Chicago's growing bourgeoisie and middle class. These were copied by high-volume, low-end producers as design styles and elements diffused through the local industry's networks. Ideas and knowledge circulated through the system when designers and skilled workers moved from one firm to another, and through formal institutions, such as the Chicago Arts and Crafts Society, the Industrial Arts League, the biannual meetings of the Chicago Furniture Exposition Association, and the trade journals.[10]

Chicago furniture firms were innovative and leaders in new technologies. The development of shaping, finishing, cutting, carving, and embossing machines led to faster speeds and lowered production costs, and enhanced Chicago's position as America's furniture center. Many local manufacturers successfully changed product lines, switching from one product mix to another, or from one design to another. Innovation was intimately tied to product and organizational innovation. This is evident in piano making when Kimball switched from organ to piano production at the end of the 1880s. Two generations later, Gulbransen, the world's largest piano maker, branched out into radio set production. The 1920s were difficult times for piano and organ firms as the industry restructured in the face of changing demand and increasing mechanization. Nationally, between 1919 and 1929, the number of firms fell dramatically, the number of employees dropped by more than a half, and output plummeted. Sensing that the piano industry had reached a plateau and that America's craze for radios was more than a passing fancy, Gulbransen switched part of its production and capital investment at its six-story factory to radios. Facilities for making radio chassis were obtained by a merger with Chicago's Well-Gardner Company.[11]

The final part of the furniture production chain was the distribution of products from the factories to the consumers. Chicago's producers had access to an enormous regional market. In the mid-1920s, more than 18,000 retail dealers of all sorts serving 50 million people lived within 500 miles of Chicago. Even though local furniture makers supplied national and regional markets, Chicago, as the second largest market in the United States, was a prime candidate for the goods spilling out of its furniture factories. In 1926 Chicago had 761 retail furniture stores with sales of almost $83 million. On the street, retail stores were not the only clients of the furniture manufacturers. Chicago, as America's mail-order center, shipped a large amount of locally made furniture throughout the country. Starting in the 1870s, a combination of the huge mail-order enterprises—Montgomery Ward and Sears, Roebuck—and smaller, more specialized ones—Hartman and Spiegel—contributed to Chicago's expanding market base. The various clients of the factories absorbed locally made furniture to the value of more than $105 million in 1925.[12]

Obviously, there is no perfect match between local output and consumption, but evidence suggests that local factories supplied local retail firms in large quantities. Take the Madison Furniture Store in the city's Northwest Side, the area lying immediately west of the Loop across the Chicago River. More than half of the number and value of its manufactured goods came from local factories (table 22). Local firms, such as Columbus Parlor Furniture and Atlas Upholstering, supplied large numbers of goods, while others had only small accounts with Madison (table 23). Even though a significant share of the retailer's products came from outside the metropolitan district, most notably from North Carolina's furniture-producing areas and the Indiana-Illinois furniture belt, all types of Chicago manufacturers sold goods to Madison Furniture.

Chicago's wholesalers played an important role in the local furniture production chain by supplying manufacturers with semi-manufactured goods and retailers with finished products. Many manufacturers sold on consignment through commission agents. The functional separation of wholesale from manufacture, however, was not as well developed in the furniture trade as it was in others. While many industries had a well-formed wholesale trade to mediate between the different segments of the production chain, in Chicago they were frequently the same. This was certainly how they were cataloged in Chicago's city directories. Reflective of the prewar industry was the 1904 directory, in which the furniture trade, other than a small number of specialty listings, was broken into two groups, "retailers"

TABLE 22. Sectoral Distribution of Madison Furniture's Creditors, 1928

			Chicago Share	
Creditors	No.	Owed ($)	No.	$
Service Industries	26	20,564.56	96.1	98.1
Finance	4	14,491.07	100.0	100.0
Wholesalers	23	7,325.72	95.7	97.5
Manufacturers	130	80,064.62	58.5	55.5
Not Known	12	9,669.03	58.3	36.7
Total	195	132,115.00	100.0	100.0

SOURCE: Records of the United States District Court for the Northern District of Illinois at Chicago, Bankruptcy Records, Act of 1898, 1898–1972; Bankruptcy Case Files, 1898–1941; Bankruptcy Case 39124, *In the Matter of Jacob Miller, Trading as Madison Furniture Company and Madison Street Furniture Company,* "Schedule A3," January 1928.

TABLE 23. Largest Creditors of Madison Furniture, 1928

Largest Creditors (City)	Owed ($)	Reason for Debt	Firm Type
Mid-City Trust and Savings	10,000.00	loan	financial
Columbus Parlor Furniture	8,439.99	parlor furniture	manufacturer
Chicago Evening American	7,721.77	advertising	newspaper
Weigle Furniture	7,121.63	parlor furniture	manufacturer
Howell Stitt (New York, NY)	5,491.53	goods	unknown
Mount Airy Chair (Mount Airy, NC)	4,273.05	chairs	manufacturer
National Printing	4,245.70	printing	manufacturer
Standard Carpet	3,951.25	carpets	wholesaler
Atlas Upholstering	3,828.21	furniture	manufacturer
Herald Examiner	2,794.00	advertising	newspaper

SOURCE: Records of the United States District Court for the Northern District of Illinois at Chicago, Bankruptcy Records, Act of 1898, 1898–1972; Bankruptcy Case Files, 1898–1941; Bankruptcy Case 39124, *In the Matter of Jacob Miller, Trading as Madison Furniture Company and Madison Street Furniture Company*, "Schedule A3," January 1928.

and "manufacturers and wholesale agents." Larger firms kept their own showrooms, usually adjoining the factory, while those with non-central factories sometimes, if they were large enough and thus could obtain some economies of scale, rented a showroom in the city center. But most were unable to do this; they needed an intermediary. Along with the wholesaler, Chicago manufacturers relied on traveling salesmen to hunt down sales from local and non-local customers. Thus, Chicago's furniture production system's distribution networks were extremely spatially diffuse.[13]

By the 1920s, however, directories and the census separated manufacturers and wholesalers. According to the 1926 census, 255 local house furniture wholesalers had Chicago sales totaling more than $58 million. But this divide between wholesale and production continued to be an arbitrary one for the larger firm. Typical of Chicago's largest manufacturers making large volume sales directly to retailers and special contract customers were S. Karpen and Brothers, with an extensive wholesale salesroom adjoining its eight-story factory on the Near South Side, and Fenske Brothers, whose sample room on South Michigan Avenue was a few blocks east of its main factory. For many other firms, however, an extensive wholesaling segment distributed furniture to customers in the metropolitan area and the rest of the nation.[14]

The glue holding this massive furniture complex together was a set of inter-firm associations. Regional trade associations and business institutions brought manufacturers together to work on industry-wide issues. In Chicago a formal institutional history can be traced back to as early as the

1870s, when proprietors organized the Furniture Manufacturers' Exchange in 1888. The exchange, which became the Chicago Furniture Manufacturers Association (CFMA), allowed manufacturers and retailers "to study trade tendencies, to interchange knowledge, to extend acquaintanceships, to renew old friendships and make new ones, and to keep in elbow touch with the progress of their fellows." In 1910 the Chicago Furniture Forwarding Company was established by the CFMA to provide coordinated rail transportation, by allowing manufacturers to build volume and reduce delivery lags of furniture destined for non-local areas.[15]

Other associations played an important part in the reproduction of Chicago's furniture complex. The National Piano Manufacturers' Association, established in the late 1800s, regulated competition, production, and distribution. As the country's largest single body, Chicago piano manufacturers played a central role in the organization. Allied industries also created associations. The Lumberman's Board of Trade and the Lumberman's Exchange, which existed from the 1870s, regulated local trade through informal face-to-face contacts and created formal committees on arbitration, appeals, inspection, and finance. In the twentieth century, local lumber firms were extremely active in the national Lumber Manufacturers Association. With almost every lumber dealer signed up, the association oversaw a variety of issues, including the promotion of trade and grademarking, the standardization of products, and finding ways to increase demand.[16]

Issues of importance to trade associations were discussed at trade meetings. Twice a year, furniture manufacturers and retailers from around the world went to Chicago. From 1895 when the Furniture Market Association (FMA) held its first showing, the industry laid out its new styles in the Near South Side exhibition halls. These shows attracted the furniture trade to Chicago because "all of the products of these [Chicago] factories are displayed in the exhibition buildings." In order to minimize the distance traveled and to maximize the clustering, the FMA built up its portfolio of exhibition buildings in the Near South Side's wholesaling district. By 1908 there were two exhibition buildings on Michigan at Van Buren, and another five within a two-block radius around Fourteenth and Michigan. Together, the seven buildings had more than a million square feet of floor space. By 1912 the meetings took place at the National Furniture Exchange of Chicago, a large central hall for furniture exhibits, built by the FMA at Twenty-second and the lakefront.[17]

Despite the new hall, the growth of the local industry continued to force wholesalers, manufacturers, and retailers to seek out new display space. By the early 1920s, the number of exhibition halls numbered 10. The ability to assemble the flow of information about prices, designs, and trade business within a central area was an obvious advantage for out-of-town and local furniture interests. The fact that many furniture factories were in the suburbs could have been disadvantageous.[18] For Chicago's furniture promoters, however, the fact that firms were spread across Chicago's vast industrial territory was not a problem. In their opinion, the central furniture mart ensured that "the out-of-town buyer is not inconvenienced by that. All of the products of these factories are displayed in the exhibition buildings."[19] The buildings provided manufacturers with a centralized site where they could show off their range of styles and bypass the dispersed geography of factories.

THE AMERICAN FURNITURE MART AND BUSINESS MARTS IN THE 1920S

A major problem was apparent by the 1920s despite Chicago's elaborate and successful furniture production chain. Continued growth created the need for better coordination of the industry. For furniture proprietors, the difficulty of having wares exhibited in many different sites was detrimental to trade. Not only did buyers have the inconvenience of traveling from hall to hall, but there was no one location where sellers and buyers from Chicago, the rest of the United States, and overseas could congregate. Cognizant of this, local furniture, financial, and political interests banded together to find a solution. In early 1922 an announcement was made about a new year-round central exhibition building or mart. For the local coalition promoting the building, a new large mart with all of the necessary facilities would have distinct advantages. Not only would the buyer be able to "save time and trouble," but "it will be possible for friends to get together without trouble or delay." As one commentator noted, "The needs of twenty years are being realized in this building."[20]

The American Furniture Mart, as the new building was called, opened in July 1924 in the Near North Side. A gigantic display center built to house local wholesalers, it replaced the National Furniture Exchange and the other buildings that previously had formed the city's marketing facilities (fig. 36). An imposing structure, it covered a city block and had 16 floors with 34 acres of rentable space and two sets of high-speed freight elevators able to handle more than 350 carloads of furniture a week. The architect, George Nimmons, worked almost exclusively for Sears, Roebuck. Trained in the Burnham and Root office, Nimmons focused on industrial buildings such as multi-story warehouses and emporiums, although he did build palatial residential homes for the executives of Sears, Roebuck, Julius Rosenwald (1903) and Richard Sears (1906). However, when he turned his attention to the mart, he brought his commercial training to bear and designed an efficient space. For the exhibits themselves, a through-flow system unloaded crated goods in the basement and transferred them by electric truck and elevator to exhibitors' spaces. The mart also moved people very quickly from the street to the exhibits. Entering into a spacious marble-lined lobby, buyers and sellers obtained information from registration clerks or from the first-floor business offices. Once they knew where they wished to go, four passenger elevators whisked them to the exhibition spaces.[21]

The mart was very successful. Within six months the building was filled to capacity and had a waiting list. In 1925 a $4.5 million addition was built, adding nine acres of floor space, four new floors, and a seven-story tower for studios and offices. More than 15,000 dealers transacted more than half of Chicago's annual furniture sales of $1.4 billion at the mart. It was the headquarters of the Furniture Club of America. Located on the building's top floor, the Furniture Club was reputed to have a membership of 4,000 manufacturers, salespeople, and retail dealers, all of whom had access to the club's dining rooms, a billiard room, a library, showers, and a banquet hall. The Furniture Club was a spatial expression of external economies constituted through manufacturing, design, credit, and commercial networks between firms. It was also an expression of the importance of the furniture industry, not only for manufacturers, wholesalers, and retailers, but also for Chicago's livelihood, wealth, and power. The building and the Furniture Club's place on the very top floor presented a powerful image of the furniture industry's metropolitan and national circuits.[22]

This was evident at the luncheon to celebrate the mart's opening in 1924. Lawrence Whiting, president of the Boulevard Bridge Bank—the mart's main financial promoter along with William Wilson, a Chicago furniture salesman—introduced those sitting at the head table. Present were the presidents of some of America's leading furniture manufacturers: Adolph Karpen of S. Karpen, D. Fennell of Pullman Couch, and A. Gorrell of A.D. Gorrell Company. Several dignitaries—including the Industrial Commissioner of the Chicago Association of Commerce, the secretary of the Chicago Furniture Manufacturers Association, the president of the Chicago Association of Commerce, and the building's construction engineers and architects—sat alongside the manufacturers. All of these people had a stake in the building and success of the mart.[23]

FIGURE 36. American Furniture Mart, 1928. Soaring above Lake Michigan and the rapidly growing Streeterville district north of the Chicago River's Main Branch, the American Furniture Mart was both a material and symbolic illustration of Chicago's dominance of furniture production and marketing in the United States. Source: *Chicago Commerce* (July 28, 1928): 28.

The mart's success was partly related to its location. While its centrality allowed it to draw together all elements of the industry's production chain, the choice of a site just outside the old business district was driven by processes that reformulated central Chicago's landscape. Chicago's preeminent place in the Midwest and its position as America's second city fueled large-scale growth and triggered intense competition between commercial, financial, retail, and other business functions for central space. After World War I, lateral expansion of the business district beyond the orbit of the elevated lines defining the Loop was tied to the recommendations of Daniel Burnham and Edward Bennett's *Plan of Chicago* (1909), the northward march of business along North Michigan Avenue, and the opening of the double-decker Wacker Drive in 1926. As early as 1917, the move of one of America's largest grocery wholesalers, John Sexton and Company, from the Loop to north of the river signaled these changes. This was reinforced by the consolidation of wholesale firms in Marshall Field's Merchandise Mart. A 25-story, $30 million building on the north side of the Chicago River, it concentrated numerous lines in one location and bolstered the trend toward large multi-story wholesaling establishments, the reorganization of Chicago's distribution networks, and the rationalization of central-city space.[24]

The American Furniture Mart propelled ancillary industries to establish warehouse and production facilities in the immediate area. As one commentator noted as early as 1924, the mart became "a magnet for the location of furniture manufacturers and distributors." This turned out to be true. The Stover Company purchased property close by and remodeled the building to accommodate exhibition rooms for household equipment. The soda fountain equipment makers Bastien-Blessing Company's decision to locate on Ontario was influenced by the desire to be close to the mart. Eagle Wabash, the nation's largest lamp-shade maker, bought a nearby seven-story property to which it added a new building and remodeled the existing one because of "the proximity of the building of the new American Furniture Mart."[25]

Chicago's long-standing position as a leading distribution center was reinforced in the 1920s by the building of other business marts. The Textile Exchange Building, serving local and out-of-town textile manufacturers and jobbers, had an office, exhibit space, a club, and overnight accommodation. The Railway Exchange Building consolidated Chicago's hold on the nation's railway supply trade. With exhibit space, a club, a swimming pool, and an auditorium, it was clearly modeled on the American Furniture Mart. Similarly, the $10 million, 40-story Jewelers' Building completed in 1926 housed many of Chicago's manufacturing and distributing jewelry firms. Completed in the same year, the Chicago Machinery Mart—an $8 million, 12-story building with 900 members, all of whom were part of the machinery trade—was managed by a board of directors chosen from the tenants.[26]

Some proposed marts simply did not materialize. In 1925, for example, it was announced that $31 million would be spent on a new agriculture mart with more than 3 million feet of rentable space. It was never built. In other cases, completed business marts did not have the impact that they thought they would. The Millinery Mart was put into operation in 1925 with the idea to centralize the millinery industry in the old "furniture row" district much closer to the Coliseum, the city's main exhibition hall (fig. 37). Since the opening of the American Furniture Mart, most of the furniture firms originally housed in the Manufacturers' Exhibition building had moved out and took residence in the new building. The Manufacturers' Exhibition's owners remodeled the building and relabeled it the Chicago Millinery Mart. At the time of remodeling, it was announced that millinery makers were "enthusiastic" about moving and that more than half of the 350,000 square feet of floor space had been let. It was further opined that once the rest of the industry realized the advantages of moving south, "Chicago's millinery center will switch from the loop to the near south side."[27]

The booster was wrong. While certain interests tried to pull the millinery trade south, wholesalers continued to be located in the old millinery district in the Loop. Three years after the mart's opening, the old district contained 81 percent of the city's 134 millinery wholesalers (table 24 and fig. 38). More than a third were housed in the newly constructed Millinery Center Building on East South Water, while a number rented space in a few buildings on Wabash, Michigan, and Randolph. This part of the Loop close to the city's major retail core also accommodated the Millinery Association of America and more than a third of the city's hat manufacturers. In other words, despite the best efforts of one group of interests, millinery wholesaling remained in its old prewar district close to its principal markets, the city's retailers and millinery manufacturers.[28]

In contrast to the one-trade business mart was the industry clearinghouse. The clearinghouse did not have the concentration of dealers, wholesalers, and manufacturers from one industry in one specially constructed space. Rather, the clearinghouse took up a small part of a building that was tenanted by other industries. Unlike the mart with its multi-functional spaces that encompassed many functions along the supply chain of one industry, the clearinghouse typically provided a simple informational and coordinating role for the industry's various elements.

MERCHANTS AND BUYERS

From Every Section Of The Country Will Come To Chicago To Attend

Chicago's Greatest Sales Event!

COLISEUM
JULY 31st
to
AUGUST 11th

300 Lines of Fall and Holiday Merchandise On Display Here For Your Convenience!

Three hundred complete sample lines assembled from the principal market centers of the country. Booth after booth of tempting merchandise which your customers will be demanding during the coming Fall and Holiday seasons. All this reliable merchandise will be PRICED VERY . . . VERY LOW because of the small selling costs to the Manufacturers, Importers and

Specialty Distributors exhibiting. On the lines you buy, your mark-up can be more generous and you can meet even your strongest competition. This Gigantic Merchandise Fair will be attended by every Merchant and Buyer who has the best interest of his store and department at heart. Only Merchants and Buyers are admitted.

Gift and Holiday Merchandise Will Be Featured!

Manufacturers, Importers and Specialty Distributors of 5¢ to $5.00 Merchandise are Invited to Visit the 8th Annual Chicago Merchandise Fair

Be sure to visit CHICAGO'S GREAT ANNUAL SALES EVENT this year. See for yourself why these 300 exhibiting concerns find it so profitable to display and sell their products here each year. If

for any reason you cannot attend the Fair this year, we will be pleased to send you full particulars regarding cost of exhibition space for next year's exhibit.

Business Hours: 8:30 A. M. to 6.00 P. M. Daily

Chicago Merchandise Fair

CHICAGO OFFICE: Pierre H. Meyer, Pres.—Walter J. Kenney, Sec'y   NEW YORK OFFICE:
1513 S. Wabash Ave.   890 Broadway

FIGURE 37. Eighth Annual Chicago Merchandise Fair, July–August 1928. Located at Wabash and Fifteenth, the Coliseum was the city's main indoor exhibition hall in the early part of the twentieth century. Source: *Chicago Commerce* (July 29, 1928): 21.

Chicago's candy trade illustrates how this worked. Although the candy industry was not an important Chicago industry, 30,000 dealers, manufacturers, and jobbers were located within one night's ride of Chicago in 1930. To capture this large number of consumers, local candy interests devised an effective system to ensure that regional buyers were better connected to

TABLE 24. Geography of Selected Wholesalers, 1928

Sector	Central No.	Central %	South Side No.	South Side %	North Side No.	North Side %	West Side No.	West Side %	Total No.
Automotive	89	62.7	17	12.0	17	12.0	14	11.3	142
Metal	197	69.1	28	9.8	25	8.8	14	4.9	285
Electrical	269	72.9	23	6.2	47	12.7	15	4.1	369
Furniture	72	77.4	3	3.2	14	15.1	1	1.1	93
Paper	99	81.1	10	8.2	7	5.7	2	1.6	122
Wood	209	82.3	18	7.1	14	5.5	7	2.8	254
Millinery	132	98.5	1	0.7	1	0.7	0	0	134
Total	1,067	76.3	100	7.1	125	8.9	53	3.8	1,399

SOURCE: *Polk's Chicago (Illinois) City Directory, 1928–1929* (Chicago: R.L. Polk, 1928).
NOTE: Row percentages do not add to 100 because "other" has been excluded.

metropolitan sellers. Learning from other trades, these interests looked to secure a better flow of goods and information between retailer, wholesaler, and manufacturer. The solution was the clearinghouse. After establishing an office in the Wrigley Building in 1921, the clearinghouse employed a small administrative staff to oversee the circulation of industry informa tion and coordination of annual meetings and exhibits. In the process, it created more centralized decision-making. According to one observer, the clearinghouse saved time for the Chicago visitor, who "would otherwise have to make long trips to plants over a wide area in order to find what he desires." As well as bringing manufacturers, dealers, and retailers together at one site close to the Loop's business offices, confectionery wholesalers, and downtown railroad depots, the clearinghouse was a place where manufacturers could exhibit goods and congregate to discuss ideas and information. It also created a milieu that generated new starts as entrepreneurs brought together local capital, expertise, and workers.[29]

In other cases, the combination of an extensive business-wholesaling complex and the centralized facilities of the candy clearinghouse attracted firms from other parts of the United States to Chicago. Typical of a well-established firm with a strong market position that moved to Chicago for these reasons was Mars Candy. In 1928 the company moved its factory (and 350 workers) from Minneapolis to Chicago's west end. Mars did not come to Chicago to gain access to direct manufacturing networks with local firms. Its raw material inputs came from a distance: chocolate from New York and Pennsylvania, sugar from Michigan, and milk from Wisconsin. Its intermediate inputs—such as packaging, boxes,

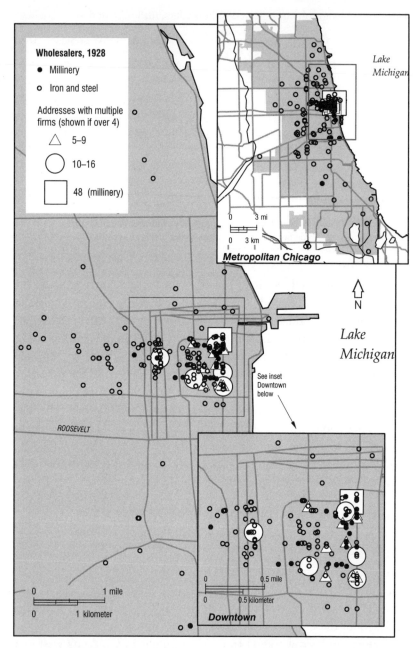

Wholesalers, 1928

- Millinery
○ Iron and steel

Addresses with multiple firms (shown if over 4)

△ 5–9

○ 10–16

□ 48 (millinery)

Lake Michigan

0   3 mi
0   3 km

Metropolitan Chicago

N

Lake Michigan

See inset Downtown below

ROOSEVELT

0   1 mile
0   1 kilometer

0   0.5 mile
0   0.5 kilometer

Downtown

FIGURE 38. The Geography of Millinery and Iron and Steel Wholesalers, 1928.
Source: *Polk's Chicago (Illinois) City Directory, 1928–1929* (Chicago: R.L. Polk, 1928).

and labels—were ubiquitous products easily supplied in most urban centers. Nor were Chicago's transportation facilities and workforce the sole inducements, although these must have been important.[30] What Mars found attractive about Chicago was the clearinghouse. Nowhere else in the Midwest, perhaps in the United States, did Mars have the same access to up-to-date market information, face-to-face business interaction with suppliers and buyers, and the latest technological and product changes. A relatively isolated west-end location was not a hindrance when you were making a national product—the five-cent Milky Way bar—and had a central node of information in the Wrigley Building.

CHICAGO'S WHOLESALERS IN THE INTERWAR PERIOD

Mars Candy came to Chicago because of numerous advantages, one of which was its major distribution and information function. It was not alone. As local business and political leaders boasted, Chicago was the "Great Central Market." They had good reason to believe this. From the last quarter of the nineteenth century, Chicago developed an elaborate distribution network that solidified its intermediary role in regional and national networks. Chicago's claim was consolidated in the 1920s with the growing importance of specialized business marts and clearinghouses. But the Great Central Market was more than the hinge in a regional and national flow of goods. Chicago's wholesalers played a decisive role in the flow of goods between Chicago firms, and in turn ensured the more effective functioning of the metropolitan production system. Despite the declining importance of urban wholesalers from the late nineteenth century, the distribution function continued to be in the hands of specialized agents; the wholesaler remained vital to the reproduction of Chicago manufacturing. Moreover, the viability of Chicago's metropolitan production chain depended on the ability of firms to transfer goods effectively. As examples from several local industries show, Chicago wholesalers remained important players in the industrial economy and decisive actors in the metropolitan complex of manufacturing linkages.

Merchants and wholesalers were critical to the development of American industrial capitalism from the late eighteenth century. They were the major intermediaries linking Europe with America, and raw material producers and manufacturers with each other and final consumers. Merchants lubricated the urban industrial economy by advancing credit and switching capital from trade into manufacturing. By investing in new productive ventures and securities, they underwrote America's burgeoning industrial economy.[31]

By the end of the nineteenth century, however, wholesaling's place in the urban economy became less important. The increasing volume of production in large corporations enabled manufacturers to construct more effective distribution networks outside the purview of the independent wholesaler. After 1880 the jobber's place in the nation's distribution system was increasingly challenged by mass retailers who bought their products directly from the manufacturer or from their own factories. By World War I a substantial share of consumer goods went from the factory floor to the final consumer via the department store, the chain store, and the mail-order house, while producer goods went from one factory to another. As manufacturers constructed their own sales organizations, obtained investment capital, and reinvested retained earnings, they became less reliant on wholesalers to provide them with the credit necessary for business operations and expansion. By 1930 two-thirds of the value of American manufactured produce bypassed the wholesaler and went directly to the retailer or to another manufacturer. Of the more than $63 billion of manufactured goods, a fifth went to retailers from manufacturers, almost a third moved from one factory to the factory of another manufacturing firm—or what the census called an industrial consumer—and a sixth was sold by the manufacturers' wholesaling outfit.[32]

Firms sold their products to retailers and other industrial producers by investing money and time in the creation of an elaborate distributional system, including salesrooms, storage facilities, administrative systems, publicity, and traveling salesmen. Despite exceptions, manufacturers' sales branches did little purchasing. Rather they focused on sales, which they performed in one of two ways. Firms either carried little stock and confined their efforts to promotion of sales and the securing of orders or operated as a typical wholesaler—buying and selling on their own account, carrying stock, and assembling goods for distribution to other dealers, other manufacturers, or retailers. Chicago's steel and meatpacking industries illustrate these points.[33]

By the interwar period, a good share of sales from Chicago's integrated steel mills went through the firms' own sales branch. In some cases, the steel mills bought out existing independent wholesaling firms. Inland Steel acquired Joseph T. Ryerson and Sons, for example. In these cases, the merger simply formalized within the boundaries of one corporation what had in fact been longtime practice. In other cases, firms created their own wholesaling system. This was part of a movement after 1900 toward increasing demands for immediate shipment of shapes, plates, and bars cut to length. Along with the steel wholesalers who increased their stock of assorted sizes, steel mills, rolling mills, foundries, and structural fitting shops added a wholesaling function to their facilities.[34]

At the end of the nineteenth century with the opening of the South Chicago and Joliet mills, Illinois Steel closed down its manufacturing operations in its original North Works and converted the space to house "varying lengths to meet local demand." By the early twentieth century, the company's wholesale function had expanded and, as one observer noted, "shipments, which in the beginning were made by wagon to the city trade only, now go to every state in the Union." With a capacity of more than 40,000 tons of beams, channels, angles, bars, and plates, the North Works of Illinois Steel in the Milwaukee Avenue area was the nation's largest warehouse before World War I. As was typical for wholesalers, orders were numerous and small in scale. On average it received more than 2,000 a month, with an average quantity of only 2.3 tons. About three-quarters of the orders were for one ton or less.[35]

The meat packers operated differently. From the 1870s perishable food companies built an elaborate sales organization to replace the wholesaler's function. By the mid-1880s perishable food distribution was centrally co-ordinated by the large, vertically integrated Chicago meatpacking firm. With an extensive branch-house system, corporations could receive and store fresh meat, which was then distributed to local customers. The branch house along with the meat packers' head offices and individual packing plants also coordinated transportation, advertising, storage, and distribution of meat products to numerous independent customers. In other words, the corporate marketing organization of the Chicago meat packers was substituted for that of the wholesaler. The internalization of a high volume of transactions within a small number of large firms not only allowed for more effective coordination between production and marketing, but also reduced unit costs of distribution and increased stock turnover.[36]

Compounding the wholesalers' position was the growth of local storage capacity. As firms grew in scale, they built warehouses at their plant to distribute goods to local and non-local markets. According to Alfred Andreas, writing in the mid-1880s, "The private warehouse and storage-room [was] a valuable accessory to commerce." By the early twentieth century, the "public" merchandise warehouse servicing general needs became more important. Butler Brothers' large public warehouse consisted of three connected buildings, each of which had 14 floors and two basements, the combined floor space reached 38 acres. These warehouses provided storage facilities for Chicago's growing regional trade. They functioned as the rail traffic pivot between the Atlantic and Pacific coasts, as well as the supply node for local firms. As the nation's Great Central Market, Chicago built a system of public warehouses on a scale far exceeding any other inland city. By the early 1920s, Chicago's 40 warehouses received, stored, and re-shipped with the most "up-to-date ideas in warehouse construction."[37]

Public warehouses were strategically scattered through the metropolitan area. Nearly all of them were located on one or more of the rail carriers coming into Chicago, while some had dockage facilities. Two of the largest were Midland Warehouse and Western Warehousing, both of which designed their warehouses to allow local manufacturers to maintain stocks of their products at the lowest cost. They permitted out-of-town firms to have a distribution system without the trouble of a branch plant, thus doing away with "the investment required for rental or ownership, and the overhead for its maintenance in the way of salaries, equipment, workmen's compensation insurance, light, heat, power, watch service, etc." In other words, they provided non-local firms with access to Chicago's manufacturers, wholesalers, and retailers "at a substantial saving."[38]

By World War I, the sales organization of the manufacturer, the mass retailer, and the public warehouse were reducing the role of the urban wholesaler. Nevertheless, the independent wholesaler remained important through the interwar period, especially in industries such as jewelry, food, and furniture, in which the large firm was rare. It was also common in regional entrepôts, such as Chicago and New York. Almost a third of manufactured goods in the United States was sold by wholesale firms. The majority of manufacturers sold through wholesalers in order to minimize their commitment to distributional overhead and to maximize their access to highly fragmented markets.[39]

Despite these historic changes, wholesaling remained important to Chicago's overall economy and to the success of the metropolitan district's production chain. According to the 1930 census, 7,237 metropolitan wholesalers employed more than 143,000 workers and had net sales of more than $6 billion. More than 90 percent of this trade—as measured in terms of firms, employees, and sales—was concentrated in the city, most of it in or close to the Loop. About half of the city's wholesale trade was in the hand of what the 1930 census labels "wholesalers only." More than 3,700 Chicago firms with almost 70,000 workers and sales of $2 billion were "engaged in the buying and selling of merchandise on their own account at wholesale." Without exception, the industrial suburbs and satellite towns ringing the city had a small wholesaling base. Gary's 38 establishments employed only 286 workers and sold just more than $10 million of goods. The same was true for Joliet. As one observer noted, "Proximity to Chicago and consequent competition of powerful firms located in that metropolis restrict the opportunity for the development of wholesale business." What was true for Joliet was true for the rest of the metropolis outside the central city.[40]

Chicago's independent wholesalers offered local manufacturers three benefits. First, they supplied them with small amounts of products, typically defined as less than a carload. While the large "mill" order was left to the large factory, the "warehouse" order was in the hands of the jobber. Second, they supplied merchandise to firms that were unwilling or unable to maintain large inventories. After the 1880s, firms, especially those experiencing rapid style and design changes, became less inclined to tie up capital in stock. The rise of these "hand-to-mouth" conditions maintained the importance of wholesalers in some industries. Third, jobbers continued to furnish manufacturers with a range of non-standard and infrequently demanded products that were extremely expensive and time-consuming to find. Wholesalers acquired and in many cases worked on semi-processed goods, especially in small lots, that large producers could not undertake. Under these conditions, the ability of manufacturers to successfully acquire inputs, much of which came from other industries, depended on wholesalers.[41] Independent wholesalers supplied semi-finished products to metropolitan, Midwest, and national manufacturers. As intermediaries, they connected a geographically dispersed set of actors and facilitated the flow of goods, information, and capital from one part of the region to another, and, of course, between Chicago and the rest of the United States and many parts of the world.

STEEL, METALWORKING, AND DISTRIBUTION

The metalworking industry illustrates the wholesalers' effectiveness in moving goods from one place to another and lubricating the metropolitan production chain. Chicago's steel industry had a few very large, vertically integrated firms. All of them sold a large share of finished steel in pre-ordered large lots (more than a carload) to regular customers. From the 1890s they also assumed the role of the wholesaler and in the process created some major problems. The most serious was the territorial struggle between mills and wholesaler. This was succinctly put by Andrew Wheeler, president of the Philadelphia jobbing house Morris Wheeler and chair of the American Iron, Steel and Hardware Association. In a 1926 speech, he stated that manufacturer-wholesaler relations were bedeviled by the inability to determine "a line of demarcation between proper mill business and proper warehouse business." Especially during the hand-to-mouth periods before World War I and in the 1920s, steel, metal goods, and machinery manufacturers increasingly encroached on the independent wholesaler's business. Long-standing trade relationships were upset as customers scaled back on their inventories and demanded immediate deliveries. Steel mills and metal makers became more willing to take over the small order so as to maintain their customers' goodwill and began to intrude upon what was traditionally the wholesalers' business. Despite plans introduced by various segments of the industry to stabilize the trade going to the manufacturer, the long-term trend was for the steel mills or the machinery maker to acquire an ever-growing share of distribution.[42]

Nevertheless, the assortment of local metalworking industries and their intimate relationship to steel producers established a byzantine system of distribution that rested on a solid jobbing basis. According to a U.S. Steel report submitted to the Temporary National Economic Committee in 1939, some 20 percent of finished steel was shipped to jobbers. In some segments of the steel industry, such as rails, the mills sold almost the entire production. In those where a large number of customers demanded all types and volumes of steel, the jobber was far more important. In this context, wholesalers played an extremely important part by distributing semi-processed products (as inputs to production) and finished products (as facilitating production and for consumption) for many industries.[43]

With hundreds of machinery shops, metalworking factories, and other metal-based firms, Chicago's market for wholesale products was insatiable. Chicago had a large and important machine tools industry, and, unlike so many of the smaller industrial centers, it did not depend upon a single large industry or a few industrial groups. As one mid-1920s survey of Chicago metalworking trades noted, wholesalers had "in Chicago and its metropolitan area thousands of machine shops and miscellaneous industries using steel products as raw materials." Moreover, "the extensive network of the metal trades and the immense diversification of manufactured products maintain a well-balanced demand for machine tools the year 'round."[44]

Chicago's metal jobbers offered advantages to its region's metalworkers. They offered specialized expertise. As one observer noted, Chicago's steel warehouses were "service institutions manned by efficient corps of employees equipped with trucks, switch track connections and every other facility requisite for rendering service." Copper and brass jobbers, machine tool wholesalers, and steel-beam distributors carried every commercial size or shape. The fact that wholesalers were close to the mills from which they took their stock only reinforced their ability to maintain a vast assortment of products. Wholesalers provided quick service, enabling local purchasers to receive goods within hours and regional buyers overnight service. Finally, the ability to buy from local distributors in small quantities on exact specifications allowed manufacturers to save on overhead expenses and to reduce inventory.[45]

In the 1920s there were hundreds of metal-related wholesalers doing business in Chicago. The largest segment in 1926 was iron and steel, whose 62 establishments had sales of $75 million. Along with this, machinery, hardware, agricultural implement, and plumbing equipment accounted for 474 firms with almost 10,000 workers and merchandise sold valued at $175 million.[46] Chicago had the largest warehouse stocks of steel products, machine tools, brass and copper, and hardware and the greatest floor space area devoted to steel stocks than other industrial centers. As one survey reported in 1925:

> Thousands of manufacturers in Chicago and the middle west exist and do business because in Chicago are large warehouse institutions operating much in the same manner as a department store from which one can purchase on a moment's notice either a pound of highly tempered spring steel or tons of steel shapes and either get an immediate delivery by truck or if the location is out of town be assured that the shipment will be on its way before night.[47]

Chicago's extensive wholesaling function met the needs of many customers.

The importance of the wholesaler is illustrated by the activities of two Chicago iron and steel firms: Joseph T. Ryerson and Sons and Scully Iron and Steel. Ryerson, Chicago's largest iron and steel wholesaler, originated in 1842. As a small enterprise providing stock for iron-working shops, it depended heavily for shipments on Pittsburgh mills. The situation changed after the Civil War as the company responded to the growing demand by Chicago's blacksmith and boilermaker trades, the increasing supply of steel from local mills, and "the uncertainty attending the prompt execution and shipment of mill orders." After incorporation in 1888, the scope of the firm's activities increased. Previously, these were limited to the handling of materials and specialties for boiler shops. But Ryerson, responding to a widening range of consumers' demands, adopted new lines. Growing scale and scope necessitated new premises adjacent to several railroads, all of which had direct connections to the warehouse through six tracks that entered the building. The new warehouse became the company's nucleus as operations were moved from the older complex and consolidated in the new one. By 1929 the plant occupied 21 acres and carried a stock of 75,000 tons in practically every type of steel. It linked the large steel mill to thousands of consumers.[48]

Scully like Ryerson, had excellent distribution facilities. Its large river-fronting warehouse had an excavated channel and unloading facilities by which boats unloaded freight directly into the warehouse, and three railroad sidings for loading and unloading cars. Chicago's manufacturers requiring small lots took up a significant share of the company's trade. In any one day, up to 30 to 40 teams of horses were busy on deliveries throughout the city. "The largest manufacturing plants in Chicago are always running out of something which they want immediately, and a part of the service of store distribution is to get the order ready and a wagon on the way with it within a very sort [sic] time after a rush order is received on the telephone." Scully serviced manufacturing firms such as Chicago Perforating, organized to manufacture perforated screens of all kinds and off all metals. As one observer noted, "Being in close touch with the iron and steel stocks of Chicago and jobbers in all classes of rolled metal," Chicago Perforating gave "quick shipment on any size and order for perforated metal."[49]

Reflecting the changing character of the steel industry, Chicago's wholesale trade experienced tremendous change after 1900. Wholesalers diversified out of iron and merchant steel to distribute products demanded by a range of industries. The shift to hand-to-mouth buying methods expanded the market for steel wholesalers such as Ryerson and Scully, and placed greater pressure on them to supply goods just-in-time. The greater volatility of styles and the faster turnover times of products forced manufacturers to cut back on inventories and to bring tied-up capital back into the production process. Under these conditions, the manufacturer's practice of carrying stocks was reduced and steel wholesalers became increasingly important.[50]

The metal wholesaling business had a two-pronged geography. A few large wholesalers such as Ryerson and Scully had huge "department store" warehouses a distance out from the downtown core. Most wholesalers, however, were to be found in the Loop and the adjoining central districts along the river and the railroads (see table 24). A central-city location was essential for metal and for other wholesalers, as it provided the most effective access to the majority of firms. Chicago's downtown area remained the nexus of financial, transportation, legal, and business producer services until after World War II. Furthermore, an effective wholesaling system required that firms minimize the time and cost of searching for goods. As Chicago became increasingly larger, busier, and more congested, the costs of acquiring goods at disparate locations became increasingly prohibitive. Clustering in the central core reduced the time and cost of searching for the appropriate materials.

In metalworking as in furniture, the linking of the components of an industry's production chain shaped the industrial landscape of metropolitan Chicago. The manner in which manufacturers gained access to raw materials, machinery and equipment, technologies, labor, information, and markets shaped the way that the industry developed and the distribution of economic agents throughout the metropolis. For example, Chicago's furniture firms—ranging from the large piano and upholstered goods makers, such as Kimball, Gulbransen, and Karpen, to the small, specialized firms producing on contract—were plugged into an array of capital, labor, and information circuits coursing through metropolitan Chicago. A well-developed furniture cluster had formed around hundreds of furniture makers and related firms producing a profusion of goods, a large wholesale complex servicing local and out-of-town interests, numerous retailers, and a well-developed institutional framework that regulated the industry's operations and attempted to minimize intra-industry conflicts. Central to this was the building of an extensive distribution system based on the institutions established by local associations. What was true for furniture and metalworking was also true for most other industries.

CHAPTER 10

Local Production
Practices and
Inter-Firm Linkages

CHICAGO'S
AUTOMOTIVE INDUSTRY,
1900–1940

Throughout the 1910s and 1920s, the *Chicago Commerce* published several reports on the local automotive industry. Frequently blurring fancy with fact, the city's boosters trotted out the rhetoric of the industry's importance for the city and how the city's unrivaled set of locational factors fueled the industry's growth. While much was hyperbole, there was truth to their proclamations. For Chicago—with its hundreds of car, taxicab, bus, truck, and motorbike assemblers and parts makers—was one of the leading contemporary automotive districts in the United States. Assorted firms produced axles, motor engines, bodies, and other vehicle components and drew on and created a unique set of scale and scope economies that operated at the metropolitan scale. However, as one commentator complained in 1925, "few" understood Chicago's important place in the automotive world.[1]

This contrasts with what is known about the economic history and historical geography of the American automotive industry. Considerable research supports the image of the vehicle industry as the exemplar of twentieth-century industrial modernism. Histories of firms and biographies of industry leaders highlight the role of, among others, Henry Ford and Alfred Sloan in creating the industry's giants, Ford Motor and General Motors.[2] Investigations of revolutions in automotive factory design demonstrate how the industry led the transformation of factory work space.[3] Labor studies emphasize work conditions and wages and the relation between the car and social life.[4] Research also looks at changes to work organization and links the rise of a modern industrial state to the emergence of mass production and mass consumption.[5]

Much is also known about the industry's geography. Automotive firms formed a triangular realm linked by Buffalo, Cincinnati, and Milwaukee, with Detroit at its center. The existence of a specialized territory in which industrial production was concentrated was common as most industries had a specific geographical division of labor. Industries clustered regionally, subregionally, and, in some cases, by metropolitan area, from the beginnings of industrial capitalism. Consider the association of early nineteenth-century New England with cotton and shoemaking, and early twentieth-century Akron and Pittsburgh with tires and steel. At the metropolitan scale, research on the automotive industry has concentrated on Detroit and the relation between increasing firm scale, vertical integration, and manufacturing decentralization.[6]

Even though few aspects of the automotive industry have not been examined, two strands of research raise several issues that complicate the picture of the industry's historical geography before World War II. The first focuses on production organization. Some writers question whether the supposed unproblematic relation between the automotive industry and mass production holds under closer scrutiny. The industry was a leader in organizational change, but what was the industry's link to mass production? For Karel Williams and his colleagues, mass production "offers not analysis but description which jumbles together results, principles and techniques."[7] This points to the gulf between mass production as an ideal type and the actual ways in which manufacturers organized work. Ford and other vehicle assemblers, for example, were not so much mass producers as repetitive manufacturers of standardized products under high-volume, short-cycle, and flexible conditions.

Industries and their constituent firms constructed different production methods, labor markets, and distribution outlets. No one "best practice" existed; industries took different organizational paths according to their particular needs. In the case of the vehicle industry, a range of firm-level production practices developed. While Ford built a vertically integrated system, firms like Studebaker fashioned their operations around flexible production strategies and niche markets. Moreover, the industry was not as vertically integrated as many commentators once thought. As Michael Schwartz and Andrew Fish recently argued, a tight relationship between suppliers and assemblers was common among Detroit assembly firms, both before World War I and during the interwar period.[8]

This research raises questions about how industrial organization relates to place. With the rise of the corporation, business transactions increasingly operated at long distances. Nevertheless, localities continued to matter, and business remained rooted in local face-to-face relations.[9] Detroit may have dominated the automotive industry, but other centers were also important. Given the different production worlds, labor conditions, and political and business cultures of urban places, it would be unrealistic to expect one industry to function the same way in every district. Production techniques were not uniformly diffused across space. The different industrial worlds that assemblers encountered outside Detroit forced them to adapt their work to local circumstances. While suppliers were beholden to assemblers, their industrial practices developed in response to the specific industrial milieus of individual cities. Moreover, the character of the industry itself differed from place to place; cities had different combinations of assemblers, suppliers, and ancillary producers. While Detroit specialized in passenger cars, Chicago was America's leading truck and bus assembly center before the Great Depression. Detroit and Chicago producers related in sharply different ways to local suppliers, credit facilities, distributors, and final markets.

Place specificity is related to the ability of a locality to attract the suppliers of an industry's commodity chain.[10] Even though the production steps of the nation's vehicle industry stretched across the Manufacturing Belt, many steps clustered within a metropolitan area. Not only did Detroit manufacturers control a significant share of the industry's production sequences; they also managed an empire of inputs within the metropolitan area itself and created extensive local networks based on price changes, long-term subcontractual relationships, innovations, market conditions, and so forth. In the automotive industry, many networks were local, and in the case of Detroit, firms had a symbiosis that stretched across the metropolitan economy as well as the region.

This was no different in Chicago. Between 1900 and 1930, a metropolitan complex of car, truck, and taxicab assemblers, parts makers, and wholesalers developed in response to the industry's production strategies and strong inter-firm relations. The local industry depended on the production practices of closely linked firms centered on a detailed division of labor between supplier and assembling plants along the metropolitan-based commodity chain. This involved the supply of vehicle parts in varying quantities to specialized assemblers and suppliers. Services received from local industries and financial institutions reinforced Chicago's automotive firms' dependence on this local network. Chicago's distinctive industrial character underpinned the development of a large industry specializing in truck and taxicab production by the interwar period.

CHICAGO'S AUTOMOTIVE INDUSTRY

Chicago developed as an important automotive production center after 1900. While not a major competitor to Detroit, the industry grew into a significant part of Chicago's industrial base and the city became one of the country's leading automotive manufacturing centers by 1930 (table 25). Detroit clearly dominated automotive production and Cleveland was always a distant second, but Chicago vied with a few other cities for third position. A major producer of automotive body and parts by 1909, Chicago grew over the following two decades.[11] Spurred by more effective production methods and increased demand for cars, trucks, and taxicabs, its 162 body, parts, assembly, and hardware firms employed 6,851 workers in 1919. Ten years later the number had almost doubled.

Most elements of the industry were evident before World War I. A large parts sector developed alongside Ford, the only passenger-car assembler. According to a 1914 trade bulletin, Chicago was the nation's second largest carburetor producer and home to the world's largest carburetor factory. Just over 10 years later, another commentator stated that "few appreciate the important role played in motorcar production by the parts and accessory makers in Chicago." He pointed out that "automotive manufacturers were quick to seize the advantages by subdividing the process necessary to produce the finished car. For instance, if two manufacturers needed axles, instead of each making his own, both would buy from a parts maker who made nothing but axles. Thus the car makers received the benefit of low costs under a production schedule neither could maintain for its own needs."[12] A major truck distributing and manufacturing center, the city boasted 10 manufacturers and 35 dealers and branch houses. Chicago's motorbike manufacturers made 100,000 bikes on the eve of World War I, accounting for about one-third of the country's output. The city's auto show, first held in 1901, exhibited the products of 87 manufacturers and 196 accessory makers in 1914.[13]

The local industry developed rapidly in the interwar period. Employment grew and the industry broadened to include firms specializing in power transmission, engines, and electrical equipment. Despite the hyperbole, a *Chicago Commerce* writer had a point when he stated that "such nationally and internationally known concerns as the Stewart-Warner Speedometer Corporation, the International Harvester Company, the

TABLE 25. Metropolitan Chicago's Automotive Industry,[a] 1909–1939

Year	Firms	Workers	Value Added
1909	41	1,725	2,232,970
1919	162	6,851	22,885,034
1921[b]	159	4,260	19,175,890
1925[b]	121	8,273	24,991,209
1929	101	12,904	56,384,945
1939	150	13,139	34,412,599

SOURCE: Various censuses.
[a]Metropolitan Chicago (or Chicago Industrial Area) as defined by the census. The definition of what could be included under the rubric of the automotive industry changed from census to census. The totals above include automobile bodies and parts, automobiles, vehicle hardware, motorcycles, transmissions, engines, automotive stampings, and electrical equipment. In all years, the numbers vastly underestimate the number of firms making products for the automotive industry.
[b]City of Chicago only.

Yellow Cab Manufacturing Company, the Vesta Battery Corporation, the Stromberg Motor Devices Company and others have contributed largely to making Chicago the industrial hub of the world, and helped to put the automotive industry up among the city's eight leading industries." The report also boasted that no other center made more parts and accessories. Although clearly untrue, the claim nevertheless captured the magnitude of the local industry, which, while centered in the city, spread through the metropolitan area (fig. 39). Automotive manufacture was also directly related to Chicago's importance as the Midwest's major distribution center. In 1915 Chicago's 97 dealers handled 100 makes of cars, and firms such as Studebaker, Ford, and General Motors serviced the region's dealers and depended on local parts manufacturers and wholesalers.[14]

Chicago's automotive industry depended on several advantages. Rapid population growth presented local manufacturers with a labor force characterized by diverse skills. America's largest railroad system linked the city with hinterland resources, moving raw materials, semi-processed goods, and finished products to and from the city in astonishing amounts. A busy entrepreneurial class—comprising manufacturers, financiers, and politicians—diverted accumulated capital into local industries, setting off multiplier effects through the local economy. Local businesses also profited from a long legacy of skills, know-how, capital, and markets. Well-established industries such as steel and farm implements came equipped with ready-made expertise, capital, and knowledge. These factors forged

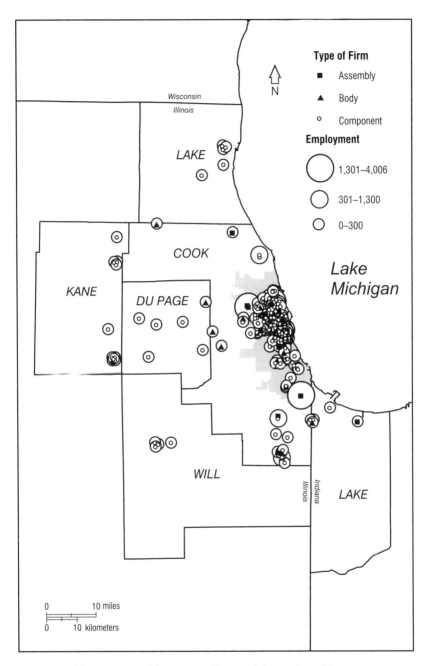

FIGURE 39. The Location of Automotive Firms in Metropolitan Chicago, 1924.
Source: *Directory of Illinois Manufacturers, 1924–1925* (Chicago: Illinois Manufacturers' Association, 1924).

the distinctive manufacturing environment in which automotive industry developed.[15]

Chicago's automotive industry grew up around two local industries. Bicycle and carriage making reconfigured their production systems and converted lines to build automobiles. The city's wagon industry, one of the nation's largest, channeled capital and information capabilities to the building of motor bodies. Similarly, the work practices and metal-stamping techniques of Chicago's large bicycle and machine-tool industries fueled the local vehicle complex's development. One mechanical engineer noted, as "the automobile industry began to exert a great deal of pressure on machine-tool manufacture and design," machine-tool makers "found it necessary to make extensive and very expensive changes in their systems of production to meet the rapidly increasing demand for cars."[16]

Several Chicago firms illustrate the contribution of these industries to the local automotive industry. In the 1860s a Prussian immigrant, William Klemm, established the E.R. Klemm Company to manufactured stone jacks for quarry work and eventually added a general line of metalworking. William's son, Emil, trained in several local metalworking establishments before joining his father's business in 1892. For the next 20 years, the firm built its reputation as a leading manufacturer of mechanical equipment and detailed machine work. In 1914 Emil turned the firm's 40 years of expertise with machines to advantage and converted a portion of the factory to motortruck production. Three years later he added 5,000 square feet in order to meet the demands of truck manufacture.[17]

International Stamping Company presents a similar story. Established by Walter Green in 1900 to make garment hangers, Green branched out into bicycle-chain guards, which he subcontracted to local manufacturers, because he had neither the machinery nor the facilities to produce them. As demand grew, he moved into larger facilities, bought machinery, secured a workforce, and began to invent and manufacture motorcycle parts. By the end of World War I, building on its technical expertise and inter-firm networks, the company expanded into automotive products, including a special type of vehicle tire carrier manufactured exclusively for Ford. In 1924, after two decades of incremental changes, the firm enjoyed a small competency in the automotive sector.[18]

Klemm and International Stamping were not alone in switching capital, skills, and knowledge to new lines. John Rath transferred capital and woodworking skills from the cooperage trade into truck-chassis manufacture, while the Albaugh-Dover Company added vehicle and tractor gears to its original line of cream separators.[19] As these examples suggest, by the end of World War I, local firms applied their traditional methods of learning by doing to the production of motor vehicle equipment. Local metal and wood producers reinvested profits and adapted the methods they established for their former product lines to the requirements of motor vehicles.

The industry grew directly out of Chicago's specialized manufacture of trucks, buses, and taxicabs. Yellow Cab Manufacturing was the largest company of its type in the world. Established in 1920 in a downtown garage, it turned out 500 cabs in its first year. Production doubled in the following year, forcing the firm to seek new premises. In 1923 the firm moved to a 35-acre plant on the city's western fringe, where it made 1,000 cabs per month. Reaching a plateau in taxicab production and looking to reap the advantages of economies of scale and scope afforded by its new plant, the firm expanded into passenger cars and motor coaches. Similarly, Diamond T Motor Car converted from passenger cars to truck manufacture in 1911, and by 1928 had $6 million in capital, employed thousands of workers, and sold trucks through factory branches and dealers in hundreds of cities.[20]

Unlike the Detroit industry, Chicago's trucks and cabs were not produced by a few large, specialized firms. Other firms designed and made a range of truck products: International Harvester built five-ton trucks, and the Harvey Motor Truck Works brought out new trucks of various sizes. The Old Reliable Motor Truck Company was organized to make dump trucks. At its factory just south of the city's largest parts-making cluster, it employed 100 workers to assemble 200 heavy-type trucks, which, as one commentator noted, "are used by many of the important industrial and commercial concerns of Chicago and other leading centers."[21] The city had an extensive parts and components sector, servicing car, truck, taxi, and motorbike manufacturers. According to a 1923–24 survey, Chicago had more 600 firms making 121 different automotive products. There were, for example, 27 automobile body makers, 22 gear manufacturers, 14 radiator producers, and 9 brake-lining makers. Thus, after 1900 a broad manufacturing developed in response to the growing demand by local automotive firms, facilitated by the willingness of Chicago's existing metal, wood, and electric equipment producers to switch lines.[22]

CHAINS, STRATEGIES, AND INTER-FIRM RELATIONS

Many scholars believe that inter-firm and geographically proximate relations declined with the rise of the large, vertically integrated corporation from the 1890s. Not only did companies internalize an increasing share of transactions, but national flows of goods and services replaced local ones. The local milieu in which business operated, however, continued to matter even though corporations operating at national (and international) scales began to litter the industrial landscape. This was the case for Chicago's automotive industry. Metropolitan Chicago offered local assemblers, manufacturers, and wholesalers most of the specialized resources necessary for vehicle manufacture. Local companies—ranging from the small, specialized parts maker to the large assembler—built a repertoire of inter-firm networks and deployed a range of productive strategies. Before turning to the strategies and relations of Chicago's automotive firms, the discussion switches briefly to the structure of the nation's automotive industry.

The automotive industry is typically cited as the exemplar of mass production, in which most of the stages along the production chain were internalized within a corporate structure. By the 1920s, if not before, the industry was an oligopoly, in which Ford and General Motors created internal economies and vertical integration on a scale that only a few other firms in a handful of industries (steel, petroleum, and cigarettes) attained. This meant that many links along the chain were internalized within the company, through either the central firm or one of its subsidiaries. Controlling an empire of raw materials, railroads, and steel mills, Ford was the classic example, although corporate control did occur in other ways. For example, the Chicago firm Borg-Warner—created in the 1928 merger of Borg and Beck, Marvel Carburetor, and Warner Gear—became one of the country's largest clutch, carburetor, and gear makers by integrating horizontally and producing in several plants. The fact that large automotive makers assembled such staggering control over many of the stages of production did not stop them from interacting outside their proprietorial boundaries. In its formative years before World War I and again in the 1920s, the industry featured an extensive system of just-in-time relations.[23] Big firms were involved with numerous suppliers, thus ensuring the externalization of activities along the production line.

There was a regional reach to these activities, as assembly firms drew upon raw materials and semi-processed products from within the automotive triangle. Certain places within this regional web captured many of these activities. Large multi-sectoral metropolitan areas—especially the motor towns of Detroit, Chicago, and Cleveland—had a diversified industrial base and considerable expertise from which automotive producers could satisfy their need for machine tools, steel, lumber, rubber, plate glass, and spark plugs. Drawing on this wide industrial base, vehicle producers built various types of inter-firm linkages. In some cases, large corporations created extensive big-firm-led relations with a host of firms that offered vehicle parts and other types of supplies. In other situations, small and medium companies developed a wide-ranging set of linkages with each other. Regardless, even though the automotive industry depended on regional inter-firm networks, it also found local suppliers to fill its production needs.

The "hand-to-mouth" (or just-in-time) system was a major reason why local networks were critical to the industry before World War II. Hand-to-mouth relations began at the new Olds Motor Works plant in 1905. The key issue facing Olds was inventories. The new assembly plant was too small to hold large stockpiles, and the pace of industrial change and market uncertainty made large stockpiles dangerous. Olds instituted a policy of providing each production station with just enough of its needed components to get them through a short production run. To provide these components, Olds established relations with suppliers located a short distance from its factory. In other words, component producers acted as Olds' inventory suppliers. This practice quickly spread to other Detroit assemblers and strongly influenced the clustering of Detroit suppliers and assemblers.[24]

The hand-to-mouth system was dropped during World War I but returned in the early 1920s. The new system was resuscitated because of Ford's cash-flow problems, which forced the firm to acquire capital by cutting back on inventory. Other assemblers quickly followed suit and implemented a policy of rapid, continuous innovation and annual model changes. The constant product changes made stockpiles too risky, while production improvements led to a downward spiral in component prices, thus adding to their expense. The construction of better transportation networks in and around Detroit made it easier to move parts between firms. It was a highly flexible production system, in which many firms produced relatively small numbers of widely varying products and rapidly developed and adapted new innovations.[25]

Chicago replicated Detroit's situation. Several commentators reported on how Chicago accessory manufacturers benefited from Ford's increased reliance on products made outside the company. According to a 1924 Motor and Accessory Manufacturers' Association estimate, 60 percent of the wholesale value of finished motorcars was produced by accessory makers. The report noted that a vehicle could be assembled in a factory where no single part or unit was actually made. Chicago assembly and parts firms subcontracted work to other local firms. A 1925 report explained that "all the component parts of the finished product are manufactured by some 600 concerns in the Chicago area" and commented that "few appreciate the important role played in motor car production by the parts and accessory makers in Chicago."[26]

The industry's importance is reflected in its share of the total national production of several sectors. In 1929 it was estimated that the industry consumed 975 million board feet of hardwood lumber (15 percent of American production), 5.6 million tons of steel (16 percent), 120 million pounds of aluminum (40 percent), 65 million square feet of plate glass (60 percent), and 814 million pounds of rubber (85 percent). Although it is impossible to say just how much of this came from local suppliers, Chicago with its diversified industrial base probably supplied a good portion of these resources. Local steel mills and foundries supplied steel and sash; veneer and planing mills easily satisfied the industry's lumber demands. Locally made rubber, aluminum, and glass products were also resources for vehicle makers.[27]

When local manufacturers failed to deliver, companies turned to the city's large wholesaling sector concentrated along the Chicago River. The 142 vehicle-accessory wholesalers listed in the 1928 *City Directory* supplied truck assemblers and parts makers with every conceivable line of product. Local firms supplying raw materials could be found elsewhere. Lumber wholesalers were a major source of material for chassis work. Corey Steel, suppliers of sheet steel and strip steel, advertised that "we specialize in Auto Body" sheets.[28] Manufacturers unable to buy in sufficient volume to satisfy the distribution channels of the steel mills turned to wholesalers for a range of steel products. Thus the industry's massive demands for raw materials and semi-processed goods created a multiplier effect that was easily satisfied within Chicago.

Several examples illustrate the relationship between metropolitan firms. American Coil Spring manufactured coil springs for local vehicle makers, dealers, and repair shops. Illinois Felt, makers of cotton felts for vehicle seats, leased warehouse space close to American Coil Spring. When Yellow Cab began passenger car manufacture in 1921, its first car, the Ambassador, was assembled from products that were made in its own factory and bought from local suppliers. The Auto Accessories Corporation of America's windshield wipers and Ford timers were sold directly to car manufacturers. United Manufacturing made an air cleaner used on more than 50 cars, trucks, and road-making machines. The Waukesha Motor Company, according to its president Harry Horning, made up to 30 models of truck engines at short notice because it was built around quick access to numerous Chicago accessory makers.[29]

A 1928 report in *Automotive Industries* noted that "a good volume of Ford business is already being handled by independent parts companies" and that the company was planning to increase outsourcing in many lines. In some cases, firms bought directly from Chicago parts makers. In others, independent companies became an extension of the car manufacturer. According to the report, "The independent source is functioning practically as the engine or clutch or whatnot department of the vehicle manufacturer, building for him a unit specifically designed and produced according to strict requirements of that individual vehicle maker."[30] Makers of parts took a share of the market because of the high quality of their parts, the savings realized by manufacturers from not having to expand their factories and inventories, and the lower costs that resulted from specialization.

Stewart-Warner, Chicago's largest manufacturer of vehicle accessories, provides a glimpse into the local world of vehicle production. By the end of the 1920s, the local company produced among other things speedometers, vacuum tanks, horns, bumpers, shock absorbers, windshields, driving lights, and electric cigarette lighters. A highly integrated firm, Stewart-Warner took advantage of the assemblers' reversion to hand-to-mouth methods. Before World War 1, according to Harry Alter, president of a local accessory wholesaler, few cars came from the factory fully equipped; car owners bought the additional equipment they required from retailers. By the 1920s, however, accessories on most cars were installed in the factory, and accessory manufacturers like Stewart-Warner changed their distribution strategy and began to sell directly to manufacturers, rather than retailers. In the process, they established a tight network of inter-firm relations with assemblers.[31]

Bankruptcy records shed more light on the interactions between Chicago's automotive makers and other local interests. Filing for bankruptcy in 1920, Chicago Standard Axle, an axle maker, had 144 creditors and debts totaling $145,288 (table 26).[32] The firm had a variety of locally based service creditors: electrical contractors, newspapers, advertisers, and messenger and towel services (fig. 40). With the exception of two trade journal subscriptions, its debts were all local. The 24 wholesaling creditors were Chicago firms, with the exception of one, a Cleveland tool producer with a Chicago office. Standard Axle received more than $70,000 in financial support from five Chicago creditors, including the Hasterlik family, owners of the Best Brewing Company, a prominent local concern. As this

TABLE 26. Sectoral Distribution of Chicago Standard Axle's Creditors, 1920

Sector	All Creditors		Chicago Creditors		
	No. of Firms	Money Owed ($)	No. of Firms	Money Owed ($)	Money Owed (%)
Service	17	1,582	15	1,185	74.9
Wholesalers	24	8,392	22	8,381	99.8
Finance	5	70,606	5	70,606	100.0
Manufacturers	89	63,565	64	44,390	69.8
Unknown	9	1,143	6	878	76.8
Total	144	145,288	112	125,440	86.3

SOURCE: Records of the United States District Court for the Northern District of Illinois at Chicago, Bankruptcy Records, Act of 1898, 1898–1972; Bankruptcy Case Files, 1898–1941; Bankruptcy Case 28646, *In the Matter of Chicago Standard Axle Company*, "Schedule A3," December 10, 1920.

case suggests, Chicago's resources ran the gamut of services, from utilities, lawyers, publishers, and city services to wholesalers, contractors, banks, and ethnic newspapers.

Some manufacturing inputs, however, were shipped in from outside the city, although local firms still supplied most of Standard Axle's goods. Eighty-nine creditors, 64 of which were Chicago based, supplied assorted manufactured goods. Five brass and several steel makers furnished metal goods, cold drawn steel, gray iron castings, steel castings, and steel tanks. Foundries, machine shops, and machinery makers produced among other things ball bearings, steel wheels, machine tools, hardware, stamped metal, springs, ornamental iron, and time recorders. Chemical plants supplied lubricants, paint, and varnish. Numerous products went into the packing and administrative stages of the business, including paper bags, wooden boxes, tags, rubber stamps, and paper.

Standard Axle depended on large loans and advances (table 27). It is likely that the Hasterlik family funneled such a large amount of money into the firm either because of a family, ethnic, or friendship connection or a carefully conceived strategy for divesting from beverage manufacture. After the establishment of Prohibition in 1919, the Hasterliks may have wished to move capital out of a problematic industry into a growing one. Regardless, the axle maker was heavily indebted to a neighborhood manufacturer: Best Brewery was just a few blocks from the axle factory. Other large creditors included well-known Chicago foundries and metal firms. The company owed a few non-local suppliers: metal wheels from Lansing,

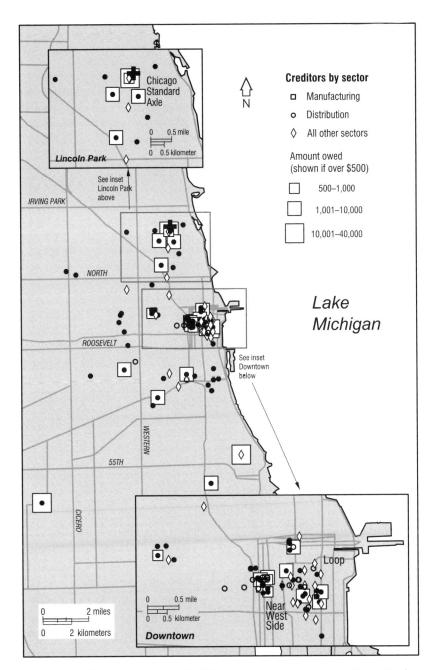

FIGURE 40. Chicago Standard Axle's Chicago Creditors, 1920. Source: Records of the United States District Court, Northern District of Illinois, Eastern Division at Chicago, Bankruptcy Records, Act of 1898, 1898–1972; Bankruptcy Case 28646, *In the Matter of Chicago Standard Axle Company*, "Schedule A(3)," filed December 10, 1920.

TABLE 27. Largest Creditors of Chicago Standard Axle, 1920

Creditor Firm	City	Products	Amount ($)
Hasterlik family	Chicago	capital	69,542.70
Interstate Foundry	Chicago	cast steel wheels	10,065.41
Albaugh Dover	Chicago	metal goods	5,751.32
Haskell & Barber Car	Chicago	railroad equipment	4,968.10
New Departure	Bristol, CT	ball bearings	4,939.75
Hills McCanna	Chicago	lubricators	4,918.11
Standard Steel & Beam	Philadelphia	steel beams	3,864.72
Hubbard Steel Foundry	East Chicago	steel	2,926.93
Warner Forge	Chicago	forgings	2,571.51
Diversey Foundry	Chicago	gray iron castings	2,533.64
La Salle Steel	Hammond	cold drawn steel	2,325.24
H. O. King	Chicago	steel products	1,767.53
McMaster Carr Supply	Chicago	mill supplies	1,761.30
Chas. H. Besly & Co.	Chicago	machine supplies	1,727.83
Electric Steel	Chicago	steel castings	1,727.27
John Hoof & Co.	Chicago	auto supplies	1,441.46
Seehausen Wehrs	Chicago	oil filters	1,305.88
Kelly Reamer	Cleveland	reamers	1,214.65
Motor Wheel	Lansing, MI	wheels	1,203.80

SOURCE: Records of the United States District Court for the Northern District of Illinois at Chicago, Bankruptcy Records, Act of 1898, 1898–1972; Bankruptcy Case Files, 1898–1941; Bankruptcy Case 28646, *In the Matter of Chicago Standard Axle Company*, "Schedule A3," December 10, 1920.

Michigan; ball bearings from Bristol, Connecticut; and steel beams from Philadelphia. Firms trying to collect from Standard Axle were located within the Manufacturing Belt: tools from Kokomo, Indiana, and Rochester, New York; machinery from Boston and Fitchburg, Massachusetts; bronze castings from Niagara Falls and Cleveland; and time recorders and journals from New York City.

Standard Axle's reliance on local suppliers reflects the importance of metropolitan Chicago's automotive inter-firm relations. Responding to the city's position as both the nation's second largest final market and a leading automobile producer, manufacturers offering an assortment of products flocked to Chicago to service the growing demand for parts. Some came to take advantage of the dense complex of inter-firm networks that had been created over time. Others hoped to benefit from the city's large and growing market. For example, U.S. Hame, a Ford truck body maker, moved to Chicago in 1923 and leased a one-story building in the heart of the city's automotive district, which it used as a convenient distribution point or, as one observer noted, "to expedite delivery of their products."[33]

Another case highlights Chicago's significance. In 1921 the business of the local firm United States Automobile Bumper was bought by Kalamazoo's C.G. Spring. The company closed the Chicago factory and moved production to its Michigan headquarters. Two and a half years later, however, the demands of the Chicago market forced Spring to open a Chicago nickeling and polishing plant and warehouse, requiring 160 workers. While the company's eastern plants serviced the Detroit and Cleveland markets, its new factory acquired semi-processed goods from local firms and made bumpers for local assemblers.[34]

Chicago companies were quick to adapt their active production practices to industry innovations. Pines Winterfront, a radiator cover maker, undertook research in conjunction with engineers from other firms. Located on the second floor of its new 1929 factory were "the offices, finishing rooms and experimental and working laboratories where winterfront engineers . . . work out the problems of built-in shutters with engineers from motor car manufacturers." Much more well-known than Pines Winterfront is Motorola, Inc. In 1928 the business of the Stewart Storage Battery Company with a factory in the Near North Side on West Superior was bought by the Galvin brothers and renamed the Galvin Manufacturing Corporation. The Galvins opened a new factory at 847 West Harrison Street right in the heart of the city's automotive cluster. By 1930, in the face of a declining market for battery eliminators, the firm switched lines and began to manufacture car radios. Success led to a move from its downtown location to a new and much larger premises on the western fringe of the city at 454 West Augusta Street. By 1940 more than 500 employees made radios for car and homes at the Augusta plant. Continued expansion and innovation during the war and the postwar period led Motorola (the company changed names in 1947) to become one of Chicago's most important producers for the automotive industry.[35]

Pines Winterfront and Motorola were not alone in being innovators. In the first four months of 1924, Pullman, which had begun making vehicle bodies to compensate for the declining demand for railroad cars, published the results of research on a steel frame that was 50 pounds lighter than an ordinary closed auto body; Buda, the engine maker, built a new six-cylinder bus engine; and Yellow Cab added a newly engineered five-seat passenger vehicle body to its production line.[36] Trade journals suggest that local firms created numerous innovative processes and new products. Even if only a small proportion turned into marketable products, the point remains that Chicago's industry was dynamic and closely knit.

Chicago was home to associations that linked manufacturers with each other and with wholesalers. In 1925 the Fenske Building was converted into the Automotive Equipment Mart. Also the home of the Automotive Manufacturers Association, the mart displayed trade products, acted as the headquarters of out-of-town buyers, provided an assortment of services, including conference rooms and messengers, and housed trade representatives. Like the American Furniture Mart and the Merchandise Mart, its purpose was to facilitate inter-firm face-to-face contacts, reduce time and costs, and create a symbolic place for its sector within the city. As one commentator put it, "Buyers will find it possible to transact in two or three days what otherwise would require several weeks time. . . . Manufacturers will find this an ideal place to meet their buyers."[37]

Manufacturers and buyers met at the Chicago Automobile Show, one of the biggest in the country, whose importance lay in the deals done, sales made, and "important meetings and conventions which took place in Chicago during auto week." Some 2,000 industry members had their "Old Timers" Club banquet and election annually in Chicago. Other groups held concurrent gatherings in which the automotive trade was "vitally interested": the National Automobile Dealers Association, the Society of Automotive Engineers, and the Automotive Electrical Service Association.[38] The local industry was connected to the national Motor and Accessory Manufacturers' Association (MAMA). Since 1905 the MAMA had functioned to protect the interests of parts makers in various ways, such as securing freight rate reductions, promoting standardization and simplification, and protecting industry from legislation and taxes. Similarly, more specialized trade associations existed, including specialized trade groups, such as a storage battery association, established in 1924.[39]

A STICKY METROPOLITAN GEOGRAPHY OF PRODUCTION

Vehicle firms were not industrial islands. They were part of a metropolitan system in which a spatial web of interactions was crucial to the functioning of the industry. Nor were these interactions randomly distributed; there was a distinct geography to these inter-firm relations. In 1924 a major automotive district with more than half of the city's firms and employees formed a crescent around the Loop made up of two main nodes (table 28 and fig. 41). One, in the Near North Side, had Stewart-Warner as its center and comprised 42 firms employing more than 5,000 workers. The

TABLE 28. The Geography of Chicago's Automotive Industry, 1924 and 1940

Zone and District	1924 Firms			1940 Firms		
	Workers	LQ	Firms	Workers	LQ	Firms
South Side	6,070	1.6	95	2,497	1.2	20
Near South	2,240	2.5	38	937	1.9	8
Near West	1,776	1.1	24	368	.4	7
Douglas	1,253	3.2	23	119	1.6	3
Lower West	801	.9	10	1,073	1.6	2
Calumet	5,652	1.2	10	4,458	2.0	6
South Chicago	4,181	3.4	2	3,015	4.5	1
Suburbs	1,471	.4	8	1,443	.9	5
North Side	5,114	2.2	42	1,172	.7	14
Lincoln Park	3,145	5.7	5	168	.4	2
West Side	1,118	1.2	21	260	.5	5
Humboldt Park	851	1.0	16	744	.9	7
North End	3,361	3.2	13	1,624	1.5	12
Portage Park	2,646	20.4	2	75	.3	1
Suburbs	393	.9	3	739	3.3	3
Lincoln Square	322	.6	8	810	1.3	8
Clearing	0	0	0	1,780	5.5	3
Other	2,512	.2	70	2,775	4	21
Total	22,709	1.0	230	14,306	1.0	76

SOURCE: *Directory of Illinois Manufacturers, 1924–1925* (Chicago: Illinois Manufacturers' Association, 1924); *Manufacturers in Chicago and Metropolitan Area* (Chicago: Chicago Association of Commerce, 1940).

other node, in the Near South Side, featured 95 small and medium firms employing more than 6,000 workers clustered along "automobile row" close to the railroad terminals, the river, and the central warehouse district. As shown by the location quotient, automotive firms and workers were concentrated in the automotive district.[40]

The industry's primary spatial feature consisted of the motor firms' link to companies that were either in the central area or geographically proximate. Bankruptcy records illustrated some of the links. Of Standard Axle's 144 creditors, 112 companies, accounting for 86 percent of the money owed, were from Chicago. While creditors came from all over the city, they tended to be concentrated in two areas. The Loop and the adjacent warehouse district accounted for 54 creditors. The second area was a function of the locale effect. More than an expected number of creditor firms were located in the north end, close to the axle maker's factory. The 30

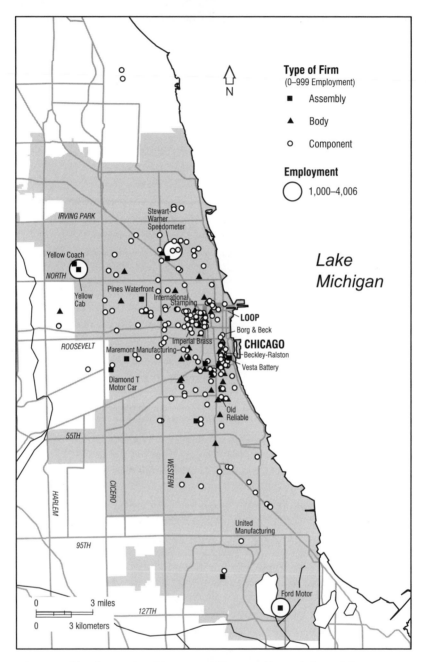

FIGURE 41. The Geography of Automotive Firms in Chicago's Central Area, 1924. Source: *Directory of Illinois Manufacturers, 1924–1925* (Chicago: Illinois Manufacturers' Association, 1924).

creditors located there accounted for more than half of the money owed by the firm. The neighborhood's tool, locks, ball bearing, brass goods, varnish, and pattern factories and warehouses had been major suppliers to Standard Axle.[41]

An examination of the creditors common to bankrupt firms provides another view of the industry's geography. While the creditors of any one firm illustrated the extent of a firm's supplier net, common creditors outlined the cohesive character of geographical networks. This was evident in the case of Sunderland Manufacturing, a tire pump manufacturer, and B&W Manufacturing, a radiator maker (table 29).[42] Even though located in different districts and dependent on neighborhood firms for a good share of their inputs (the locale effect), they shared 22 firms in common, ranging from utilities (Chicago Telephone, Commonwealth Edison, and Peoples Gas) to services (*Chicago Daily News, Chicago Tribune*, Chicago Towel, and Fuller-Morrison). Most creditors were local wholesalers and manufacturers, and some outside manufacturers, which fulfilled most of the semi-processed requirements necessary for making tire pumps and radiators. The majority of creditors were located in the area surrounding the Loop and the two major automotive districts.

Mohawk Automobile Equipment and Unexcelled Automobile Products also shared a significant number (98) of creditors.[43] The firms they had in common accounted for 43 and 31 percent, respectively, of the firms owed money by the two bankrupts. Chicago firms supplied them with services and inputs, from telephone and printing to radio tubes, battery terminals, lamps, and malleable iron. The commonality of inter-firm linkages is further illustrated by the large number of non-Chicago firms (45) the two companies had in common. Chicago's automotive districts, however, remained important, as almost half of the non-local companies had Chicago warehouses, mostly located in the Near South Side. Not only was Chicago a major producer; it was also a major distribution site for manufacturers and wholesalers from other parts of the country. In almost all cases, these firms established a branch in Chicago's automotive district.

The other major center of automotive employment in 1924 was the Calumet district. Here, in the far southern reaches of the metropolis, 10 firms employed 5,652 workers. Unlike the other districts with their multitude of small and medium-size firms, the Calumet district was dominated by Ford. Located at Hegewisch, an area close to the Indiana border, Ford with more than 4,000 workers accounted for more than 80 percent of the district's employment. What is interesting about Ford's move to Hegewisch in 1923 is that it had little impact on the geography of the rest of

TABLE 29. Common Creditors of B&W Manufacturing and Sunderland Manufacturing, 1920

Creditor	District	Product	City	Money Owed to Creditors ($)	
				B&W	Sunderland
Baltimore Tube	—	tubing	Baltimore	1,030	1,041
Chicago Daily News	Loop	publishing	Chicago	8	12
Chicago Telephone	Loop	telephone	Chicago	40	20
Chicago Towel	Douglas	towel service	Chicago	5	5
Chicago Tribune	Loop	publishing	Chicago	49	27
Commonwealth Edison	Loop	power	Chicago	153	366
Dallas Brass & Copper	Lincoln Park	brass goods	Chicago	3,904	683
Fuller-Morrison	West Side	druggists	Chicago	9	5
Harris, Samuel	West Side	machine supplies	Chicago	17	428
Hollis & Duncan	West Side	paper bags	Chicago	15	20
Horders Stationery	Loop	rubber stamps	Chicago	17	11
Illinois Nail	Milwaukee	nails	Chicago	254	50
International Time Recorder[a]	—	time recorder	New York	1	6
Johns-Manville[b]	—	asbestos	Manville, NJ	5	83
L'Hommedieu, Chas.	Far West Side	plating supplies	Chicago	14	11
Machinist Supply	Near West	factory supplies	Chicago	111	83
Norton & Co.[c]	—	grinding wheels	Worcester	16	28
Peoples Gas	Loop	power	Chicago	282	30
Reeves Pulley	Near West	pulley	Chicago	22	11
Standard Oil	Near South	oil goods	Chicago	71	297
Steel Sales	Near West	wire rope	Chicago	433	215
Stevens, Chas.	Near West	metal goods	Chicago	na	52

SOURCE: Records of the United States District Court for the Northern District of Illinois at Chicago; Bankruptcy Records, Act of 1898, 1898–1972; Bankruptcy Case Files, 1898–1941; Bankruptcy Case 28657, *In the Matter of Sunderland Manufacturing Company*, "Schedule A3," October 15, 1920; and Bankruptcy Case 28534, *In the B & W Manufacturing Company*, "Affidavit of Mailing," December 20, 1920.

[a]International Time Recorder had a branch office on West Madison.
[b]Johns-Manville kept a branch office and warehouse on South Michigan.
[c]Norton kept a branch office and warehouse on the North West Side.

Chicago's automotive-related firms. Even the city's loss of its major producer and only car assembler left the centrally located automotive districts intact. Existing locational assets in the centrally located automotive districts continued to play a decisive part in the geography of the industry.

Ford began its strategy of branch plant assembly in 1909 with the construction of a plant at Kansas City. Others quickly followed, and by 1917 the company had built 18. One of these was in Chicago. The original plant in the Central Manufacturing District was quickly replaced, however, as its small size and outmoded production methods forced Ford to seek larger premises. Ford bought the Hegewisch land in 1920, began construction in 1921, and moved into the facility two years later. A year later the remaining stock and equipment were moved to Hegewisch, thus ensuring that all of Ford's Chicago operations were under one roof. The Chicago plant was sold to a real estate broker for $1,113,000.[44]

The Hegewisch plant did not become an industrial anchor for other firms. Located south of the South Chicago steel mills, the area was named after Achilles Hegewisch, a local manufacturer. Concerned about the consequences of George Pullman's move to Chicago in the 1880s, Hegewisch platted a community to "outrival Pullman's." In 1883 he moved his plant to Hegewisch and created a land development company that bought 1,500 acres close to the plant. Fueled by a vision of developing industrial property on the empty prairie, Hegewisch believed that harnessing and connecting the city's southern waterways would make it a transportation and industrial center.[45] Despite his ambitions, the planned model city fell through. When Hegewisch died in the 1890s, the town "was nothing more than an industrial plant with a handful of nearby workingmen's cottages."[46] Growth continued to lag. Some development occurred after the South Shore electric line was built in the 1890s and after construction of the Calumet Sag Channel began in 1903. Hegewisch's car plant became Pressed Steel Car and then the ordnance division of U.S. Steel. Despite Ford's 1923 move, which inspired a real estate bubble that went bust, the area only had a population of 7,890 in 1930. Over the next 20 years, the population declined, and by 1950 it was still "largely an area of prairie and underdeveloped land."[47]

Because of its scale and close relations with many Chicago supplier firms, Ford had the potential to exert considerable influence on the character of Chicago's automotive geography. It seemed likely that a portion of the industry would be drawn southward following the company's move to Hegewisch, but this never happened. According to the trade journals, most new starts, additions to existing premises, and moves occurred within the older automotive districts. Certainly by 1940 the industrial directory reported few changes in the location 16 years later. Some related firms appeared in the Calumet district, other than Harvey's Buda Company (truck engines) and Gary's Standard Steel Spring (automotive and tractor springs), which were established before Ford moved south.[48]

Several possibilities explain this. One is that Ford may have had little contact with the wider set of assembler-supplier relationships that characterized the city's production practices. However, this was not the case. While company records describing Ford's suppliers do not exist, scraps of information from other sources portray a range of linkages existing between Ford and Chicago firms. As one observer noted in 1925 of the new plant, "Cars are not manufactured there from the ground up, [while] a number of small parts are manufactured in Chicago."[49] Some of the city's leading automotive part suppliers did business with Ford: Stewart-Warner furnished speedometers and signal horns, C.G. Spring and Bumper supplied bumpers, and Chicago Rivet and Machine made rivets.[50]

Another possible explanation is that the Hegewisch plant received much of its equipment from suppliers located elsewhere in the automotive realm. Ford plants in St. Paul and Hamilton, Ohio, for example, sent glass and steering wheels, respectively, while the River Rouge plant furnished other parts. As Michael Schwartz and Andrew Fish's work on just-in-time relationships among Detroit firms suggests, some material destined for Hegewisch could have come from supplier firms in and around Detroit. In that event, Ford's new site at the end of Lake Michigan would have reduced transport time. The appeal of such a location for smaller Chicago firms whose clients were not as far-flung, however, was much weaker.[51]

The fact that Chicago was America's truck center contributed to the city's stability. Ford was the largest single automotive firm, but local truck, taxi, and tractor assemblers were more influential than a single firm, despite its large size. After the Great War, truck and cab makers moved to the city's western fringes. By 1924 Diamond T was located in Cicero, while the industrial district to the north was home to Yellow Cab. Thus, if local geographic proximity was of any consequence, and the difficulty and cost of intra-urban transportation at that time suggests that it must have been, the growing demand for parts and equipment by the likes of Yellow Cab and Diamond T would have enticed suppliers to locate centrally or in the city's western environs.[52] These firms, and the cluster that they formed by the early 1920s, acted as a counterweight to whatever pull that Ford exerted.

Despite the importance of the large assemblers, Chicago's production practices by the 1920s ensured that suppliers were not only close to each other but also had immediate access to raw materials and semi-finished products. In a world of lumber, textile, rubber, and metalworking firms that produced goods for various industries, the greatest need was access to materials, design capabilities, and know-how. Take chassis and panel construction, a major Chicago trade. Until the mid-1930s, when steel became the major element, a vehicle's body was composed of steel sheets attached to a wooden frame. Even though the time needed to build a body fell drastically in the interwar period, these firms, building on skills inherited from carriage makers, continued to make bodies from traditional materials. Vehicle manufacturers such as Maremont Manufacturing, Martin-Parry, Met-L-Wood, and Able Auto Body, for example, all clustered near Chicago's lumber district in order to be close to the wood products needed for chassis and side-panel construction.[53]

By the 1920s the Calumet district had firmly cemented its reputation as an industrial producer of steel and ancillary services. Large, high-volume producers like Ford, which drew on many suppliers and served a national market, might have realized the benefits of this particular business climate. Smaller supplier firms dependent on local companies—practicing specialized production practices and subject to the changing climates of demand from larger and more independent firms—would not have been so attracted to Calumet's charms. For them, the embedded world of specialist firms in the existing automotive districts far outweighed the dubious fortunes of close geographic proximity to a firm, however important, that was located at the farther reaches of Chicago's industrial territory. Space matters, and in this case, the space was a downtown crescent composed of a thick set of interrelated firms covering all aspects of the automotive industry's production chain.

On the eve of the Great Depression, Chicago was America's second industrial city. With the exception of New York, Chicago, by almost any measure, outdistanced all other urban centers. With a population of more than 3.3 million in the city and another 1.2 million in the surrounding suburbs, the district's economy encompassed nearly every aspect of the manufacturing, service, transportation, construction, commercial, and professional world in the United States. Metropolitan Chicago was either the leader in or among the top few producers of railroad equipment, agricultural implements, meat, steel, machinery, printed goods, clothing, and furniture. Territorially, Chicago was massive, stretching 30 to 40 miles from the Loop, with hundreds of suburbs, towns, and satellite cities forming an endless combination of specialized social and economic spaces. Wealthy white districts such as the Gold Coast, Riverside, and Kenilworth were scattered throughout the city and suburbs. In contrast, large swaths of the metropolis such as Packingtown, the Black Belt, and Cicero were home to concentrations of poor minorities and immigrants. Many residential areas—with a small selection of retailers, professionals, and workshops—offered few local employment opportunities and differed in fundamental ways from the steel-producing districts of the South Side or the bustling industrial districts encircling the Loop. Large in scale, differentiated in composition, Chicago was a teeming and dynamic metropolis.

This was the point made time after time by those taking the prewar industrial inspection tours. One guest of a July 1912 tour noted that a "city council committee which understands the relations of the factors of production and distribution in a workshop of the size of Chicago comes to its view its duties broadly, and legislates not for a ward, or to reciprocate favors as between aldermen, but for a vast community, the focal point of the earth's richest valley."[1] But the city wasn't the only concern of policy makers and economic interests, as was made clear a year later by W. Buchanan, vice president of the Chicago Association of Commerce's Civic Industrial Committee. To an audience of industrialists, railway executives, politicians, and developers, he outlined the committee's purpose, stating that it was not a parochial group, obsessed with small, local matters and institutions. Rather, its mandate was metropolitan: "The kind of questions that are referred to this committee from the industrial side have to do with almost every kind of an institution which is looking for a new location in or about the Chicago. That would mean as far as Waukegan, Joliet or any other place within the Manufacturing zone of Chicago."[2]

Chicago was not simply Chicago. The economic center of the earth's richest valley, Chicago consisted of central city, suburbs, and satellite cities, spilling over municipal lines and involving the interests of numerous economic, social, and political institutions and actors. The decentralization of industry out of the city core was a major force driving the development of a multi-nucleated metropolis. Metropolitan Chicago's manufacturing geography, with its incremental decentralization and continued specialization of factory districts, was based on a cyclical process of economic change involving an assortment of industries that can be traced back to the mid-nineteenth century.[3] The formation of factory districts on the urban fringe has always been part of Chicago's growth. In his study of Chicago's employment, John McDonald notes that "from the very beginning of industrial capitalism, firms have moved to the accessible periphery of the metropolitan area to gain space and capture economies of scale."[4] Industry was not alone. As Graham Taylor reminds us in his 1915 study of industrial suburbs in Chicago, St. Louis, and Cincinnati, "households, small stores, saloons, lodges, churches, [and] schools" accompanied the huge plants, which were "uprooting themselves bodily from the cities."[5]

In other words, Chicago was not simply a city; it was a complex metropolitan district composed of several business centers and factory districts, hundreds of municipalities, and thousands of residential areas. As William Cronon and others have made clear, metropolitan districts such as Chicago were ensconced in national flows of material, capital, information, and people.[6] As this book shows, the metropolitan scale is important for understanding the industrial and urban histories of the United States. Probing the internal dynamics of manufacturing change and interchange in metropolitan areas unravels the relationship of capital investment and industrial change, reasserts the importance of the metropolis as the central coordinating entity of American society, and demonstrates the autonomy of industrial relations within the metropolis.

Adding to the metropolitan complexity was the speed at which Chicago changed. This was made abundantly clear by city alderman Albert Fisher. In June 1913 he told members of an industrial inspection tour that it "is almost a crime for a committee that has any responsibility to think of Chicago as Chicago of today. In a week Chicago will be a vastly different proposition than it is today and we cannot look at Chicago today as being the measure of our responsibility."[7] But here was the rub. The planning and control of an economic machine the size of Chicago was a forbidding task. Growth—oftentimes anarchic, always disorderly—had to be reconciled with order and regulation. Tensions between the twin necessities of order and disorder had to be resolved. Planning and control was necessary, for a capitalist industrial economy needs constant attention, maintenance, and coordination. The issue facing Fisher and others was how was this to be done and who was going to do it? The answer was growth through territorial expansion and order through the rules and regulations laid down by elites.

LOCATIONAL ASSETS AND THE MAKING OF THE METROPOLIS

As the industrial inspection tours made quite clear, Chicago's growth as an industrial metropolis was predicated on the ability of local capitalists to incorporate vacant land and existing settlements on the urban fringe into its territorial frame. To do this they created what I have termed locational assets—packages of infrastructures, institutions, and practices vital to the expansion of Chicago's industrial base, population growth, and metropolitan form. Creating these assets was not an easy task as the ability of elites to build effective coalitions was neither assured nor automatic. Even though they were required for successful economic and urban growth, the ability of self-interested groups to act collectively was shaky at best. Nevertheless, elites did form local alliances, and in the process built new areas and restructured existing parts of the metropolis to accommodate capitalist industrialization. Even though locational assets could inhibit growth or turn it down pathways that were far from optimal, the success of rapidly expanding industrial districts depended on building effective infrastructures and institutions.

The unspoken purpose of the industrial tours was to forge effective coalitions out of different metropolitan interests. Participants were working to change "unimproved" land into factory districts and to construct new and consolidate existing business networks. These networks, it was hoped, would create the material and social circuits to underpin Chicago's next bout of industrial development. For urban promoters, the present was not enough. The past had to be manufactured into the future. A shifting set of class-based local alliances refashioned a past and present Chicago into a future one. While speculators and developers sought out property-based profits, other interests rallied behind the booster flag and actively created locational assets. Railroads built the iron paths linking firms to the outside world. Local politicians passed bylaws, constructed long-lasting infrastructures, and zoned the city to accommodate industry's needs. Local professional and financial groups facilitated economic growth in its various forms. Despite their own specific interests, these groups formed coalitions, sometimes loose and for a specific project, other times coming together for a long period of time.[8]

This was evident in those city and the suburban areas where class coalitions worked together to convert empty prairie into industrial sites. An alliance of railroad interests, manufacturers, and politicians, for example, formed locational assets in order to transform the Central Manufacturing District into productive industrial land. Before World War I, the district was turned from vacant railroad land into prime industrial property by a local business and political coalition. In June 1913 Dr. W. Evans of the Chicago Association of Commerce's Civic Industrial Committee averred that previously there had been "practically nothing in this district—there was no development here. It was about as desolate and unpromising a part of the city that we could find in close range of what we were undertaking." Five year later Francis Harman, assistant manager of the Chicago Junction Railway, opined that "this tract of some three hundred acres of land was unknown except as a barren prairie. The ground was a mere swamp, without the semblance of a passable road."[9] Class coalitions turned this empty land into one of Chicago's fastest-growing interwar factory districts.

The Central Manufacturing District was not unique. The planned district was just one of the hundreds of areas, from Hammond to Cicero, Packingtown and the West Side, where class alliances transformed land into industrial space from the middle of the nineteenth century. As Evans told the members of an industrial inspection tour, Chicago

> is spreading to those outlying districts which only a few years ago we called prairies. These same prairies now are occupied by busy factories, warehouses and foundries. Heavily loaded trains carry the enormous amount of freight turned out by these "flowers of the prairie" to all parts of the world.

Urban space was restructured through the building of industrial locational assets. And, as Evans made quite clear, local alliances were instrumental in the creation of these assets and in making distant empty lands the home of industry. "Not long ago these same industries scoffed at the idea of establishing their plants 'way out there.' Now, however, they resent any insinuation that they are located away from the center of Chicago."[10]

The building of locational assets depended on earlier paths of metropolitan development. Shaped by the procedures, routines, and practices inherited from the past, each round of industrial growth built on preexisting attributes and in the process created new economic, social, and institutional ones. As a result of these place-based endowments, firms faced constraints and opportunities in the way they behaved and the choices they made. In part, these derived from a firm's history. Past decisions regarding plant investment, product lines, technologies, and strategic alliances occurred over space, thus shaping the choices firms made. Just as importantly, decisions were rooted in the particular histories of place. The accumulated economic, social, and institutional place-based "fixed capital" created over time established path dependency.[11]

Locational assets were magnets of urban growth. They anchored firms in place. Capital, labor, land, institutions, and infrastructures were place-specific and channeled the locational choices of firms and industries. The existing social, political, and economic structure of a place established a self-reinforcing and cumulative process that framed new rounds of economic growth. Firms sought spatial proximity to other firms, transportation facilities, services, and workers. The uneven development of this search and the building of locational assets produced specific parts of the metropolis. Assets also anchored workers in place. In Chicago, as elsewhere, the city's social geography was strongly shaped by the siting of industries and their associated locational assets. Ethnic, racial, and class-based neighborhoods were built in tandem with manufacturing development across the metropolitan landscape.[12] As elsewhere, the industrial metropolis required the increased scale and speed of money, labor, information, and commodity circulation over an ever-widening urban landscape.

Locational assets established the necessary components for building on the urban frontier and in the process stretched the perimeters of what was considered the industrial district. It was incumbent upon local elites to establish the territory under their control. The development of suburban factory districts—a dense network of transportation facilities, working-class residential districts, and inter-firm linkages—had a tremendous impact on what constituted the district. The stretching of the geographic limits of the industrial city and the assembling of locational assets in suburban districts forced the U.S. Census to formally acknowledge industrial districts as metropolitan areas. First recognized in the 1900 census, the census compilers by 1910 were providing data on metropolitan districts, which they defined as "not only the area within the corporate limits of the city but also parts of the surrounding territories which may in a general way be regarded as closely associated with the development of the city." However, the notion that locational assets combined with population and economic growth were taking root in a profusion of suburbs, market towns, and satellite cities was already part of the way in which nineteenth-century Chicagoans and their contemporaries framed the idea of their rapidly growing industrial centers. By the interwar period, the idea of metropolitan areas was firmly on the agenda of groups seeking to better read capitalist urban industrialization. The metropolis was central to understanding urban growth, locational assets, and factory districts.[13]

Locational assets also linked the metropolis with exogenous opportunities, capital, and resources. A place was dependent on the way it was inserted into worlds outside the place itself. Firms, institutions, and infrastructures were all connected to networks that were fixed, locally and non-locally. Any place was composed of a multitude of local, regional, national, and international links. These take a variety of forms from private and state capital to build new docks and factories to the development of a range of social and economic signals, such as stock exchange prices, fashion statements, and technological innovations. The opportunities available to any firm in a place were related to its ability to combine non-local and local locational assets. Chicago's specialized wholesaling community, for example, linked local firms to other localities. Machinery wholesalers provided Chicago's large local machine shops, foundries, and factories with machines of all sorts from the entire manufacturing world, both within and outside the United States. The Chicago industrial district was home to locational assets that allowed manufacturing interests to function across various spatial scales, and to a capitalist class that manipulated and controlled the networks holding the district together.

DIVISION OF LABOR, WORK, AND GEOGRAPHIC SCALES OF INDUSTRY

No firm, place, or person is an island. From the smallest suburban workshop to a corporation's downtown office, business networks formed the basis of the industrial system. Companies acquired production inputs from and sold processed goods to other enterprises because they were able to forge functional links with other firms, agencies, and individuals. Without this complex and intersecting array of connections, the business system could not function. From the small industrial suburb perched on the edge of a small industrial city to the vibrant and bustling office district of a massive metropolis, places were linked through an almost infinite set of linkages. Many of these were everyday and unremarkable; companies acquired heat, water, and power from utility companies, workers from local labor pools, and materials from other firms. In some cases, however, economic relations—ranging from the development of new industrial sites to workplace conflict, the development of new technologies, or the implementation of new productive practices—were distinctive. Regardless, all were part of a broadly knit industrial world that focused on the metropolis.

The production chain is of vital importance for understanding Chicago's metropolitan development. Production chains linked industry, place, and people, and in the process created specialized metropolitan geographies. Production chains were not abstract and placeless, but were embedded in place-making, business networks, and working-class neighborhoods.[14] The counter flows of city and rural goods within the Manufacturing Belt are well-known. Non-urban raw materials—such as Mesabi Range iron ore, Pennsylvania coal, and Corn Belt hogs—traveled by train and ship to cities. Similarly, finished products, such as farm implements and furniture, moved from city to city. Urban flows were critical. The manufacturing world of Chicago reached out of individual firm boundaries to span the neighborhood, the city, the rest of the metropolis, and the larger region. In the process, conduits of capital, information, and trust were opened up, linking nodes of production within the wider flows of industrial capitalism. Place mattered, in both its absolute and relational senses. Chicago was Chicago because its economic actors, its business entrepreneurs, local boosters, politicians, and working families were able to forge spaces that were linked to the broader regional capitalist space economy.

Urban places, of course, were central organizing nodes in the production chain. From large multi-sectoral metropolitan areas such as Chicago to smaller industrial centers such as Detroit and Grand Rapids, firms were embedded in regional, national, and international production relations. Urban relations were hierarchically structured. Economic agents operating out of the large metropolises, most notably New York and Chicago before 1940, controlled a substantial share of the flows of goods, finance, labor, and information that underpinned the development of regional production chains. The industries that dominated smaller, lower-order, specialized industrial cities—such as Pittsburgh, Providence, and Lowell—linked the industrial centers into wider webs of production sequencing and placed them into well-defined sets of relationships along the regional production chain. Detroit's automotive firms, for example, assembled an assortment of inputs collected from throughout the automotive realm surrounding the Motor City. Rubber came from Akron, steel from Pittsburgh and Cleveland, and speedometers from Chicago.[15]

Production chains also had important effects within urban areas. All urban places, from the multipurpose metropolis to the specialized industrial town, depended on local economic agents to maintain production chain links. Material flows, knowledge transfers, trust networks, and symbiotic relationships were organized at the urban level. From Pullman and the Calumet district to the city's automotive and instrument districts, local firm interaction—sometimes at the neighborhood level, sometimes not—formed the production chain that stretched across the urban economy. The larger the place, the more likely it contained a greater number of steps along the production chain. Nevertheless, a distinct metropolitan geography was formed through the networks constructed by Providence's jewelry producers, Pittsburgh's steelmakers, and Grand Rapids' furniture manufacturers. Within a metropolis, different areas offered advantages to firms: the central business district offered financial and legal services, while factory districts supplied semi-processed inputs, a labor force, and other commodities. The ability of manufacturers, wholesalers, bankers, and workers to sustain the interconnected character of the metropolis was made possible by the development of a distinctive transportation, energy, and communicative network that spanned across the metropolis and linked the various parts together.

Firms were embedded within a wider world of local innovation, interaction, client-supplier networks, financing, and servicing. Coexisting with, and linked to, the production practices of large firms was an elaborate set of smaller, specialized firms that participated in a local industry's production chain. From the provision of services through the supply of capital, insurance, and semi-processed products, metropolitan firms generated resources for local industry to build upon. This, in turn, formed an intricate body of networks between firms. Chicago's network-based economy facilitated collective action and constant adjustment, both inside and outside the local industry. This was in turn directly related to Chicago's metropolitan geography. As the bankruptcy records and other evidence indicate, firms were stuck in place and linked to central and neighborhood suppliers. This did not mean that the firms were backward or stagnant. In fact, the reverse was true. The existence of strong geographic links indicated that industries belonged to factory districts with a well-established set of local connections.

Factory districts also reached outside the metropolitan area as manufacturers sought raw materials, semi-processed goods, information, capital, and expertise from other industrial districts. Together, the need to maintain the viability of the production chain forced firms to build up a dense inter-firm network spanning several spatial scales. From Packington and East Chicago to Philadelphia's Kensington or New York's Hoboken factory districts, a firm's functioning economic world embraced local business connections that included national and international networks. These places were not unique. The manufacturing bases of all industrial urban centers—from multi-sectoral metropolitan areas such as Chicago and New York to the specialized city such as Grand Rapids or Lowell—were built upon an array of factory districts. These districts tended to be larger and more numerous in the multi-sectoral metropolis than in the single-industry city. Regardless, specialized areas in industrial districts as different as Chicago, Boston, Detroit, Paterson, and Hartford provided external economies and other opportunities that would not have been otherwise available.

The scale and character of production chains were a product of the industrial division of labor. Different production systems, markets, and cost structures created different inter-firm and inter-industry linkages. Chicago's mass and bulk producers, which tended to be vertically integrated and multi-unit, functioned in a quite different world of inputs and outputs from the city's batch and custom producers. Routinized producers such as U.S. Steel and Campbell Soup internalized many production steps and often led a hierarchically based network of subcontracting, strategic alliances, and input-output relations. In contrast, smaller firms functioned in various modes, from independent entities receiving subcontracts from larger firms to captive subsidiaries that were spin-offs of a vertically integrated corporation. In Chicago, as elsewhere, batch and custom producers, such as furniture makers and printers, had a much greater dependence on linkages external to the firm itself.[16]

In Chicago the division of labor and external economies produced spatial relations that contained firm boundaries. This took several territorial forms. The Chicago industrial district was an extremely effective spatial container for industrial production. The Windy City, like all other industrial centers, offered firms the advantages of geographic proximity. This had several important effects, most notably reducing the friction of space; lessening the costs and troubles of moving inputs, outputs, and personnel; assembling resources in an area; and providing access to locational assets. As bounded spaces, metropolitan areas had well-developed political and economic boundaries, which limited and defined industrial movement, social interaction, and social relations. In short, firms were linked together across metropolitan space in a functional and hierarchical productive relationship that was built on a division of labor and external economies.[17]

This metropolitan hierarchy varied according to the degree of urban specialization. There were few places that matched metropolitan Chicago's broad industrial base before 1940. New York and Philadelphia certainly did; Los Angeles by 1940 may have, although this is unlikely. The range of industry ensured that all four metropolitan areas had an elaborate industrial geography.[18] From an early date, industries spilled over the borders of the four cities, setting up new factory districts and adding to the metropolitan industrial agglomeration. Industrial diversity ensured that the factory districts scattered throughout the metropolis were well linked and that firms had strong linkages. In New York, factory districts in Brooklyn, Queens, and eastern New Jersey, which specialized in the entire gamut of industrial production in America at the time, formed an elaborate interlinked industrial geography centered on Manhattan. The array of metropolitan options could not be surpassed; from extensive central-city services and financial functions to the dazzling array of specialized input producers, transportation choices, markets, and workers, these four metropolitan areas had the largest industrial choices in the United States.

In contrast to these metropolitan giants were the specialized economies of other industrial centers. Smaller in scale and with a much more truncated industrial spectrum, places such as Pittsburgh, Providence, and Portland could not match the services, linkages, and networks offered by the large multi-sectoral centers. Nevertheless, every industrial center, regardless of scale and diversity, forged a distinctive array of coexisting production practices. It is important to appreciate that even though Cleveland, Cincinnati, and Chicopee had their own distinctive industrial structure, these places had elements in common with the large multi-sectoral metropolitan areas. While lacking the full range of elements available to the industrial firm, these places were marked by the geographies of the division of labor and production chains. In Chicago as elsewhere, the building of metropolitan collective economies forged a range of metropolitan geographies.

Production centers differed from one another. As the automotive industry illustrates, Chicago was not Detroit. Local differences in scale and composition mattered. Chicago's dense set of steel, lumber, metal, furniture, and wholesaling firms presented opportunities that were available to few other centers, including Detroit. Chicago's distinctive industrial structure provided a strong foundation that was difficult to match. Similarly, the city's specialized automotive industry facilitated a unique structure of local and outside inter-firm relations. Chicago's relatively large supplier and parts industry, and its concentration of truck and taxicab assemblers, formed an array of links that set it apart from those in other centers, such as Detroit, Flint, and Akron. Moreover, the city's late start as an automotive producer forced it to pursue avenues outside large-scale passenger car manufacture. Accordingly, a unique industry, centered on truck and taxicab assembly and featuring an extremely diversified parts manufacture, flourished in Chicago before World War II.

INDUSTRIAL DISTRICTS AND SOCIAL SPACE

In his study of Chicago's Polish immigrant neighborhoods between 1880 and 1922, Dominic Pacyga states that the South Chicago and Back of the Yards industrial areas "were based on a firm economic/symbolic base rooted in the steel and meatpacking industries."[19] He is not alone in pointing to the symbiotic relationship between industrial districts and working-class residential neighborhoods found in American urban places before World War II. From the built-up West Side to the industrial suburb of Cicero and industrial satellite city of Joliet, metropolitan Chicago was a veritable checkerboard of industrial plants and residential districts.[20] The relationship of the home-work link and urban growth has a long history. As a 1942 report spelled out, "The potentialities of suburban industrial development as a factor in population decentralization cannot be overlooked. . . . [I]n view of lower housing costs and greater living amenities in suburban regions, the tendency is inevitable for those workers whose places of employment are well outside Chicago likewise to change their residences to outlying points, thus diverting a considerable portion of their family expenditures."[21] Industrial and residential movements were intimately linked.

The strong home-work relationship in Chicago was not unique. In rapidly expanding Detroit, working-class suburbanization after 1880 "took place simultaneously with the diffusion of factories toward the city's periphery." Moreover, this connection between home and work "was no mere coincidence."[22] Metropolitan Pittsburgh's ethnically diverse working-class districts such as Homestead and Vandergrift were built around factories and mills, both large and small.[23] The Omaha stockyards "proved to be a major nucleus of residential growth in spite of the fact that stockyards and packing plants probably are among the least attractive industries in respect to encouraging nearby residential development."[24] Birmingham, Alabama, was home to "new plants on the fringes of the old city that attracted thousands of workers. By 1920 most iron and steel workers had moved away from the center of the city to neighborhoods closer to work."[25] In the western industrial cities of Los Angeles and San Francisco, factory and residential development went hand in hand.[26] In urban America, working-class homes and industrial workplaces were joined at the hip.

The character of factory districts shaped the everyday lives of the workers who were employed and lived in them. Metropolitan factory district growth was related to housing markets, transit, and working-class lifeways. As Chicago Heights, Hammond, Bridgeport, and many other factory districts show, these were tied to investment in manufacturing. Scholars have pointed out that a graduated set of working-class housing districts—typically along the lines of cost, race, and ethnicity—were built around factories. Housing provision took many forms, from company housing to small speculators who constructed batches of new houses to workers who built their own. The extension of the metropolitan transportation systems, from train and streetcar lines to highways, was often driven by the need to facilitate the journey to work, to allow firms to secure their necessary inputs, and to ensure that finished goods were able to make it to market. Place-based investment in manufacturing produced workplace conflict. Depending on contingencies, this conflict could range from minor to serious. Regardless, all factory districts, both in the central city and elsewhere, gave rise to a distinct set of industrial relations that involved not only workers, managers, and their institutions (unions and trade organizations), but other local agents, such as local politicians and land developers.[27]

The place-based locational assets so assiduously created by ever-shifting coalitions of capital and labor formed the social, political, economic, and physical worlds where workers labored and lived. Metropolitan business networks forged between manufacturers, wholesalers, service providers, land developers, and politicians framed the everyday context of working-class life. These worlds differed greatly. The ethnic and racial composition of a neighborhood shaped the internal occupational hierarchy of the workplace, which in turn shaped the rewards—income, status, promotion—that workers had and could deploy back into the neighborhood. According to a 1935 survey of East Chicago, for example, the seven housing areas of the industrial city were defined by their income and ethnicity, and the presence or not of industry. The poorer areas such as the "New Addition" were populated by recent immigrants and African Americans who worked as laborers in the steel mills, while the "good class of working people and professionals" lived away from the heavy industrial districts.[28] In all other manner of ways—from factory labor and gender politics to the ethnic and racial composition of the workforce—the difference in workplaces mattered and played themselves out in the locational worlds, both outside and inside the factory.

Inside the factory, the elaborate hierarchy of the workforce in large producers such as U.S. Steel, International Harvester, Standard Oil, and Ford had profound implications for workers' interactions within the workplace. Occupation, race, gender, and ethnicity shaped the distribution of income and the ability of unions to create a collective consciousness. These differed in important ways from the conditions of most labor-intensive batch and specialized producers, such as those in the machinery and furniture industries, which tended to have less elaborate internal occupational hierarchies and stronger paternalistic relations. These internal dynamics were molded by the geography of workplaces.

Outside the factory, large corporations wielded considerable more power at the local level than the smaller batch and custom firms. The ability of the corporation to shape local events was reduced in a large city where manufacturing executives had to compete for political space with many other powerful interests, most notably trade unions, land developers, and utilities. Nevertheless, a few large corporations did sway local decisions in ways that the thousands of batch and custom producers in their workshops, sweatshops, and small factories could not even begin to contemplate. In some cases, they shaped decisions specific to the firm, such as the zoning of a parcel of land and of the reduction of property taxes, while in others they had decisive input into citywide matters, such as public housing, public transit, and Prohibition.

A different situation was to be found in the suburbs, however, as corporate executives had unrivaled control over political decisions about local locational assets. Few executives of large corporations such as Pullman Palace Car (Pullman), U.S. Steel (Gary), and Western Electric (Cicero) were municipal politicians. They did not have to be, as most local merchants, property developers, and professionals had few ideological disagreements with corporate managers. Most important suburban political actors, from local councillors and businesspeople to union executives, realized that their own interests were at risk if they did not concede to the demands or follow the interests of the largest and sometimes only major employer of the suburb.

Workers had different experiences of work and home depending on where they worked and lived. For working-class families tied to an anchor industry—such as Poles in South Chicago's steel mills or Packingtown's packing plants—work and home were often shared with ethnic members and were geographically circumscribed. Everyday routines of family, religion, and leisure were rooted to industrial neighborhoods. In other cases, industry attracted workers from a distance and in the process stretched the perimeters of the everyday. While thousands of women who worked in the downtown garment factories walked to work from adjacent residential districts, others, especially unmarried women, traveled from more distant working-class residential areas. In other cases, manufacturing corporations, financial institutions, and other office-related companies in their search for cheap and disposable labor employed an increasing number of young women in their downtown offices as typists and low-grade clerical workers. Difficult to organize, many saw the Loop's bright lights as an escape from the rigors of life in the city's factory districts.

The character of workplace conflict was shaped by the concentration of industries in different parts of the metropolis and the ability of local employer associations to coordinate activities in Chicago's factory areas. Labor and business conflict over working conditions, wages, and hours operated at various spatial scales. Clusters of industry—such as clothing and printing downtown, steel in the Calumet district, and electrical appliances in Cicero—were home to actions linked to neighborhood locals, city associations, and national unions. The concentration of workers in specific neighborhoods ensured that the intricate set of workers' political activities in their unions were place-based. Likewise, the power of business associations operating out of Chicago had important implications for labor. According to Josiah Currey, in 1917 there were about 100 commercial organizations in the industrial zone encompassed by Waukegan, Gary, and the Outer Belt Railroad. The Chicago Association of Commerce with more than 4,000 members was the central organizational node of metropolitan Chicago's capitalist class. Through its different divisions and committees the association provided information, advice, and resources to its members.[29]

The actions of Chicago's unions—such as the Amalgamated Clothing Workers of America, the International Ladies' Garment Workers' Union, the Chicago Typographical Union, and the Amalgamated Association of Iron and Steel Workers—individual employees, and business associations were shaped by the politics of the local workplace. In turn, these actions linked the local workplace to city, regional, and national union movements. These nested labor relations ranged from informal sabotage and work-to-rule tactics to the one-firm, industry-wide, or general strike. While strikes mobilized the local place, linking workers to a world larger than their everyday, they also unleashed employer assaults. Using a mix of firings, blacklisting, open shop, ethnic and gender divisions, and the threat of moving to another city, managers resisted and contested the actions of place-based labor and their nested relations at the neighborhood, urban, and national levels.[30]

Although common workplace interests brought workers together, ethnic, ideological, and political differences both inside and outside the factory gate led to conflict and dissent. Strong ties within a factory district could work both ways. In some cases, tight neighborhood networks laid the basis for strong worker resistance, while in others these ties hindered intergroup and intra-class cooperation. Even though worker actions were contingent, the support found in clustered industrial areas was stronger than that in industries where factories were scattered across the metropolitan area. The ability of workers to forge effective networks of association and support was greater in a factory neighborhood than among dispersed and isolated plants. Local, place-based community networks mattered. While labor's political strength as manifested in local, state, and national labor organizations was responsible for legislation protecting workers, this was based in local neighborhoods and the links workers made in the workplace, home, church, and saloon. Success was frequently dependent upon the intersecting politics of home, neighborhood, and workplace.

These politics were embedded within the world outside the local. The war between the meat packers and the workers' union illustrates this point. The workers' defeat in the 1904 and 1921 strikes and the destruction of the union by the packing companies were linked to a set of relations that flowed from corporate offices and from neighborhood homes. Managerial decisions made in the business districts of Chicago, New York, and London framed the reasons for, the direction of, and the outcomes of these strikes. Packing company executives decided when the strikes would take place and devised a strategy that struck at the very heart of neighborhood politics—the use of immigrant and African American scabs. Interracial conflict and tensions were inflamed and played out in the homes, streets, saloons, and shops of Packingtown and the surrounding residential districts, stymieing the building of a united front and leading to eventual defeat. The geography of factory location was a politics of place.

METROPOLITAN CHICAGO: THE INDUSTRIAL DISTRICT

The politics of place, the economics of factory location, and imperatives of capital investment established Chicago as an industrial district. Shifting coalitions of financiers, manufacturers, politicians, and land developers built an elaborate system of inter-firm networks, formed a distinctive production system, molded a range of economic and social geographies, and, in the process, created an extensive metropolitan industrial district. Chicago's industrial prominence between the Civil War and World War II rested on its ability to function as a territorially rooted industrial district. A multitude of inter-firm linkages operating over various spatial scales established the industrial and geographic parameters of metropolitan growth. This almost infinite set of networks that created and held the industrial district together was controlled by a small elite. While workers and small businesspeople were critical to the success of capitalist industrial-urban growth, the decisions were in the hands of a select capitalist class who controlled the powerful coalitions.

Manufacturers operated within interlinked regional and local industrial worlds and in the process created a set of specialized industrial spaces. Firms were tightly linked with one another across the metropolitan landscape. The coordinating node of the Chicago industrial district was the city core. Flows of capital, information, people, and materials were controlled from the downtown offices of the large corporations, financial institutions, law and accounting firms, politicians, and developers. The central factory districts were home to thousands of manufacturing, wholesaling, and service firms that provided a substantial share of the urban economy's everyday and episodic transactions. In this way, firms from the rest of the metropolitan industrial district stretching out from the Loop were tied to the central districts and their distinctive array of inputs. Manufacturing firms of all sorts—whether they were located in the city's industrial neighborhoods, industrial suburbs, or satellite cities—were connected to the dense set of downtown networks. By providing inputs that were not generated within the factory, this close-packed set of eternal economies allowed large integrated firms access to the benefits of the urban fringe, most notably a captive labor force, a pliant local government, and cheap land. Even the small firms clustered in factory districts throughout the metropolis drew upon the advantages of the downtown business and industrial districts. All manufacturing firms were deeply embedded within the stretched boundaries of Chicago's industrial district.

The Chicago industrial district was not cast in the mold of the European one. Nineteenth- and early twentieth-century American industrial districts were similar in some ways to their European counterparts, such as the English Midlands or the French Lyonnais. All had an intricate set of inter-firm relations, specialized workforces, and coordinating elites that bound the district together as a coherent economic body. But there were differences. The American industrial districts were much more tightly controlled than European districts by large corporations and a continental market. The organization of geographic relations was shaped by the distinctive industrial structure of American capitalism. This ensured that the American industrial district had a different set of spatial relations than its European counterparts. To an extent, this was the product of the much shorter history of urban development in the United States. Just as importantly, industrial districts such as Chicago were nested in a geographic hierarchy, dependent on a huge number of central production facilities, the formation of a specialized spatial division of labor, an elaborate degree of inter-firm flows, and strong regional relations that tied local firms, not only to each other, but to the industrial centers of the Manufacturing Belt. Proximity mattered, but inter-firm relations were stretched across space. American industrial districts were centered on the large, diverse, multi-functional metropolitan areas such as Chicago that pulled in resources from a broad region reaching hundreds of miles from the city, not on a regional complex of strongly interlinked small industrial cities characteristic of Europe.

The building of an industrial district resulted from differential flows of capital investment over several generations. Competition between sites was tremendous as manufacturers sought out areas within metropolitan Chicago that would provide them with what they believed were the best returns. Although rational decisions concerning the best location for their firms were impossible to make under the conditions that managers and owners operated, a capitalist logic drove firms to locate where they did. Managers of large multi-unit corporations and proprietors of custom firms had a good sense of the advantages and disadvantages of the old districts close to the city core, the first ring of suburban districts 10 miles from State and Washington, and the newer greenfield sites 20 or 30 miles from the Loop. Intra-industry relations facilitated this spatial sorting out over long periods of time. Place and capital were mutually constitutive. As the prewar industrial tours of the empty prairie made quite clear, capitalist control of industry-specific expertise, specialized inputs and outputs, and booster institutions created and maintained Chicago's industrial district.

Bankruptcy Records,
1872–1928

This study rests on information taken from 97 bankrupt firms and their 7,196 creditors. The firms were selected from four periods: 1872–1878, 1898–1901, 1920, and 1928 (table 30). Bankruptcy law underwent several important changes in this period but was not effectively systematized until the bankruptcy law of 1898. Bankruptcy laws were enacted in 1800, 1841, and 1867, only to be repealed in 1803, 1842, and 1878, respectively. The modern bankruptcy system was put into place with the 1898 Bankruptcy Act. This legislation—the product of intense and drawn-out wrangling by different business interests, regions, and social groups—created what John Cover calls the "federal legal process by which a debtor is declared insolvent, his assets seized and distributed among his creditors, after which he is formally discharged of further liability."[1] While the 1867 and 1898 bankruptcy acts covered all types of firms, the acts were typically used by small and medium-size firms. Large firms typically sought out corporate reorganization through equity receivership. Nevertheless, the sample includes a range of firms by size and is representative of the vast majority of small- and medium-scale firms found in Chicago during this period.[2]

Creditor information varies by bankruptcy laws. In general, the later cases are more complete because of the stricter controls and more routinized nature of proceedings imposed by the 1898 act. Case files also differed by firm. There is no indication why, but some case files had much less information than others. Assuming that the law was evenly enforced, the

TABLE 30. Summary of Bankruptcy Cases, 1872–1928

	1870s	1890s	1920s	1928	All Cases
Firm (No.)	19	25	24	29	97
Creditors (No.)	841	1,426	2,359	2,570	7,196
Creditors Linked (No.)	765	1,269	2,186	2,380	6,600
Creditors Linked (%)	91.0	88.9	92.7	92.6	91.7
Chicago Creditors (%)	76.0	66.3	77.2	70.3	72.3
Non-Chicago Creditors (%)	24.0	33.7	22.4	29.7	27.7

SOURCE: Compiled from 97 cases taken from Record Group 21, Records of the United States District Court, Northern District of Illinois, Eastern Division at Chicago, Bankruptcy Records, Act of 1867, 1867–1878; Bankruptcy Case Files, 1867–1878; and Bankruptcy Records, Act of 1898, 1898–1972; Bankruptcy Case Files, 1898–1946.

reason probably is the variation in the records upon which the creditor lists were made; some firms kept extremely good records, while others did not. Nevertheless, all bankruptcy cases contained a list of creditors with several pieces of information. The most important for my study were the creditor's name and street address, the service or good supplied by the creditor, and the amount of money owed to the creditor. In nearly all cases, the dollar amount, creditor name, and address were provided, although only the city and not the street address were sometimes given.

The lack of a street address did not present a problem with non-Chicago firms, as the city was the appropriate scale for firms outside of the metropolitan area. But it was an obstacle to locating a firm's street address within Chicago. Another problem related to the product or service supplied by the creditor. In some cases, the information was simply lacking. But even in those cases where the product or service delivered by the creditors was given, the description lacked specificity and, with a few exceptions, was extremely vague. Rather than a specific description, such as machine tool, wooden crates, or legal fees, the most common labels given were "goods," "merchandise," or "open account." As this study seeks to determine the character of inter-firm networks, it was necessary to ascertain what these vague terms referred to. I was able to obtain the product and street address, rather successfully it so happens, by linking the creditor listings with the creditors' products, services, and addresses. The two major sources were the *Chicago City Directory* (1870–1928) and the *Thomas Register of American Manufacturers* (1906–1930). The links between the bankruptcy records and these sources enabled me to specify the product or service in 90 percent of all cases and addresses in 94 percent of all cases.

Another issue concerning the usefulness of bankruptcy records for delineating the character of nineteenth and early twentieth-century firm networks is the idea that failed firms are not representative of ones that did not go bankrupt, and that therefore the case files tell us very little about firm interactions. The idea may have some basis. Firms may have failed because of the number and quality of interactions. William Cronon, for example, argues that as bankruptcy is not a random event, it hits those who are already in economic trouble. Thus, the "relations between bankrupt debtors and creditors are probably atypical of the population as a whole."[3] In his study of nineteenth-century Chicago's regional linkages, Cronon dealt with this problem by only analyzing firms that filed for bankruptcy during a national economic crisis. The rationale was that these firms were more likely to be typical than firms that went bankrupt during good economic times.

While there is a germ of truth to what Cronon has to say, the bankrupt records provide very useful material for understanding the urban economy. As William Mollard notes in his study of the upper Illinois industrial corridor, despite the unevenness of their information, bankruptcy records do present a representative view of a regional economy. Other studies reinforce Mollard's point. Bankruptcy was not that unusual—most firms went out of business, whether as a bankrupt or not. According to Edward Balleisen, "At least one in three and as perhaps as many as one in two eventually succumbed to an unsupportable load of debt" in antebellum America. A century later a 1933 study showed that firms failed in large numbers and for a variety of reasons, from "discernible errors in management" (the most common) to environmental conditions over which they had no control to personal and family reasons (such as illness).[4]

The suggestion that failed firms are representative—not atypical—of industrial capitalist is supported elsewhere. In his discussion of automobile parts manufacture, one writer made the point that the nature of a capitalist economy ensured that most firms would not be successful. Referring to the early 1920s economic depression, the writer noted that "there is apprehension lest some smaller automobile companies may fail to weather the storm. . . . There is a considerable number of companies . . . which are virtually in the hands of their creditors. Every effort will be made to save them, but some have reached the limit of their resources."[5] While the number of bankrupts may increase during economic downturns, there are always firms that, despite their best efforts, fail because they are squeezed out of the industrial market and cannot survive the rigors of a competitive environment. This does not make them atypical; rather, they are typical of most industrial enterprises. In other words, the bankruptcy records, while not presenting the gamut of firms at any one time, especially the large firms, do provide a workable cross section of the urban economy.

A more practical reason for using the bankruptcy records is that they are, as far as I know, the best extant source for delineating business linkages. Historians typically use company records to understand how manufacturing firms operate. These, however, are not as useful as bankruptcy records for a study of metropolitan business linkages and geography. While unrivaled in their insight of an individual firm, there are few company records for any one city. Nor can company records, unlike bankruptcy ones, yield a systematic picture of the inter-firm links operating in one urban setting. Even if such company records were to exist, it would be a herculean task to collect and analyze the same number of business records from individual firms as is possible from the bankruptcy cases. Moreover, unlike the bankruptcy records, the typical company record rarely provides a summary amount of money owed to creditors. Even if an acceptable number of company records were to exist, the work to collate the hundreds if not thousands of invoices for any one firm into a manageable whole would be too time-consuming. This is the position taken by Mollard, who states that "in weighing the drawbacks of bankruptcy records as reliable sources of information against the wealth of data the cases present, the latter emerges as being well worth the investigation."[6] If the intent is to portray a broad picture of one urban place's business linkages, then the bankruptcy files are superior to company records.

INTRODUCTION

1. "City's Industrial Centers Visited by Council and Association Committees in First Inspection Trip of the Year," *Chicago Commerce* (July 21, 1911): 11.

2. Ibid.

3. Ibid., 11–12.

4. Sam Bass Warner, *The Private City: Philadelphia in Three Periods of Its Growth* (Philadelphia: University of Pennsylvania, 1968); Robert Fogelson, *The Fragmented Metropolis: Los Angeles, 1850–1930* (Cambridge, MA: Harvard University Press, 1967); Olivier Zunz, *The Changing Face of Inequality: Urbanization, Industrial Development, and Immigrants in Detroit, 1880–1920* (Chicago: University of Chicago Press, 1982); Edward Muller and Paul Groves, "The Changing Location of the Clothing Industry: A Link to the Social Geography of Baltimore in the Nineteenth Century," *Maryland Historical Magazine* 71 (1976): 403–20.

5. John Bodnar, *The Transplanted: The History of Immigrants in Urban America* (Bloomington: University of Indiana Press, 1987); David Ward, *Cities and Immigrants* (New York: Oxford University Press, 1971); David Montgomery, *Workers' Control in America: Studies in the History of Work, Technology, and Labor Struggles* (New York: Cambridge University Press, 1979). For Chicago, see James Barrett, *Work and Community in the Jungle: Chicago's Packinghouse Workers, 1894–1922* (Urbana: University of Illinois Press, 1987); Lizabeth Cohen, *Making a New Deal: Industrial Workers in Chicago, 1919–1939* (New York: Cambridge University Press, 1990); Perry Duis, *Challenging Chicago: Coping with Everyday Life, 1837–1920* (Urbana: University of Illinois Press, 1998); and Richard Schneirov, *Labor and Urban Politics: Class Conflict and the Origins of Modern Liberalism in Chicago, 1864–1897* (Urbana: University of Illinois Press, 1998).

6. Henry Binford, *The First Suburbs: The Residential Communities on the Boston Periphery, 1815–1860* (Chicago: University of Chicago Press, 1985); Sam Bass Warner, *Streetcar Suburbs: The Process of Growth in Boston, 1870–1900* (Cambridge, MA: Harvard University Press, 1962); Richard Harris, *Unplanned Suburbs: Toronto's American Tragedy, 1900 to 1950* (Baltimore: John Hopkins University Press, 1996); Alexander Von Hoffman, *Local Attachments: The Making of an American Urban Neighborhood, 1850 to 1920* (Baltimore: John Hopkins University Press, 1994).

7. Stanley Buder, *Pullman: An Experiment in Industrial Order and Community Planning, 1880–1930* (New York: Oxford University Press, 1967); Dominic Pacyga, *Polish Immigrants and Industrial Chicago: Workers on the South Side, 1880–1922* (Columbus: Ohio State University Press, 1991); Joseph Bigott, *From Cottage to Bungalow: Houses and the Working Class in Metropolitan Chicago, 1869–1929* (Chicago: University of

Chicago Press, 2001); Robert Lewis, *Manufacturing Montreal: The Making of an Industrial Landscape, 1850 to 1930* (Baltimore: John Hopkins University Press, 2000); Philip Scranton, *Proprietary Capitalism: The Textile Manufacture at Philadelphia, 1800–1885* (Cambridge: Cambridge University Press, 1983) and *Figured Tapestry: Production, Markets, and Power in Philadelphia Textiles, 1885–1941* (Cambridge: Cambridge University Press, 1989).

8. See the case studies in Robert Lewis, ed., *Manufacturing Suburbs: Building Work and Home on the Metropolitan Fringe* (Philadelphia: Temple University Press, 2004). For Chicago, see Buder, *Pullman*; Pacyga, *Polish Immigrants*; and Mary Beth Pudup, "Model City? Industry and Urban Structure in Chicago," in *Manufacturing Suburbs*, ed. Lewis, 53–75.

9. Barrett, *Work and Community*; Buder, *Pullman*; Cohen, *Making a New Deal*; Harold Mayer, *The Port of Chicago and the St. Lawrence Seaway* (Chicago: University of Chicago, 1957); Raymond Mohl and Neil Betten, *Steel City: Urban and Ethnic Patterns in Gary, Indiana, 1906–1950* (New York: Holmes and Meier, 1986); Harold Platt, *The Electric City: Energy and the Growth of the Chicago Area, 1880–1930* (Chicago: University of Chicago Press, 1991); Max White, *Water Supply Organization in the Chicago Region* (Chicago: University of Chicago Press, 1934).

10. David Meyer, "Emergence of the American Manufacturing Belt: An Interpretation," *Journal of Historical Geography* 9 (1983) 145–74; William Cronon, *Nature's Metropolis: Chicago and the Great West* (New York: Norton, 1991); Gordon Winder, "The North American Manufacturing Belt in 1880: A Cluster of Regional Industrial Systems or One Large Industrial District?" *Economic Geography* 75 (1999): 71–91, and "Building Trust and Managing Business over Distance: A Geography of Reaper Manufacturer D. S. Morgan's Correspondence, 1867," *Economic Geography* 77 (2001): 95–121.

11. John Appleton, *The Iron and Steel Industry of the Calumet District: A Study in Economic Geography* (Urbana: University of Illinois Press, 1927); Barrett, *Work and Community*; Buder, *Pullman*; Robert Cramer, *Manufacturing Structure of the Cicero District, Metropolitan Chicago* (Chicago: University of Chicago, Department of Geography Research Paper No. 27, 1952); Homer Hoyt, *One Hundred Years of Land Values in Chicago: The Relationship of the Growth of Chicago to the Rise of Its Land Values, 1830–1933* (Chicago: University of Chicago Press, 1933); William Mitchell, *Trends in Industrial Location in the Chicago Region since 1920* (Chicago: University of Chicago Press, 1933); Pacyga, *Polish Immigrants*; Thomas Philpott; *The Slum and the Ghetto: Neighborhood Deterioration and Middle-Class Reform, Chicago, 1880–1930* (New York: Oxford University Press, 1978); Graham Taylor, *Satellite Cities: A Study of Industrial Suburbs* (New York: Appleton, 1915); Louise Carroll Wade, *Chicago's Pride: The Stockyards, Packingtown, and Environs in the Nineteenth Century* (Urbana: University of Illinois Press, 1987).

12. Harold Mayer and Richard Wade, *Chicago: Growth of a Metropolis* (Chicago: University of Chicago Press, 1969); Ann Durkin Keating, *Building Chicago: Suburban Developers and the Creation of a Divided Metropolis* (Columbus: Ohio State University Press, 1988) and *Chicagoland: City and Suburbs in the Railroad Age* (Chicago: University of Chicago Press, 2005).

13. Barrett, *Work and Community*; Buder, *Pullman*; Cohen, *Making a New Deal*; Duis, *Challenging Chicago*; Rick Halpern, *Down on the Killing Floor: Black and White Workers in Chicago's Packinghouses, 1904–1954* (Urbana: University of Illinois Press, 1997); Robert Ozannne, *A Century of Labor: Management Relations at McCormick and International Harvester* (Madison: University of Wisconsin Press, 1967); Pacyga, *Polish Immigrants*; Schneirov, *Labor and Urban Politics*; Wade, *Chicago's Pride*.

14. Alfred Chandler, *The Visible Hand: The Managerial Revolution in American Business* (Cambridge, MA: Belknap, 1977) and *Scale and Scope: The Dynamics of Industrial Capitalism* (Cambridge, MA: Belknap, 1990).

15. David Gordon, "Capitalist Development and the History of the American Cities," in *Marxism and the Metropolis: New Perspectives in Urban Political Economy*, ed. William Tabb and Larry Sawers (New York: Oxford University Press, 1978), 25–63; Allen Scott, "Locational Patterns and Dynamics of Industrial Activity in the Modern Metropolis," *Urban Studies* 19 (1982): 111–42. Also see Michael Conzen, "The Progress of American Urbanism, 1860–1930," in *North America: The Historical Geography of a Changing Continent*, ed. Robert Mitchell and Paul Groves (Totowa, NJ: Rowman and Littlefield, 1987), 363–64; Raymond Fales and Leon Moses, "Land-Use Theory and the Spatial Structure of the Nineteenth-Century City," *Papers of the Regional Science Association* 28 (1972): 49–80; and Richard Walker, "The Transformation of Urban Structure in the Nineteenth Century and the Beginnings of Suburbanization," in *Urbanization and Conflict in Market Societies*, ed. Kevin Cox (Chicago: Maaroufa Press, 1979), 165–212.

16. David Hounshell, *From the American System to Mass Production, 1800–1932: The Development of Manufacturing Technology in the United States* (Baltimore: John Hopkins University Press, 1984); John Ingham, *Making Iron and Steel: Independent Mills in Pittsburgh, 1820–1920* (Columbus; Ohio State University Press, 1991); Robert Lewis, "Productive and Spatial Strategies in the Montreal Tobacco Industry, 1850–1918," *Economic Geography* 70 (1994): 370–89; Charles Sabel and Jonathan Zeitlin, "Historical Alternatives to Mass Production: Politics, Market and Technology in Nineteenth-Century Industrialization," *Past and Present* 108 (1985): 133–76; Philip Scranton, *Endless Novelty: Specialty Production and American Industrialization, 1865–1925* (Princeton, NJ: Princeton University Press, 1997); Richard Walker, "The Geographical Organization of Production-Systems," *Environment and Planning D* 6 (1988): 377–408; John Brown, *The Baldwin Locomotive Works, 1831–1915: A Study in American Industrial Practice* (Baltimore: John Hopkins University Press, 1995).

17. David Meyer, "Formation of Advanced Technology Districts: New England Textile, Machinery and Firearms, 1790–1820," *Economic Geography* Special Issue (1998): 31–45, and *The Roots of American Industrialization* (Baltimore: John Hopkins University Press, 2003); Scranton, *Proprietary Capitalism*.

18. Edward Muller, "Industrial Suburbs and the Growth of Metropolitan Pittsburgh, 1870–1920," in *Manufacturing Suburbs*, ed. Lewis, 124–42; Fred Viehe, "Black Gold Suburbs: The Influence of the Extractive Industry on the Suburbanization of Los Angeles, 1890–1930," *Journal of Urban History* 8 (1981): 3–26; Michael Schwartz and Andrew Fish, "Just-in-Time Inventories in Old Detroit," *Business History* 40 (1998): 48–71; Howell John Harris, "Getting It Together: The Metal Manufacturers'

Association of Philadelphia, c. 1900–1930," in *Masters to Managers: Historical and Comparative Perspectives on American Employers*, ed. Sanford Jacoby (New York: Columbia University Press, 1991), 111–31; Philip Scranton, "Webs of Productive Association in American Industrialization: Patterns of Institution-Formation and Their Limits, Philadelphia, 1880–1930," *Journal of Industrial History* 1 (1998): 9–34.

19. The quote is from Bennett Harrison, "Industrial Districts: Old Wine in New Bottles?" *Regional Studies* 26 (1992): 471. Also see Ash Amin, "Industrial Districts," in *A Companion to Economic Geography*, ed. Eric Sheppard and Trevor Barnes (Oxford: Blackwell, 2000), 149–68; Manuel Castells and Peter Hall, *Technopoles of the World: The Making of 21st Century Industrial Complexes* (London: Routledge, 1994); Giacomo Becattini, "The Marshallian Industrial District as Socio-Economic Notion," in *Industrial Districts and Inter-Firm Cooperation in Italy*, ed. Frank Pyke, Giacomo Becattini, and Werner Sengenberger (Geneva: International Institute for Labour Studies, 1990), 37–51; Bennett Harrison, *Lean and Mean: The Changing Landscape of Corporate Power in the Age of Flexibility* (New York: Basic Books, 1994); and Michael Enright, "Organization and Coordination in Geographically Concentrated Industries," in *Coordination and Information: Historical Perspectives on the Organization of Enterprise*, ed. Naomi Lamoreaux and Daniel Raff (Chicago: University of Chicago Press, 1995), 103–42.

20. Mia Gray, Elyse Golub, and Ann Markusen, "Big Firms, Long Arms, Wide Shoulders: The 'Hub-and-Spoke' Industrial District in the Seattle Region," *Regional Studies* 30 (1996): 651–66; Ann Markusen, "Sticky Places in Slippery Space: A Typology of Industrial Districts," *Economic Geography* 72 (1996): 293–313; Castells and Hall, *Technopoles of the World*; Sabel and Zeitlin, "Historical Alternatives to Mass Production"; AnnaLee Saxenian, *Regional Advantage: Culture and Competition in Silicon Valley and Route 128* (Cambridge, MA: Harvard University Press, 1994); Margaret O'Mara, *Cities of Knowledge: Cold War Science and the Search for the Next Silicon Valley* (Princeton, NJ: Princeton University Press, 2005).

21. Meyer, "Emergence of the American Manufacturing Belt"; Cronon, *Nature's Metropolis*; Winder, "The North American Manufacturing Belt in 1880" and "Building Trust and Managing Business."

22. The quote is from Winder, "The North American Manufacturing Belt," 89. Brian Page and Richard Walker, "From Settlement to Fordism: The Agro-Industrial Revolution in the American Midwest," *Economic Geography* 67 (1991): 281–315; Cronon, *Nature's Metropolis*; Meyer, "Emergence of the American Manufacturing Belt"; David Meyer, "The National Integration of Regional Economies, 1860–1920," in *North America*, ed. Mitchell and Groves, 321–46.

23. Muller, "Industrial Suburbs."

24. The production chain is also known as the commodity chain and the supply chain. Most studies of these chains focus on their workings at the global scale. Few, if any, consider how they operate within metropolitan areas. Gary Gereffi and Miguel Korzeniewicz, eds., *Commodity Chains and Global Capitalism* (Westport, CT: Praeger, 1994); Jeffrey Henderson, Peter Dicken, Martin Hess, Neil Cole, and Henry Wai-Chung Yeung, "Global Production Networks and the Analysis of Economic Development," *Review of International Political Economy* 9 (2002): 436–64; Deborah

Leslie and Suzanne Reimer, "Spatializing Commodity Chains," *Progress in Human Geography* 23 (1999): 401–20.

25. Michael Storper and Richard Walker, *The Capitalist Imperative: Territory, Technology, and Industrial Growth* (New York: Blackwell, 1989), 53–54; Scranton, *Endless Novelty*; Harrison, *Lean and Mean*.

26. Storper and Walker, *The Capitalist Imperative*, 79, 128–32; Harrison, *Lean and Mean*; John Parr, "Missing Elements in the Analysis of Agglomeration Economies," *International Regional Science Review* 25 (2002): 151–68.

27. Gernot Grabher, "Rediscovering the Social in the Economics of Interfirm Relations," in *The Embedded Firm*, ed. Gernot Grabher (London: Routledge, 1993), 1–31; Peter Maskell and Anders Malmberg, "The Competitiveness of Firms and Regions: 'Ubiquitification' and the Importance of Localized Learning," *European Urban and Regional Studies* 6 (1999): 9–25. For a discussion of locational assets in a historical context, see Lewis, *Manufacturing Montreal*.

28. Platt, *The Electric City*; White, *Water Supply Organization*; Mayer, *The Port of Chicago*; James Putnam, *The Illinois and Michigan Canal: A Study in Economic History* (Chicago: University of Chicago Press, 1918).

29. Keating, *Building Chicago* and *Chicagoland*; Bigott, *From Cottage to Bungalow*; Hoyt, *One Hundred Years*; Christine Meisner Rosen, *The Limits of Power: Great Fires and the Process of City Growth in America* (Cambridge: Cambridge University Press, 1986); Michael Ebner, *Creating Chicago's North Shore: A Suburban History* (Chicago: University of Chicago Press, 1988); Mohl and Betten, *Steel City*.

30. "Representatives of Association and City Council Inspect Factory Districts—See Great Future for City," *Chicago Commerce* (July 28, 1911): 19.

CHAPTER ONE

1. Elmer Riley, "The Development of Chicago and Vicinity as a Manufacturing Center Prior to 1880" (Ph.D. diss., University of Chicago, 1911), 132.

2. Harvey Land Association, *The Town of Harvey, Illinois: Manufacturing Suburb of Chicago* (Chicago: Harvey Land Association, 1892), 6.

3. Michael Storper and Richard Walker, *The Capitalist Imperative: Territory, Technology, and Industrial Growth* (New York: Blackwell, 1989).

4. Local businesspeople and politicians typically defined metropolitan Chicago before World War II as consisting of six countries, five in Illinois (Cook, Du Page, Kane, Lake, and Will) and one in Indiana (Lake). But definitions varied, usually based on the interests of the parties involved. The *Plan of Chicago* of 1909 defined Chicago as extending into Indiana and Wisconsin. The Chicago Regional Association defined the metropolis as the surrounding "tributary" area of 15 counties within 50 miles of the Loop, while another group defined it as being made up of 204 cities and villages, 280 special districts, 350 police forces, and 978 school districts. Regardless, Chicago's major functions that previously had operated as city-based ones increasingly became metropolitan ones. Helen Monchow, *Seventy Years of Real Estate Subdividing in the Region of Chicago* (Evanston, IL: Northwestern University, Studies in the Social Sciences, No. 3, 1939), 7; Charles Merriam, Spencer Parratt, and Albert

Lepawsky, *The Government of the Metropolitan Region of Chicago* (Chicago: University of Chicago Press, 1933).

5. The central city was also the home to the mass of Chicago's wholesaling employment. The vast majority of the metropolitan area's 7,200 wholesaling establishments, 143,381 workers, and $6 billion sales in 1930 were in the central city. United States, Bureau of the Census, *Fifteenth Census of the United States: 1930 Distribution*, (Washington, DC: Government Printing Office, 1933), 2:13, 19–20.

6. Michael Conzen, "The Progress of American Urbanism, 1860–1930," in *North America: The Historical Geography of a Changing Continent*, ed. Robert Mitchell and Paul Groves (Totowa, NJ: Rowman and Little, 1987), 347–70; David Meyer, "The National Integration of Regional Economies, 1860–1920," in *North America*, ed. Mitchell and Groves, 321–46; Allan Pred, *City-Systems in Advanced Economies* (New York: Halsted Press, 1977).

7. William Cronon, *Nature's Metropolis: Chicago and the Great West* (New York: Norton, 1991); Ann Durkin Keating, *Building Chicago: Suburban Developers and the Creation of a Divided Metropolis* (Columbus: Ohio University Press, 1989), 180–87.

8. Lizabeth Cohen, *Making a New Deal: Industrial Workers in Chicago, 1919–1939* (New York: Cambridge University Press, 1990); Cronon, *Nature's Metropolis*; Stephen Adams and Orville Butler, *Manufacturing the Future: A History of Western Electric* (Cambridge: Cambridge University Press, 1999); James Barrett, *Work and Community in the Jungle: Chicago's Packinghouse Workers, 1894–1922* (Urbana: University of Illinois Press, 1987); Stanley Buder, *Pullman: An Experiment in Industrial Order and Community Planning, 1880–1930* (New York: Oxford University Press, 1967).

9. Louise Carroll Wade, *Chicago's Pride: The Stockyards, Packingtown, and Environs in the Nineteenth Century* (Urbana: University of Illinois Press, 1987); Kenneth Warren, *The American Steel Industry, 1850–1970: A Geographical Interpretation* (Pittsburgh: University of Pittsburgh Press, 1973), 205; Kenneth Warren, *Big Steel: The First Century of the United States Steel Corporation, 1910–2001* (Pittsburgh: University of Pittsburgh Press, 2001), 358; U.S. Department of Commerce, *Biennial Census of the Manufacturers: 1937*, part I (Washington, DC: Government Printing Office, 1939), 1553–63.

10. Philip Scranton, *Endless Novelty: Specialty Production and American Industrialization, 1865–1925* (Princeton, NJ: Princeton University Press, 1997); Conzen, "The Progress of American Urbanism"; Meyer, "The National Integration of Regional Economies."

11. Scranton, *Endless Novelty*; John Ingham, *Making Iron and Steel: Independent Mills in Pittsburgh, 1820–1920* (Columbus: Ohio State University Press, 1991), 1–11; Paul Hirst and Jonathan Zeitlin, "Flexible Specialization versus Post-Fordism: Theory, Evidence and Policy Implications," *Economy and Society* 20 (1991): 1–56; Richard Walker, "The Geographical Organization of Production-Systems," *Environment and Planning D* 6 (1988): 377–408.

12. *Directory of Illinois Manufacturers, 1924–1925* (Chicago: Illinois Manufacturers' Association, 1924).

13. Cronon, *Nature's Metropolis*; David Hounshell, *From the American System to Mass Production, 1800–1932: The Development of Manufacturing Technology in the United*

States (Baltimore: John Hopkins University Press, 1984); Keating, *Building Chicago*; David Meyer, "Emergence of the American Manufacturing Belt: An Interpretation," *Journal of Historical Geography* 9 (1983): 145–74; Meyer, "The National Integration of Regional Economies"; Brian Page and Richard Walker, "From Settlement to Fordism: The Agro-Industrial Revolution in the American Midwest," *Economic Geography* 67 (1991): 281–315; Storper and Walker, *The Capitalist Imperative*.

14. A. T. Andreas, *History of Chicago: From the Earliest Period to the Present Time*, 3 vols. (Chicago: Andreas Company, 1885), vol. 2; Josiah Currey, *Manufacturing and Wholesale Industries of Chicago*, 3 vols. (Chicago: Thomas Poole, 1918), 3:154–58; James Grossman, Ann Durkin Keating, and Janice Reiff, eds., *The Encyclopedia of Chicago* (Chicago: University of Chicago Press, 2004), 911–12, 940–43, 951–52.

15. United States, Bureau of the Census, *Fourteenth Census of the United States: 1920, Population* (Washington, DC: Government Printing Office, 1922), 3:261–69, 297–99.

16. Louis Cain, "The Sanitary District of Chicago: A Case Study of Water Use and Conservation" (Ph.D. diss., Northwestern University, 1969); Keating, *Building Chicago*; Harold Mayer, *The Port of Chicago and the St. Lawrence Seaway* (Chicago: University of Chicago, 1957); James Putnam, *The Illinois and Michigan Canal: A Study in Economic History* (Chicago: University of Chicago Press, 1918); Max White, *Water Supply Organization in the Chicago Region* (Chicago: University of Chicago Press, 1934); Harold Mayer, "The Railway Pattern of Metropolitan Chicago" (Ph.D. diss., University of Chicago, 1943).

17. "Industries Form Groups as City Grows," *Chicago Commerce* (April 14, 1923): 15.

18. Ann Durkin Keating, *Chicagoland: City and Suburbs in the Railroad Age* (Chicago: University of Chicago Press, 2005), 65–92; Graham Taylor, *Satellite Cities: A Study of Industrial Suburbs* (New York: Appleton, 1915). Also see various chapters in Robert Lewis, ed., *Manufacturing Suburbs: Building Work and Home on the Metropolitan Fringe* (Philadelphia: Temple University Press, 2004).

19. Homer Hoyt, *One Hundred Years of Land Values in Chicago: The Relationship of the Growth of Chicago to the Rise of Its Land Values, 1830–1933* (Chicago: University of Chicago Press, 1933), 50, 65–66; Andreas, *History of Chicago*, 2:673–98.

20. The data for the 1881 analysis is taken from *Lakeside Annual Business Directory for 1881* (Chicago, 1881). The directory gives the name and address of all businesses listed alphabetically by product. The tables show the firms listed under product headings, which suggested that the firms were actually manufacturers. Companies from headings that were obviously wholesale or retail were not included. In some cases, it can be very difficult to decipher whether a firm was a manufacturer or not. Industrial material from city or business directories fall victim to the plague of residential directories: smaller firms are more likely to be missing than larger ones. Finally, the directories only provide information about products and address. They tell us nothing about firms' scale, capital inputs, raw materials, and employment structure.

21. Christine Meisner Rosen, *The Limits of Power: Great Fires and the Process of City Growth in America* (Cambridge: Cambridge University Press, 1986), 152–61; Raymond Fales and Leon Moses, "Land-Use Theory and the Spatial Structure of

the Nineteenth-Century City," *Papers of the Regional Science Association* 28 (1972): 49–80.

22. Hoyt, *One Hundred Years*, 95–96; Rosen, *The Limits of Power*, 141–60; S. Schoff, *The Industrial Interests of Chicago* (Chicago: Knight and Leonard, 1873); Wade, *Chicago's Pride*; Dominic Pacyga and Ellen Skerrett, *Chicago: City of Neighborhoods* (Chicago: Loyola University Press, 1986); Andreas, *History of Chicago*, 2:673–98.

23. Keating, *Chicagoland*; Hoyt, *One Hundred Years*, 65–66, 95–96, 201–19, 230; Schoff, *The Industrial Interests of Chicago*; Wade, *Chicago's Pride*; Pacyga and Skerrett, *Chicago*; Andreas, *History of Chicago*, vols. 2 and 3; Monchow, *Seventy Years of Real Estate Subdividing*, 100–103; Taylor, *Satellite Cities*.

24. Keating, *Chicagoland*; Harry Jebsen, "Blue Island, Illinois: The History of a Working-Class Suburb" (Ph.D. diss., University of Cincinnati, 1971); E. C. Alft, *Elgin: An American History, 1835–1985* (Elgin: Daily-Courier, 1984); United States Census, *Report of the Social Statistics of Cites* (Washington, DC: Government Printing Office, 1887), 477–80, 521–24; Michael Conzen, "The Making of an Industrial Corridor," in *The Industrial Revolution in the Upper Illinois Valley*, ed. Michael Conzen, Glenn Richard, and Carl Zimring (Chicago: University of Chicago, Studies on the Illinois and Michigan Canal Corridor, No. 6, 1993), 1–16; Teddy Kim, "The Industrialization of Joliet: From Stone to Steel, 1850–1920," in *The Industrial Revolution*, ed. Conzen, Richard, and Zimring, 63–70.

25. Robert Lewis, "The Changing Fortunes of American Central-City Manufacturing, 1870–1950," *Journal of Urban History* 28 (2002): 580–85.

26. Hoyt, *One Hundred Years*, 201–19, 230; Barrett, *Work and Community*; Buder, *Pullman*; Taylor, *Satellite Cities*.

27. Alft, *Elgin*; U.S. Census, *Report of the Social Statistics of Cites*; Conzen, "The Making of an Industrial Corridor"; Kim, "The Industrialization of Joliet."

28. Monchow, *Seventy Years of Real Estate Subdividing*; Dominic Pacyga, *Polish Immigrants and Industrial Chicago: Workers on the South Side, 1880–1922* (Columbus: Ohio State University Press, 1991), 63–64; Raymond Mohl and Neil Betten, *Steel City: Urban and Ethnic Patterns in Gary, Indiana, 1906–1950* (New York: Holmes and Meier, 1986).

29. Keating, *Chicagoland*, 7; Hoyt, *One Hundred Years*, 137–38.

30. On annexation, see Kenneth Jackson, *Crabgrass Frontier: The Suburbanization of the United States* (New York: Oxford University Press, 1985); and Jon Teaford, *City and Suburb: The Political Fragmentation of Metropolitan America, 1850–1970* (Baltimore: John Hopkins University Press, 1979). For Chicago, see Lewis, "The Changing Fortunes of American Central-City Manufacturing"; and Harold Mayer and Richard Wade, *Chicago: Growth of a Metropolis* (Chicago: University of Chicago Press, 1969), 177.

31. For the quote, see Monchow, *Seventy Years of Real Estate Subdividing*, 89. Michael Ebner, *Creating Chicago's North Shore: A Suburban History* (Chicago: University of Chicago Press, 1988); Hoyt, *One Hundred Years*; Keating, *Building Chicago*; Barbara Posadas, "Suburb into Neighborhood: The Transformation of Urban Identity on Chicago's Periphery—Irving Park as a Case Study, 1870–1910," *Journal of the Illinois State Historical Society* 76 (1983): 163–76; Edward Greer, *Big Steel: Black Politics and*

Corporate Power in Gary, Indiana (New York: Monthly Review Press, 1979); Mohl and Betten, *Steel City.*

32. For the social reformers, see Edith Abbott, *The Tenements of Chicago, 1908–1935* (Chicago: University of Chicago Press, 1936); and Robert Hunter, *Tenement Conditions in Chicago: Report by the Investigating Committee of the City Homes Association* (Chicago: City Homes Association, 1901). Other important studies include Barrett, *Work and Community*; Joseph Bigott, *From Cottage to Bungalow: Houses and the Working Class in Metropolitan Chicago, 1869–1929* (Chicago: University of Chicago Press, 2001); Pacyga, *Polish Immigrants*; Buder, *Pullman*; Thomas Philpott, *The Slum and the Ghetto: Neighborhood Deterioration and Middle-Class Reform, Chicago, 1880–1930* (New York: Oxford University Press, 1978); Harvey Zorbaugh, *The Gold Coast and the Slum: A Sociological Study of Chicago's Near North Side* (Chicago: University of Chicago Press, 1929).

33. David Bensman and Roberta Lynch, *Rusted Dreams: Hard Times in a Steel Community* (New York: McGraw-Hill, 1987); Keating, *Chicagoland*; Posadas, "Suburb into Neighborhood"; Pacyga, *Polish Immigrants*; Barrett, *Work and Community*; Buder, *Pullman*; Wade, *Chicago's Pride*; Philpott, *The Slum and the Ghetto*; Zorbaugh, *The Gold Coast and the Slum.*

34. John Appleton, *The Iron and Steel Industry of the Calumet District: A Study in Economic Geography* (Urbana: University of Illinois Press, 1927), 92–94; Bigott, *From Cottage to Bungalow*; Dominic Candelero, "Suburban Italians: Chicago Heights, 1890–1975," in *Ethnic Chicago*, ed. Peter Jones and Melvin Holli (Grand Rapids, MI: W.B. Eerdmans, 1981), 180–209; Bensman and Lynch, *Rusted Dreams*; Keating, *Chicagoland*; Jebsen, "Blue Island," 116–48; Monchow, *Seventy Years of Real Estate Subdividing*, 88–118; Pacyga, *Polish Immigrants*; Buder, *Pullman*; Mohl and Betten, *Steel City.*

35. Bigott, *From Cottage to Bungalow*; James Walker, "Planning for the Future of East Chicago, Indiana: General Survey of Its Social and Economic Problems" (East Chicago: Chamber of Commerce of East Chicago, 1926); Candelero, "Suburban Italians"; Keating, *Chicagoland*; Jebsen, "Blue Island"; Monchow, *Seventy Years of Real Estate Subdividing*, 88–118; Pacyga, *Polish Immigrants*; Buder, *Pullman*; Mohl and Betten, *Steel City.*

36. Zorbaugh, *The Gold Coast and the Slum*; Ebner, *Creating Chicago's North Shore*; Jebsen, "Blue Island"; Cohen, *Making a New Deal*; Robert Cramer, *Manufacturing Structure of the Cicero District, Metropolitan Chicago* (Chicago: University of Chicago, Department of Geography Research Papers No. 27, 1952); Candelero, "Suburban Italians"; Philpott, *The Slum and the Ghetto.*

CHAPTER TWO

1. George Plumbe, *Chicago, the Great Industrial and Commercial Center of the Mississippi Valley* (Chicago: Chicago Association of Commerce, 1912), 48; William Cronon, *Nature's Metropolis: Chicago and the Great West* (New York: Norton, 1991).

2. "Civic-Industrial Committee and Local Industries Committee of City Council Make Eleventh Industrial Excursion," *Chicago Commerce* (July 19, 1912): 15–17.

3. For quote, see "Representatives of Association and City Council Inspect Factory Districts—See Great Future for City," *Chicago Commerce* (July 28, 1911), 19. Henry Abbott, *Historical Sketch of the Confectionery Trade of Chicago* (Chicago: Jobbing Confectioners' Association, 1905), 72; Chicago Park District, "Beilfuss Playlot Park," www.chicagoparkdistrict.com (accessed August 16, 2007).

4. Harry Jebsen, "Blue Island, Illinois: The History of a Working-Class Suburb" (Ph.D. diss., University of Cincinnati, 1971); Powell Moore, *The Calumet Region: Indiana's Last Frontier* (Indianapolis: Indiana Historical Bureau, 1959); Phyllis Bate, "The Development of the Iron and Steel Industry of the Chicago Area, 1900–1920" (Ph. D. diss., University of Chicago, 1948).

5. *Our Suburbs: A Resumé of the Origin, Progress and Present Status of Chicago's Environs*, 3, reprint from *Chicago Times*, May 1873, Special Collections, Chicago Public Library.

6. For quote, see *Year Book of the Commercial, Banking and Manufacturing Interests of Chicago with a General Review of Its Business Progress, 1885–1886* (Chicago: Howe, 1885), 128–29. Harvey Land Association, *The Town of Harvey, Illinois: Manufacturing Suburb of Chicago* (Chicago: Harvey Land Association, 1892), 8, 10.

7. Cronon, *Nature's Metropolis*; Everett Hughes, *The Growth of an Institution: The Chicago Real Estate Board* (Chicago: Society of Social Research of the University of Chicago, 1931), series II, monograph no. 1; Helen Monchow, *Seventy Years of Real Estate Subdividing in the Region of Chicago* (Evanston, IL: Northwestern University Press, Studies in the Social Sciences, No. 3, 1939), 119–58. For a discussion of the relations between factory district growth and the real estate industry, see Richard Walker and Robert Lewis, "Beyond the Crabgrass Frontier: Industry and the Spread of North American Cities, 1850–1950," *Journal of Historical Geography* 27 (2001): 3–19.

8. Joseph Bigott, *From Cottage to Bungalow: Houses and the Working Class in Metropolitan Chicago, 1869–1929* (Chicago: University of Chicago Press, 2001), 59–67, 151–56; Monchow, *Seventy Years of Real Estate Subdividing*, 105–7; Ann Durkin Keating, *Building Chicago: Suburban Developers and the Creation of a Divided Metropolis* (Columbus: Ohio State University Press, 1988); Joseph Arnold, "Riverside, IL," in *The Encyclopedia of Chicago*, ed. James Grossman, Ann Durkin Keating, and Janice Reiff (Chicago: University of Chicago Press, 2004), 712.

9. Hughes, *The Growth of an Institution*; Monchow, *Seventy Years of Real Estate Subdividing*; Keating, *Building Chicago*.

10. Quote from Raymond Mohl and Neil Betten, *Steel City: Urban and Ethnic Patterns in Gary, Indiana, 1906–1950* (New York: Holmes and Meier, 1986), 24. Also see James Lane, *City of the Century: A History of Gary, Indiana* (Bloomington: Indiana University Press, 1978); and Graham Taylor, *Satellite Cities: A Study of Industrial Suburbs* (New York: Appleton, 1915).

11. *Year Book of the Commercial, Banking and Manufacturing Interests*, 130–31, 144–46; Monchow, *Seventy Years of Real Estate Subdividing*; Keating, *Building Chicago*; Carl Abbott, "Necessary Adjuncts to Its Growth: The Railroad Suburbs of Chicago, 1854–1875," *Journal of the Illinois State Historical Society* 73 (1988): 117–31; Richard Harris, "Chicago's Other Suburbs," *Geographical Review* 84 (1994): 394–410; Michael

Ebner, *Creating Chicago's North Shore: A Suburban History* (Chicago: University of Chicago Press, 1988); Homer Hoyt, *One Hundred Years of Land Values in Chicago: The Relationship of the Growth of Chicago to the Rise of Its Land Values, 1830–1933* (Chicago: University of Chicago Press, 1933); 137–38; Ann Durkin Keating, *Chicagoland: City and Suburbs in the Railroad Age* (Chicago: University of Chicago Press, 2005); Barbara Posadas, "Suburb into Neighborhood: The Transformation of Urban Identity on Chicago's Periphery—Irving Park as a Case Study, 1870–1910," *Journal of the Illinois State Historical Society* 76 (1983): 163–76.

12. *Year Book of the Commercial, Banking and Manufacturing Interests*, 130–31, 144–46; Monchow, *Seventy Years of Real Estate Subdividing*; Keating, *Building Chicago*; Abbott, "Necessary Adjuncts to Its Growth"; Harris, "Chicago's Other Suburbs"; Ebner, *Creating Chicago's North Shore*; Hoyt, *One Hundred Years of Land Values*, 137–38; Keating, *Chicagoland*.

13. *Year Book of the Commercial, Banking and Manufacturing Interests*, 130–31, 144–46; Monchow, *Seventy Years of Real Estate Subdividing*; Keating, *Building Chicago*; Bigott, *From Cottage to Bungalow*, 149–62; Abbott, "Necessary Adjuncts to Its Growth"; Harris, "Chicago's Other Suburbs"; Ebner, *Creating Chicago's North Shore*; Hoyt, *One Hundred Years of Land Values*, 137–38; Keating, *Chicagoland*; Posadas, "Suburb into Neighborhood."

14. Bigott, *From Cottage to Bungalow*; Keating, *Building Chicago*; Abbott, "Necessary Adjuncts to Its Growth"; Posadas, "Suburb into Neighborhood."

15. Harold Platt, *The Electric City: Energy and the Growth of the Chicago Area, 1880–1930* (Chicago: University of Chicago Press, 1991).

16. Keating, *Chicagoland*, 7–9, 64, 91; Harold Mayer, "The Railway Pattern of Metropolitan Chicago" (Ph.D. diss., University of Chicago, 1943); Mary Beth Pudup, "Model City? Industry and Urban Structure in Chicago," in *Manufacturing Suburbs: Building Work and Home on the Metropolitan Fringe*, ed. Robert Lewis (Philadelphia: Temple University Press, 2004), 53–75.

17. Mayer, "The Railway Pattern," 27–41, 67–68; Committee on Co-ordination of Chicago Terminals, *The Freight Traffic of the Chicago Terminal District* (Chicago: privately published, 1927), 18–19.

18. Harold Mayer, *The Port of Chicago and the St. Lawrence Seaway* (Chicago: University of Chicago, 1957), 10–18; Chicago Harbor Commission, *Report to the Mayor and Alderman of the City of Chicago* (Chicago: Adair, 1909), 19–23.

19. Pudup, "Model City?"; Monchow, *Seventy Years of Real Estate Subdividing*, 119–58; Walker and Lewis, "Beyond the Crabgrass Frontier"; Platt, *The Electric City*.

20. Newspaper Clippings on the History of South Chicago, p. 11, Chicago Public Library, Special Collections, South Chicago Community Collection, box 1, file 19; box 3, file 16; and box 4, file 9; Ellen Weiss, "Americans in Paris: Two Buildings," *Journal of the Society of Architectural Historians* 45 (1986): 166–67; *The Story of Chicago Regional Port District, Port of Chicago, Lake Calumet Harbor: Mid-America Water-Rail-Truck Link with the World: "A City within a City"* (n.p., n.d. [CA, 1940s]); Calumet National Bank, *South Chicago: Its History and Progress* (Chicago: Calumet National Bank, 1927). Hoyt, *One Hundred Years of Land Values*, 108, 134–35; Monchow, *Seventy Years of Real Estate Subdividing*, 126; Dominic Pacyga, *Polish Immigrants and*

Industrial Chicago: Workers on the South Side, 1880–1922 (Columbus: Ohio State University Press, 1991), 30.

21. Calumet National Bank, *South Chicago*; Pacyga, *Polish Immigrants*, 30–31; Harold Mayer and Richard Wade, *Chicago: Growth of a Metropolis* (Chicago: University of Chicago Press, 1969), 186–92; Hoyt, *One Hundred Years of Land Values*, 134–35; Jim Martin, "South Chicago History Is Older than City," *Daily Calumet*, March 1, 1982, Chicago Public Library, Special Collections, South Chicago Community Collection, box 4, file 3; Mary Faith Adams, "Present Housing Conditions in South Chicago, South Deering and Pullman" (MA thesis, University of Chicago, 1926), 5–9.

22. Moore, *The Calumet Region*, 83–113; Chicago Harbor Commission, *Report to the Mayor*, 45–50; Committee on Co-ordination of Chicago Terminals, *The Freight Traffic of the Chicago Terminal District*, 15–27; Keating, *Chicagoland*; Mayer, "The Railway Pattern of Metropolitan Chicago," 27–41, 61–68; Mayer, *The Port of Chicago*, 13, 183–213; Henry Lee, "The Calumet Region as an Industrial Center," *Chicago, the Great Central Market Magazine* (July 1908): 71.

23. "Representatives of Association and City Council," 19.

24. Quote from Moore, *The Calumet Region*, 83. Also see Keating, *Building Chicago*, 11–32.

25. George Roberts, "History of West Pullman" (typescript, 1941), Chicago Public Library, Special Collections, Calumet Region Community Collection, box 6, file 7; Stanley Buder, *Pullman: An Experiment in Industrial Order and Community Planning, 1880–1930* (New York: Oxford University Press, 1967); Janice Reiff, "'His Statements . . . Will Be Challenged': Ethnicity, Gender, and Class in the Evolution of the Pullman/Roseland Area of Chicago, 1894–1917," *Mid-America* 74 (1992): 231–52; Carl Smith, *Urban Disorder and the Shape of Belief: The Great Chicago Fire, the Haymarket Bomb, and the Model Town of Pullman* (Chicago: University of Chicago Press, 1995).

26. Monchow, *Seventy Years of Real Estate Subdividing*, 119–58.

27. William Rowan, "South Chicago's History for First Hundred Years," *Daily Calumet*, June 6, 1926, Chicago Public Library, Special Collections, South Chicago Community Collection, box 4, file 5; "A Chicago District Steel Plant," *Iron Age* (May 16, 1912): 1226–27; Monchow, *Seventy Years of Real Estate Subdividing*.

28. Newspaper Clippings on the History of South Chicago, Chicago Public Library, Special Collections, South Chicago Community Collection, box 3, file 16.

29. Hughes, *The Growth of an Institution*, 18, 41–42.

30. "South Austin Problems," *Austinite*, March 14, 1924, in "Austin: Its People and History," vol. 1, Chicago Public Library, Special Collections, Austin Community Collection, box 4, file 6.

31. Paul Bourget, *Outre-Mer Impressions of America* (New York: Scribner, 1895), quoted in *As Others See Chicago: Impressions of Visitors, 1673–1933*, ed. Bessie Louise Pierce (Chicago: University of Chicago Press, 1933), 387.

32. George Steevens, *The Land of the Dollar* (New York: Dodd, Mead, 1897), quoted in *As Others See Chicago*, ed. Pierce, 399–400. For takes by other visitors, see Harold

Platt, *Shock Cities: The Environmental Transformation and Reform of Manchester and Chicago* (Chicago: University of Chicago Press, 2005).

33. Ernest Bicknell, "Problems of Philanthropy in Chicago," *Annals of the American Academy of Political and Social Science* 21 (1903): 379–88; George Hooker, "Congestion and Its Causes in Chicago," *Proceedings of the Second National Conference on City Planning* (Boston, 1910), 42–57; Louise Wade, *Graham Taylor, Pioneer for Social Justice, 1851–1938* (Chicago: University of Chicago Press, 1964); William Stead, *If Christ Came to Chicago!* (London: London Review of Books, 1894). Sophinisba Breckinridge and Edith Abbott made several surveys of Chicago's housing problems. For example, "Chicago's Housing Problem: Families in Furnished Rooms," *American Journal of Sociology* 16 (1910): 289–308; "Housing Conditions in Chicago, III: Back of the Yards," *American Journal of Sociology* 16 (1911), 433–68; "Chicago Housing Conditions, IV: The West Side Revisited," *American Journal of Sociology* 17 (1911): 1–34; and "Chicago Housing Conditions, V: South Chicago at the Gates of the Steel Mills," *American Journal of Sociology* 17 (1911), 145–76. Also see Edith Abbott, *The Tenements of Chicago, 1908–1935* (Chicago: University of Chicago Press, 1936).

34. For quote, see "Representatives of Association and City Council," 19. Recent studies of these issues include Maureen Flanagan, "Gender and Urban Political Reform: The City Club and the Women's City Club of Chicago in the Progressive Era," *American Historical Review* 95 (1990): 1032–50; Emily Gilbert, "Naturalist Metaphors in the Literatures of Chicago, 1893–1925," *Journal of Historical Geography* 20 (1994): 283–304; Thomas Philpott, *The Slum and the Ghetto: Neighborhood Deterioration and Middle-Class Reform, Chicago, 1880–1930* (New York: Oxford University Press, 1978); Harold Platt, "Creative Necessity: Municipal Reform in Gilded Age Chicago," in *The Constitution, Law, and American Life: Critical Aspects of the Nineteenth-Century Experience*, ed. Donald Nieman (Athens: University of Georgia Press, 1992), 162–90; and Harold Platt, "Jane Addams and the Ward Boss Revisited: Class, Politics and Public Health in Chicago, 1890–1930," *Environmental History* 5 (2000), 194–222. For the Chicago school, see Robert Park, Ernest Burgess, and Roderick McKenzie, eds., *The City* (Chicago: University of Chicago Press, 1925); and Harvey Zorbaugh, *The Gold Coast and the Slum: A Sociological Study of Chicago's Near North Side* (Chicago: University of Chicago Press, 1929).

35. Quote from Hoyt, *One Hundred Years of Land Values*, 50. Marcel De Meirleir, *Manufactural Occupance in the West Central Area of Chicago* (Chicago: University of Chicago, Department of Geography, Research Paper No. 11, 1950). Park, Burgess, and McKenzie, *The City*.

36. Thomas Bender, *Toward an Urban Vision: Ideas and Institutions in Nineteenth-Century America* (Lexington: University of Kentucky Press, 1975); Paul Boyer, *Urban Masses and Moral Order in America, 1820–1920* (Cambridge, MA: Harvard University Press, 1978); Allen Davis, *Spearheads for Reform: The Social Settlements and the Progressive Movement, 1890–1914* (New York: Oxford University Press, 1967); John Fairfield, *The Mysteries of the Great City: The Politics of Urban Design, 1877–1937* (Columbus: Ohio State University Press, 1993); Christopher Mele, "The Materiality of

Urban Discourse: Rational Planning in the Restructuring of the Early Twentieth-Century Ghetto," *Urban Affairs Review* 35 (2000): 628–48; Max Page, *The Creative Destruction of Manhattan, 1900–1940* (Chicago: University of Chicago Press, 1999); David Ward, *Poverty, Ethnicity and the American City, 1840–1925: Changing Conceptions of the Slum and the Ghetto* (New York: Cambridge University Press, 1989); Morton White and Lucia White, *The Intellectual versus the City: From Thomas Jefferson to Frank Lloyd Wright* (Cambridge, MA: Harvard University Press and MIT Press, 1962), 139–78.

37. "Executive Committee Approving Report of Housing Committee, Recommends Housing Survey of Chicago," *Chicago Commerce* (May 23, 1913): 9–11; "Housing Conditions in Chicago which Demand a Citywide Survey as Preliminary to a Reform Program," *Chicago Commerce* (June 13, 1913): 31–34. Also see Robert Hunter, *Tenement Conditions in Chicago: Report by the Investigating Committee of the City Homes Association* (Chicago: City Homes Association, 1901); Abbott, *The Tenements of Chicago*; and Philpott, *The Slum and the Ghetto.*

38. For quotes, see Daniel Burnham and Edward Bennett, *Plan of Chicago* (Chicago: The Commercial Club, 1909), 1, 4. Also see Art Institute of Chicago, *The Plan of Chicago: 1909–1979* (Chicago: Art Institute of Chicago, 1979); Daniel Bluestone, *Constructing Chicago* (New Haven, CT: Yale University Press, 1991), 194–204; and Fairfield, *The Mysteries of the Great City*, 119–24.

39. For quote, see James Ford, "Residential and Industrial Decentralization," in *City Planning*, ed. John Nolen (New York: Appleton, 1922), 337. *Proceedings of the Second National Conference on City Planning and the Problems of Congestion* (Boston, 1910); Edward Pratt, *Industrial Causes of Congestion of Population in New York City* (New York: Columbia University Studies in the Social Sciences, No. 109, 1911); Taylor, *Satellite Cities*. A recent study that understands suburban growth as a search for a "spatial fix" is Smith, *Urban Disorder*, 197–200.

40. For quote, see Robert Haig and Roswell McCrea, "Major Economic Factors in Metropolitan Growth and Arrangement," in *Regional Survey* (New York: Regional Plan of New York and Its Environs, 1927), 18. Fairfield, *The Mysteries of the Great City*; Edward Muller, "The Pittsburgh Survey and 'Greater Pittsburgh': A Muddled Metropolitan Geography," in *Pittsburgh Surveyed: Social Science and Social Reform in the Early Twentieth Century*, ed. Maurine Greenwald and Margo Anderson (Pittsburgh: University of Pittsburgh Press, 1996), 69–87; Anne Mosher, "'Something Better than the Best': Industrial Restructuring, George McMurty and the Creation of the Model Industrial Town of Vandergrift, Pennsylvania, 1883–1901," *Annals of the Association of American Geographers* 85 (1995): 84–107; Boyer, *Urban Masses and Moral Order*. For studies of how the machine metaphor permeated other aspects of urban life, see Lindy Biggs, *The Rational Factory: Architecture, Technology, and Work in America's Age of Mass Production* (Baltimore: John Hopkins University Press, 1996); and Terry Smith, *Making the Modern: Industry, Art, and Design in America* (Chicago: University of Chicago Press, 1993).

41. For quote, see Ford, "Residential and Industrial Decentralization," 337. Pratt, *Industrial Causes of Congestion*, 204. Also see Buder, *Pullman*; Mohl and Betten, *Steel City*; and Taylor, *Satellite Cities.*

42. Hooker, "Congestion and Its Causes," 55; Ruth Crocker, *Social Work and Social Order: The Settlement Movement in Two Industrial Cities, 1889–1930* (Urbana: University of Illinois Press, 1992); Davis, *Spearheads for Reform*; Rivka Lissak, *Pluralism and Progressives: Hull House and the New Immigrants, 1880–1919* (Chicago: University of Chicago Press, 1989); Kathleen McCarthy, *Noblesse Oblige: Charity and Philanthropy in Chicago, 1849–1929* (Chicago: University of Chicago Press, 1982).

43. The comments can be found in "City's Industrial Centers Visited by Council and Association Committees in First Inspection Trip of the Year," *Chicago Commerce* (July 21, 1911): 12.

CHAPTER THREE

1. Allen Scott, "Locational Patterns and Dynamics of Industrial Activity in the Modern Metropolis," *Urban Studies* 19 (1982): 111–42; David Gordon, "Capitalist Development and the History of the American Cities," in *Marxism and the Metropolis: New Perspectives in Urban Political Economy*, ed. William Tabb and Larry Sawers (New York: Oxford University Press, 1978), 25–63; Stanley Buder, *Pullman: An Experiment in Industrial Order and Community Planning, 1880–1930* (New York: Oxford University Press, 1967); Graham Taylor, *Satellite Cities: A Study of Industrial Suburbs* (New York: Appleton, 1915); Ann Durkin Keating, *Chicagoland: City and Suburbs in the Railroad Age* (Chicago: University of Chicago Press, 2005); Christine Meisner Rosen, *The Limits of Power: Great Fires and the Process of City Growth in America* (Cambridge: Cambridge University Press, 1986); John F. McDonald, *Employment Location and Industrial Land Use in Metropolitan Chicago* (Champaign, IL: Stipes, 1984).

2. Rosen, *The Limits of Power*, 3–85.

3. Albert Dickens, "Armor Square, Fuller Park, McKinley Park, Bridgeport, and New City," in Chicago Plan Commission, *Forty-four Cities in the City of Chicago* (Chicago: Chicago Plan Commission, 1942), 56–58; Dominic Pacyga, "Bridgeport," in *The Encyclopedia of Chicago*, ed. James Grossman, Ann Durkin Keating, and Janice Reiff (Chicago: University of Chicago Press, 2004), 92–93; Dominic Pacyga and Ellen Skerrett, *Chicago: City of Neighborhoods* (Chicago: Loyola University Press, 1986), 452–58; David Solzman, *The Chicago River* (Chicago: Wild Onion, 1998), 226–29; Louise Carroll Wade, *Chicago's Pride: The Stockyards, Packingtown, and Environs in the Nineteenth Century* (Urbana: University of Illinois Press, 1987), 15–17, 66–69.

4. Dickens, "Armor Square," 56–58; Pacyga and Skerrett, *Chicago*, 452–58; Solzman, *The Chicago River*, 226–29; Wade, *Chicago's Pride*, 15, 17, 66–69; A. T. Andreas, *History of Chicago: From the Earliest Period to the Present Time*, 3 vols. (Chicago: Andreas Company, 1886), 3:482, 492.

5. Andreas, *History of Chicago*, 3:368; Homer Hoyt, *One Hundred Years of Land Values in Chicago: The Relationship of the Growth of Chicago to the Rise of Its Land Values, 1830–1933* (Chicago: University of Chicago Press, 1933), 50, 65–66, 95–96; Bessie Louise Pierce, *A History of Chicago*, 3 vols. (New York: Alfred Knopf, 1957), 3:92–93; "Membership Engineers Survey Industrial Sections of Great Town Which

Association Strives to Upbuild," *Chicago Commerce* (July 24, 1914): 17–18; Pacyga and Skerrett, *Chicago*, 239–67.

6. Hines (Edward) Lumber Co., *50 Years of Edward Hines Lumber Co.* (Chicago: Edward Hines, 1942); Andreas, *History of Chicago*, 3:368; Hoyt, *One Hundred Years*, 50, 65–66, 95–96; Pierce, *A History of Chicago*, vol. 3; "Membership Engineers Survey Industrial Sections of Great Town"; Gabriela Arredondo, "Lower West Side," in *The Encyclopedia of Chicago*, ed. Grossman, Keating, and Reiff, 494–95.

7. Andreas, *History of Chicago*, 3:368; Pierce, *A History of Chicago*, 3:92–93.

8. William Cronon, *Nature's Metropolis: Chicago and the Great West* (New York: Norton, 1991), 159–80; Hines Lumber, *50 Years of Edward Hines Lumber Co.*

9. H. Thomas, "Re-Building of Chicago," in Andreas, *History of Chicago*, 3:54. Also see Andreas, *History of Chicago*, 3:365–66.

10. Don Fehrenbacher, *Chicago Giant: A Biography of Long John Wentworth* (Madison, WI: American Historical Research Center, 1957); Harold Mayer and Richard Wade, *Chicago: Growth of a Metropolis* (Chicago: University of Chicago Press, 1969), 29, 38–40; Donald Miller, *City of the Century: The Epic of Chicago and the Making of America* (New York: Simon and Schuster, 1996), 77–78, 98–101, 131; Pierce, *A History of Chicago*, , 2:14, 143, 150, 413; Wade, *Chicago's Pride*, 14, 19.

11. Edith Abbott, *The Tenements of Chicago, 1908–1935* (1936; repr., New York: Arno Press, 1970); James Barrett, *Work and Community in the Jungle: Chicago's Packinghouse Workers, 1894–1922* (Urbana: University of Illinois Press, 1987); Sophonisba Breckinridge and Edith Abbott, "Housing Conditions in Chicago, III: Back of the Yards," *American Journal of Sociology* 16 (1911): 433–68; Charles Bushnell, "Some Social Aspects of the Chicago Stock Yards," *American Journal of Sociology* 7 (1901): 289–330; Chicago Union Stock Yard and Transit Co., *History of the Yards* (n.p., n.d.); Dickens, "Armor Square"; J. Kennedy et al., *Wages and Family Budgets in the Chicago Stock Yards District* (Chicago: University of Chicago Press, No. 3 of the Chicago Stock Yards Community Study, 1914); Alice Miller, "Rents and Housing Conditions in the Stock Yards District of Chicago, 1923" (MA thesis, University of Chicago, 1923); Pacyga and Skerrett, *Chicago*, 451–77; Mary Beth Pudup. "Model City? Industry and Urban Structure in Chicago," in *Manufacturing Suburbs: Building Work and Home on the Metropolitan Fringe*, ed. Robert Lewis (Philadelphia: Temple University Press, 2004), 53–75; Robert Slayton, *Back of the Yards: The Making of a Local Democracy* (Chicago: University of Chicago Press, 1986); Wade, *Chicago's Pride*; Howard Wilson, *Mary McDowell, Neighbor* (Chicago: University of Chicago Press, 1928).

12. Breckinridge and Abbott, "Housing Conditions in Chicago, III," 435; Chicago Union Stock Yard, *History of the Yards*, 4, 12; Pudup, "Model City?" and "Packers and Reapers, Merchants and Manufacturers: Industrial Restructuring and Location in an Era of Emergent Capitalism" (MA thesis, University of California, Berkeley, 1983), 70–71.

13. For quote, see Chicago Union Stock Yard, *History of the Yards*, 12. Pacyga and Skerrett, *Chicago*, 464; Wade, *Chicago's Pride*, 47–60.

14. For quote, see Chicago Union Stock Yard, *History of the Yards*, 12. Keating, *Chicagoland*; Pudup, "Model City?" 68–73; Pacyga and Skerrett, *Chicago*, 464.

15. Pudup, "Packers and Reapers," 71–74; Wade, *Chicago's Pride*, 47–60.

16. Wade, *Chicago's Pride*, 98–101; Pudup, "Packers and Reapers," 80–81; Chicago Union Stock Yard, *History of the Yards*, 18.

17. Wade, *Chicago's Pride*, 98–101; Pudup, "Packers and Reapers," 80–81; Chicago Union Stock Yard, *History of the Yards*, 18.

18. Andreas, *History of Chicago*, 2:674–77, 3:480; Ulf Beijbom, *Swedes in Chicago: A Demographic and Social Study of the 1846–1880 Immigration* (Stockholm: Scandinavian University Books, 1971), 127–44; Pacyga and Skerrett, *Chicago*, 165–97; Elmer Riley, "The Development of Chicago and Vicinity as a Manufacturing Center Prior to 1880" (Ph.D. diss., University of Chicago, 1911), 99–109.

19. Andreas, *History of Chicago*, vols. 2 and 3; Beijbom, *Swedes in Chicago*; Pacyga and Skerrett, *Chicago*, 165–97; Christaine Harzig, "Chicago's German North Side, 1880–1900: The Structure of a Gilded Age Ethnic Neighborhood," in *German Workers in Industrial Chicago, 1850–1910: A Comparative Perspective*, ed. Harmut Keil and John Jentz (Dekalb: Northern Illinois University Press, 1983), 127–44.

20. Andreas, *History of Chicago*, 3:478–79; Josiah Currey, *Manufacturing and Wholesale Industries of Chicago*, 3 vols. (Chicago: Thomas Poole, 1918), 2:151–52.

21. "Crucible Steel Casting Co.," *Chicago Journal of Commerce* (May 16, 1888): 12; "Soft, Strong, Crucible Steel Castings," *Chicago Journal of Commerce* (August 15, 1888): 1; Beijbom, *Swedes in Chicago*, 127–44; Harzig, "Chicago's German North Side."

22. Kennedy et al., *Wages and Family Budgets*, 2–4; Wade, *Chicago's Pride*, 198–99.

23. Kennedy et al., *Wages and Family Budgets*, 2–4; *Manufacturers in Chicago and Metropolitan Area* (Chicago: Chicago Association of Commerce, 1940).

24. Powell Moore, *The Calumet Region: Indiana's Last Frontier* (Indianapolis: Indiana Historical Bureau, 1959), 131–39.

25. Moore, *The Calumet Region*, 131–39.

26. Cronon, *Nature's Metropolis*, 183–98, 203–205; "Cigar Box Industry Shifts to New Basis," *Chicago Commerce* (September 15, 1923): 15–16; Pierce, *A History of Chicago*, 3:92–93; "Membership Engineers Survey Industrial Sections of Great Town."

27. Gilbert Lacher, "Acme Steel Expands Cold Strip Capacity," *Iron Age* (January 31, 1924): 353; Simon Dekker, "The History of Roseland and Vicinity," (typescript, c. 1935), 108–10; "Acme Steel Company Organized in 1880," Newspaper Clippings on the History of South Chicago, ca. 1936, Chicago Public Library, Special Collections, South Chicago Community Collection, box 3, file 18.

28. U.S. Census, *Compendium of the Tenth Census (June 1, 1880)*, vol. 11 (Washington, DC: Government Printing Office, 1883); *Directory of Illinois Manufacturers, 1924–1925* (Chicago: Illinois Manufacturers' Association, 1924).

29. For quotes, see George Hooker, "Congestion and Its Causes in Chicago," in *Proceedings of the Second National Conference on City Planning* (Boston, 1910), 54; and William Mitchell, *Trends in Industrial Location in the Chicago Region since 1920* (Chicago: University of Chicago Press, 1933), 51, 62. Also see Taylor, *Satellite Cities*; Dickens, "Armor Square"; Moore, *The Calumet Region*, 241–44; Irving Cutler, *The Chicago-Milwaukee Corridor: A Geographic Study of Intermetropolitan Coalescence* (Evanston, IL: Department of Geography, Studies in Geography, No. 9, Northwestern University, 1965); James Kenyon, *The Industrialization of the Skokie Area*

(Chicago: University of Chicago, Department of Geography, Research Paper No. 33, 1954); Leo Reeder, "Industrial Location Trends in Chicago in Comparison to Population Growth," *Land Economics* 30 (1954): 177–82; Leo Reeder, "Industrial Deconcentration as a Factor in Rural-Urban Fringe Development," *Land Economics* 31 (1955): 275–80; and Martin Reinemann, "The Pattern and Distribution of Manufacturing in the Chicago Area," *Economic Geography* 36 (1960): 139–44.

30. For Crane, see Currey, *Manufacturing and Wholesale Industries of Chicago*, 2:258; and "Fiftieth Anniversary of the Crane Co.," *Machinery* 11 (August 1905): 620. For factory design principles, see Lindy Biggs, *The Rational Factory: Architecture, Technology, and Work in America's Age of Mass Production* (Baltimore: John Hopkins University Press, 1996); and Robert Lewis, "Redesigning the Workplace: The North American Flexible Factory in the Interwar Period," *Technology and Culture* 42 (2001): 665–84.

31. Andreas, *History of Chicago*, 2:680–81; Currey, *Manufacturing and Wholesaling Industries*, 2:253–59; "Fiftieth Anniversary of the Crane Co."

32. Andreas, *History of Chicago*, 2:680–81; Currey, *Manufacturing and Wholesaling Industries*, 2:253–59; "Fiftieth Anniversary of the Crane Co."

33. Andreas, *History of Chicago*, 2:680–81; Currey, *Manufacturing and Wholesaling Industries*, 2:253–59; "Fiftieth Anniversary of the Crane Co."

34. Moore, *The Calumet Region*, 178–215; Emmett Dedmon, *Challenge and Response: A Modern History of Standard Oil Company (Indiana)* (Chicago: Mobium, 1984), 7; "3,000 Mile Pipe Makes Chicago Oil Center," *Chicago Commerce* (February 10, 1923): 7–8, 12.

35. Currey, *Manufacturing and Wholesale Industries*, 2:10–13; Helen Monchow, *Seventy Years of Real Estate Subdividing in the Region of Chicago* (Evanston, IL: Northwestern University Press, Studies in the Social Sciences, No. 3, 1939), 100–103; Michael Conzen, "The Making of an Industrial Corridor," in *The Industrial Revolution in the Upper Illinois Valley*, ed. Michael Conzen, Glenn Richard, and Carl Zimring (Chicago: University of Chicago, Studies on the Illinois and Michigan Canal Corridor, No. 6, 1993), 1–16; Teddy Kim, "The Industrialization of Joliet: From Stone to Steel, 1850–1920," in *The Industrial Revolution*, ed. Conzen, Richard, and Zimring, 63–70.

36. Alft, *Elgin*; Monchow, *Seventy Years of Real Estate Subdividing*, 100–103; Mark Wilson with the assistance of Steven Porter and Janice Reiff, "Elgin National Watch Co.," in *The Encyclopedia of Chicago*, ed. Grossman, Keating, and Reiff, 922.

37. Robert Cramer, *Manufacturing Structure of the Cicero District, Metropolitan Chicago* (Chicago: University of Chicago, Department of Geography Research Paper No. 27, 1952); Frank Findlay, *A Survey of the Town of Cicero, Illinois* (Chicago: Department of Statistics and Research of the Council of Social Agencies of Chicago, 1937); Evelyn Kitagawa and Karl Taeuber, *Local Community Fact Book: Chicago Metropolitan Area, 1960* (Chicago: Chicago Community Inventory, University of Chicago, 1963), 178–79; Monchow, *Seventy Years of Real Estate Subdividing*, 113–15.

38. The quote is from Abbott, *The Tenements of Chicago*, 129. Also see Breckinridge and Abbott, "Housing Conditions in Chicago, III"; Chicago Plan Commission, *Residential Chicago* (Chicago: Chicago Plan Commission, 1942); and Joseph Bigott,

From Cottage to Bungalow: Houses and the Working Class in Metropolitan Chicago, 1869–1929 (Chicago: University of Chicago Press, 2001).

39. For quote, see Bushnell, "Some Social Aspects of the Chicago Stock Yards," inset between 290–91, 289. Craig Colten, "Industrial Waste in Southeast Chicago: Production and Disposal," *Environmental Review* 10 (1986): 95–105, and "Chicago's Waste Lands: Refuse Disposal and Urban Growth, 1840–1990," *Journal of Historical Geography* 20 (1994): 124–42; Abbott, *The Tenements of Chicago*; Breckinridge and Abbott, "Housing Conditions in Chicago, III"; Chicago Plan Commission, *Residential Chicago*; Ernest Bicknell, "Problems of Philanthropy in Chicago," *Annals of the American Academy of Political and Social Science* 21 (1903): 379–88.

40. Miller, "Rents and Housing Conditions," 15.

41. Charles Wacker, "Shows How All Business Is Intimately Related to the Chicago Plan," *Chicago Commerce* (November 19, 1921): 28.

42. Jerome Wilcox, "Charles Henry Wacker," in *Dictionary of American Biography*, ed. Dumas Malone (New York: Scribner's, 1936), 19:298.

43. Miller, "Rents and Housing Conditions," 1.

44. For quotes, see Wacker, "Shows How All Business"; and Barrett, *Work and Community*, 71. To get a sense of how important the yards were to Chicago's image before the Great Depression, see Bessie Louise Pierce, ed., *As Others See Chicago: Impressions of Visitors, 1673–1933* (Chicago: University of Chicago, 1933), especially the pieces by Sir John Leng, Rudyard Kipling, Paul Bourget, and Julian Street.

CHAPTER FOUR

1. Gilbert Lacher, "Planned for Expansion in Three Directions," *Iron Age* (January 10, 1924): 141–46; "From the Shops," *American Machinist* (June 11, 1891): 4–5; "A–Z List of Leading Manufacturers," *Thomas' Register of American Manufacturers, 7th Edition, October 1915* (New York: Thomas Publishing, 1915), 582.

2. "Studies of Migration of Industry," *Iron Age* (October 24, 1929), 1089.

3. William Mitchell, *Trends in Industrial Location in the Chicago Region since 1920* (Chicago: University of Chicago Press, 1933); Chicago Plan Commission, *Industrial and Commercial Background for Planning Chicago* (Chicago: Chicago Plan Commission, 1942), 26.

4. Allen Scott, "Locational Patterns and Dynamics of Industrial Activity in the Modern Metropolis," *Urban Studies* 10 (1982): 111–42; Graham Taylor, *Satellite Cities: A Study of Industrial Suburbs* (New York: Appleton, 1915); Edward Pratt, *Industrial Causes of Congestion of Population in New York City* (New York: Columbia University Studies in the Social Sciences, No. 109, 1911); Ernest Burgess, "The Growth of the City," in *The City*, ed. Robert Park, Ernest Burgess, and Roderick McKenzie (Chicago: University of Chicago Press, 1925), 47–62.

5. The data are taken from two lists of plant investments and moves for 1923 and 1929. The first are the amount firms spent on existing and new plants, as published in the *Chicago Commerce*, a weekly magazine put out by the Chicago Association of Commerce. The 1923 and 1929 data are taken from "Industrial Building Sets

Great Record," *Chicago Commerce* (January 5, 1924): 13; "Big Industrial Buildings of 1930," *Chicago Commerce* (January 11, 1930): 122; and "Industrial Building Figures Record Number of Small Structures," *Chicago Commerce* (January 18, 1930): 10, 29–31. The second source is the weekly "Machinery News" section of *Iron Age*, which lists plant expenditures and firm moves. The information was cross-checked with the "Industrial News" section of the *Chicago Commerce*, various Illinois industrial directories (1920, 1924, 1928), the *Chicago City Directory* (1928), and the *Thomas' Register* (1920, 1928). "Machinery News" informed the construction industry and machinery manufacturers of new markets. "Industrial News" supplied information on the weekly goings-on among firms in the Chicago industrial district. There are problems with the data: the coverage is centered on the city, and there is a strong likelihood that the listings, like city directories, are biased against small firms. Set against this, however, is that additions and new plants as small as $5,000 were included in the list, and the association's industrial department claims that the listings were compiled from building permits, implying that all (legal) work was included in the tabulations. Another problem is that while the data provide ample information about the building of an industrial base through the positive actions taken by firms (in terms of investment to plant), they do not provide evidence for firms that go out of business. In other words, while the data set provides an excellent basis for mapping the geography of manufacturing expansion, this is not true for the geography of manufacturing decline.

6. The number of cases for each of the four decisions are as follows: 167 new plants, 368 additions to existing factory plant, 193 new starts, and 270 moves.

7. Michael Storper and Richard Walker, *The Capitalist Imperative: Territory, Technology, and Industrial Growth* (New York: Blackwell, 1989).

8. Peter Maskell and Anders Malmberg, "Localised Learning and Industrial Competitiveness," *Cambridge Journal of Economics* 23 (1999): 167–85.

9. "Studies of Migration of Industry," 1089–90.

10. Most non-local capital that entered Chicago probably did so through the takeover of existing Chicago firms and factories.

11. "Machinery News," *Iron Age* (March 22, 1923): 867, and (July 12, 1923): 95; *Directory of Illinois Manufacturers, 1924–1925* (Chicago: Illinois Manufacturers' Association, 1924); Larry Sypolt and Bruce Seely, "Youngstown Sheet and Tube Company," in *Iron and Steel in the Twentieth Century*, ed. Bruce Seely (New York: Bruccoli Clark Layman, 1994), 501.

12. "Machinery News," *Iron Age* (July 12, 1923): 114. Halligan was formerly employed by Federal Pipe Company and Western Pipe and Supply, and Shoff worked at George Limbert and Company.

13. Josiah Currey, *Manufacturing and Wholesale Industries of Chicago*, 3 vols. (Chicago: Thomas Poole, 1918), 2:18–21; "Latest News of Chicago Industrial Development Plans," *Chicago Commerce* (February 24, 1923): 32; "Machinery News," *Iron Age* (August 1, 1929): 330; *Directory of Illinois Manufacturers, 1924–1925*.

14. This analysis does not include the numerous small changes made to the Calumet mills. Louis Cain and Robert Aduddell, "Inland Steel Corporation," in *Iron and Steel*, ed. Seely, 220–21.

15. "Youngstown Purchase of Steel and Tube Co.," *Iron Age* (January 11, 1923): 167–68; "Machinery News," *Iron Age* (July 12, 1923): 95; Sypolt and Seely, "Youngstown," 501; William Hogan, *Economic History of the Iron and Steel Industry in the United States*, 5 vols. (Lexington, MA: D.C. Heath, 1971), 3:982–90.

16. "Youngstown Purchase of Steel and Tube Co."; Sypolt and Seely, "Youngstown," 501; Hogan, *Economic History*, 3:982–90.

17. "Flexibility Increased by New Foundry," *Iron Age* (March 12, 1925): 751.

18. Lindy Biggs, *The Rational Factory: Architecture, Technology, and Work in America's Age of Mass Production* (Baltimore: John Hopkins University Press, 1996); Reyner Banham, *A Concrete Atlantis: U.S. Industrial Building and European Modern Architecture, 1900–1925* (Cambridge, MA: MIT Press, 1986); Robert Lewis, "Redesigning the Workplace: The North American Flexible Factory in the Interwar Period," *Technology and Culture* 42 (2001): 665–84. For the Jones Foundry, see Roger Fiske, "Relocated Equipment Boots Output," *Iron Age* (February 6, 1930): 438–42.

19. "Industrial Development Plans," *Chicago Commerce* (February 23, 1929): 22. For more detail on Clearing and the city's other major planned industrial district, the Central Manufacturing District, see chapter 7.

20. The concentration index captures the degree to which an activity is (or is not) concentrated or clustered in a particular area. An index of greater than 1.0 means that an activity is concentrated, while an index less than 1.0 signifies the reverse.

21. Glenn McLaughlin, *Growth of American Manufacturing Areas* (Pittsburgh: Bureau of Business Research, University of Pittsburgh, 1938), 186–88; Charles Colby, "Centrifugal and Centripetal Forces in Urban Geography," *Annals of the Association of American Geographers* 23 (1933): 4–7.

22. Currey, *Manufacturing and Wholesale Industries*, 2:29–31. In 1918 the Chicago plant was only one of five plants operated by Chicago Railway Equipment. The other four were a rolling mill at Franklin, Pennsylvania; malleable iron plants at Grand Rapids, Michigan, and Marion, Indiana; and a forging plant at Detroit.

23. Ibid., 2:155–57, 332–34.

24. Colby, "Centrifugal and Centripetal Forces," 5, 8–9.

25. "Latest News of Chicago Industrial Development Plans," *Chicago Commerce* (March 24, 1923): 32, and (March 31, 1923): 38.

26. "Latest Industrial Development Plans," *Chicago Commerce* (April 23, 1927): 27.

27. William Mitchell and Michael Jucius, "Industrial Districts of the Chicago Region and Their Influence on Plant Location," *Journal of Business* 6 (1933): 142; "Interpretative Digital Essay: Water in Chicago," in online Chicago Encyclopedia at www.encyclopedia.chicagohistory.org/pages/300044.html, accessed August 16, 2007. According to its website, Van Vlissingen and Co. continues to be one of the largest commercial and industrial real estate companies in the Chicago area.

28. Irving Cutler, *The Chicago-Milwaukee Corridor: A Geographic Study of Intermetropolitan Coalescence* (Evanston, IL: Department of Geography, Studies in Geography, No. 9, Northwestern University, 1965), 193–252; Mitchell, *Trends in Industrial Location*; Chicago Plan Commission, *Industrial and Commercial Background*; James Kenyon, *The Industrialization of the Skokie Area* (Chicago: University of Chicago, Department of Geography, Research Paper No. 33, 1954).

1. "Factors in Calumet Region Development," *Calumet Region Today* (December 24, 1925): 61, Chicago Public Library, Special Collections, Calumet Region Community Collection, box 4, file 18.

2. The chapter's analysis centers on the examination of the case files of 44 Chicago firms that went bankrupt between 1872–1878 and 1898–1901. These case files, which failed under the 1867 and 1898 bankruptcy acts, allow me to document the spatial extent and the business interactions of Chicago firms.

3. Gordon Winder suggested that a city-centered industrial district is where "perhaps 70 percent of the suppliers and suppliers of establishments . . . might be sourced from the immediate industrial region." With 76 percent and 66 percent of the bankrupt firms' creditors in 1878–1882 and 1898–1901, respectively, coming from the metropolitan area, Chicago can be defined as city-centered. See Gordon Winder, "The North American Manufacturing Belt in 1880: A Cluster of Regional Industrial Systems or One Large Industrial District?" *Economic Geography* 75 (1999): 75. The decline in the Chicago share of creditors was not a trend. In 1920 and 1928, the numbers were 77 and 70 percent.

4. Records of the United States District Court, Northern District of Illinois, Eastern Division at Chicago, Bankruptcy Records, Act of 1867, 1867–1878, Bankruptcy Case Files, 1867–1878 [hereafter Act of 1867]; Bankruptcy Case File 4172, *In the Matter of the Petition of Richards Iron Works*, "Schedule" (April 26, 1878) and "Printed List of Creditors and Debt Amounts" (n.d.).

5. Records of the United States District Court, Northern District of Illinois, Eastern Division at Chicago, Bankruptcy Records, Act of 1898, 1898–1972, Bankruptcy Case Files, 1898–1946 [hereafter Act of 1898], Bankruptcy Case File 3114, *In the Matter of Allen J. White*, "List of Creditors" (n.d.) and "Schedule" (March 12, 1900).

6. Act of 1867, Bankruptcy Case File 3946, *In the Matter of the Schureman and Hand Mantel Company*, "Notice" (April 4, 1880) and "Schedules" (February 20, 1878).

7. Act of 1867, Bankruptcy Case File 3349, *In the Matter of Henry Bennett*, "Schedules" (October 9, 1876).

8. *In the Matter of Henry Bennett*, "Schedules"; Act of 1867, Bankruptcy Case File 2649, *In the Matter of Warren Proprietary Medicine Company*, "Schedules A and B" (May 16, 1874); Act of 1898, Bankruptcy Case File 317, *In the Matter of August Keil, Albert Polley and Frank Thomas*, "List of Creditors" (January 16, 1900).

9. Act of 1898, Bankruptcy Case File 5012, *In the Matter of Hall Steel Tank Company*, "List of Creditors" (June 24, 1901); Act of 1898, Bankruptcy Case File 2069, *Pratt and Lambert et al. vs. Atlas White Lead and Color Company*, "List of Creditors" (November 1899); Act of 1898, Bankruptcy Case File 164, *In the Matter of Buchanan and Reens*, "Affidavit of Mailing Notices of Final Meeting of Creditors" (December 5, 1900); Act of 1867, Bankruptcy Case File 2507, *In the Matter of Bright Side Company*, "Schedules" (January 7, 1874). For example, Hall Steel Tank, Atlas White Lead, and Buchanan and Reens owed money for loans to Emma Hall, Mrs. Minnie Minehardt, and J. Buchanan, respectively, while the printing firm Bright Side was

supported by Mrs. Sherwood of Massachusetts and Mary Paine of Ravenswood, a Chicago suburb.

10. Act of 1867, Bankruptcy Case File 2143, *In the Matter of Union Screw and Bolt Company*, "Bankrupts Inventory and Schedule" (November 2, 1872).

11. Raymond Fales and Leon Moses, "Land-Use Theory and the Spatial Structure of the Nineteenth-Century City," *Papers of the Regional Science Association* 28 (1972): 49–80; Benjamin Chinitz, *Freight and the Metropolis: The Impact of America's Transport Revolutions on the New York Region* (Cambridge, MA: Harvard University Press, 1960).

12. Gordon Winder, "Building Trust and Managing Business over Distance: A Geography of Reaper Manufacturer D. S. Morgan's Correspondence, 1867," *Economic Geography* 77 (2001): 95–121.

13. Act of 1867, Bankruptcy Case 3812, *In the Matter of the Lakeside Publishing and Printing Company*, "Petition" (December 6, 1877); *Pratt and Lambert et al. vs. Atlas White Lead*, "List of Creditors"; *Buchanan and Reens*, "Affidavit."

14. *In the Matter of Allen J. White*, "List of Creditors" and "Petition"; Edgar Hoover, *Location Theory and the Shoe and Leather Industries* (Cambridge, MA: Harvard University Press, 1937); Philip Scranton, *Proprietary Capitalism: The Textile Manufacture at Philadelphia, 1800–1885* (Cambridge: Cambridge University Press, 1983).

15. The five firms are Artemis Plating, maker of bicycle parts and fittings in a factory capitalized at $20,000 on South Clinton Street; Woltz Wheel Works (Jasper and Frank), a wheel-making outfit in a small workshop at Racine and Twelfth; Northern Cycle and Supply, a bicycle wheel maker and wholesaler with a factory and storage space at Lake close to Clark; Hero Cycle, a bicycle maker on Ontario Street; and Chicago Tube Company, which constructed bicycle frames on Washington at Market. Information about the firms is from Act of 1898, Bankruptcy Case File 1995, *In the Matter of Northern Cycle and Supply Company*, "Petition" (November 1, 1899) and "Affidavit of Mailing Notices of Final Meeting of Creditors" (July 8, 1901); Act of 1898, Bankruptcy Case File 2451, *In the Matter of the Artemis Plating and Manufacturing Company*, "Petition" (February 8, 1900) and "Affidavit of Mailing Notices of First Meeting of Creditors" (February 9, 1900); Act of 1898, Bankruptcy Case File 2579, *In the Matter of the Hero Cycle Company*, "Schedule" (February 8, 1900); Act of 1898, Bankruptcy Case File 3433, *In the Matter of Woltz Wheel Works*, "Petition" (April 17, 1900); Act of 1898, Bankruptcy Case File 4008, *In the Matter of Chicago Tube Company*, "Schedules" (July 14, 1900). Unless otherwise indicated, firm information is taken from these sources.

16. Bruce Epperson, "Failed Colossus: Strategic Error of the Pope Manufacturing Company, 1878–1900," *Technology and Culture* 41 (2000): 300–320; Stephen Goddard, *Colonel Albert Pope and His American Dream Machine* (Jefferson, NC: McFarland, 2000); David Hounshell, *From the American System to Mass Production, 1800–1932: The Development of Manufacturing Technology in the United States* (Baltimore: John Hopkins University Press, 1984), 188–215; Glen Norcliffe, *The Ride to Modernity: The Bicycle in Canada, 1869–1900* (Toronto: University of Toronto Press, 2001), 89–119.

17. The information for the August purchases comes from *In the Matter of Chicago Tube Company*, "Statement of Expenses of Running Business" (August 2, 1900), "Report of Trustee" (August 15, 1900; August 17, 1900; and August 23, 1900).

18. Act of 1898, Bankruptcy Case File 490, *In the Matter of Bennett Box Manufacturing Company*, "List of Creditors and Debt Amounts" (February 8, 1899); *The Lakeside Annual Directory of the City of Chicago, 1874–1875* (Chicago: Williams, Donnelley, 1874); *The Lakeside Annual Business Directory of the City of Chicago, 1881* (Chicago: Chicago Directory Company, 1881); *The Lakeside Annual Directory of the City of Chicago, 1895* (Chicago: Chicago Directory Company, 1895).

19. David Skeel, *Debt's Dominion: A History of Bankruptcy Law in America* (Princeton, NJ: Princeton University Press, 2001).

20. Hoffman was forced into bankruptcy twice. I have not been able to ascertain why this was the case. Regardless of the fact that each case was initiated by a different set of creditors, the records of both cases are almost identical. See Act of 1898, Bankruptcy Case File 2317, *In the Matter of J.G. Hoffman Company*, "Lists of Creditors and Debts Amounts" (February 1, 1900) and "Debtors Sheet" (November 27, 1900); and Act of 1898, Bankruptcy Case File 2338, *In the Matter of J.G. Hoffman Company*, "Petition" (January 23, 1900).

21. Ann Durkin Keating, *Building Chicago: Suburban Developers and the Creation of a Divided Metropolis* (Columbus: Ohio State University Press, 1988), 20–22, 25–27; Powell Moore, *The Calumet Region: Indiana's Last Frontier* (Indianapolis: Indiana Historical Bureau, 1959); Chicago Sanitary District, Real Estate Development Committee, *The Manufacturing Possibilities of the Land along the Sanitary and Ship Canal* (Chicago: Chicago Sanitary District, Real Estate Development Committee, 1916); Stanley Buder, *Pullman: An Experiment in Industrial Order and Community Planning, 1880–1930* (New York: Oxford University Press, 1967); Raymond Mohl and Neil Betten, *Steel City: Urban and Ethnic Patterns in Gary, Indiana, 1906–1950* (New York: Holmes and Meier, 1986).

22. Act of 1898, Bankruptcy Case File 818, *In the Matter of Hill Cart and Carriage Company*, "Petition" (April 19, 1899) and "List of Creditors and Debt Amounts" (July 5, 1899); Act of 1898, Bankruptcy Case File 6099, *In the Matter of the Hammond Rolling Mill Company*, "Petition" (July 6, 1901), "Report of Referee" (August 19, 1901), and "Schedule of Creditors" (n.d.).

23. Dominic Candelero, "Suburban Italians: Chicago Heights, 1890–1975," in *Ethnic Chicago*, ed. Peter Jones and Melvin Holli (Grand Rapids, MI: W.B. Eerdmans, 1981), 181; Alfred Meyer and Paul Miller, "Manufactural Geography of Chicago Heights, Illinois," *Proceedings of the Indiana Academy of Science* 66 (1956): 209–29; Helen Monchow, *Seventy Years of Real Estate Subdividing in the Region of Chicago* (Evanston, IL: Northwestern University Press, Studies in the Social Sciences, No. 3, 1939), 103–5; Jerome Wilcox, "Charles Henry Wacker," in *Dictionary of American Biography*, ed. Dumas Malone (New York: Scribner's, 1936), 19:298.

24. Quote from E. Palma Beaudette, *Chicago Heights, Illinois* (n.p., 1914), 34. Candelero, "Suburban Italians"; Meyer and Miller, "Manufactural Geography of Chicago Heights"; Monchow, *Seventy Years of Real Estate Subdividing*.

25. The Chicago Steel Company was not a great success at first. Only open for six months, it was revived in 1890 and remained in business under one or another set of owners for several years. Moore, *The Calumet Region*, 141–77, 160–66, 221–24; Joseph Bigott, *From Cottage to Bungalow: Houses and the Working Class in Metropolitan Chicago, 1869–1929* (Chicago: University of Chicago Press, 2001), 57–86.

26. Quote from Beaudette, *Chicago Heights*, 27. Arthur Longini, *Chicago–Chicago Heights Industrial Economic Blueprint* (Chicago: Chicago and Eastern Illinois Railroad, 1957), 340; Meyer and Miller, "Manufactural Geography of Chicago Heights," 209–11; Bigott, *From Cottage to Bungalow*; Moore, *The Calumet Region*, 173–76; Alfred Meyer and Nora Mitchell, "Manufactural Geography of Hammond, Indiana: A Study in Geographic Anomaly," *Proceedings of the Indiana Academy of Science* 72 (1962): 194–96.

27. Moore, *The Calumet Region*, 160–66.

28. Longini, *Chicago–Chicago Heights*; Meyer and Miller, "Manufactural Geography of Chicago Heights"; Moore, *The Calumet Region*; Meyer and Mitchell, "Manufactural Geography of Hammond."

29. Dominic Pacyga, *Polish Immigrants and Industrial Chicago: Workers on the South Side, 1880–1922* (Columbus: Ohio State University Press, 1991); Meyer and Miller, "Manufactural Geography of Chicago Heights," 220. For the journey to work elsewhere, see Richard Harris and A. Victoria Bloomfield, "The Impact of Industrial Decentralization on the Gendered Journey to Work," *Economic Geography* 73 (1997): 94–117.

30. Moore, *The Calumet Region*, 160–66.

31. Scranton, *Proprietary Capitalism*; Olivier Zunz, *The Changing Face of Inequality: Urbanization, Industrial Development, and Immigrants in Detroit, 1880–1920* (Chicago: University of Chicago Press, 1982).

CHAPTER SIX

1. "Chicago Now Leads in Iron and Steel," *Chicago Commerce* (June 13, 1925): 7–8; "Chicago to Lead as World Steel Center," *Chicago Commerce* (July 26, 1924): 9.

2. Colonel A. A. Sprague, "Explains Growth of Calumet Industries," *Chicago Commerce* (June 20, 1925): 24.

3. Arend Van Vlissingen, "Why the Calumet District Draws Manufacturers" (Typescript, n.d. [c. 1910]), 1, Chicago Public Library, Special Collections, Calumet Region Community Collection [hereafter CRCC], box 4, file 18.

4. For quote, see George Plumbe, *Chicago, the Great Industrial and Commercial Center of the Mississippi Valley* (Chicago: Chicago Association of Commerce, 1912), 43. According to Phyllis Bate, "Mill locations were in close connection with the railroad lines entering Chicago, either directly or by means of belt roads, as well as with consuming fabricating industries which soon built their factories in the area surrounding the mills." She also noted, quoting *Iron Age*, "Gary will eventually be surrounded by a chain of iron and steel working factories, attracted thither by convenient transportation facilities, suitable sites and proximity to an ample

supply of raw material." More recently, Ann Markusen has shown that Southeast Chicago in the mid-1970s was a tightly bound production complex consisting of five interdependent industrial sectors. Phyllis Bate, "The Development of the Iron and Steel Industry of the Chicago Area, 1900–1920" (Ph.D. diss., University of Chicago, 1948), 14, 196–97; Ann Markusen with Josh Lerner, Wendy Patton, Jean Ross, and Judy Schneider, *Steel and Southeast Chicago: Reasons and Opportunities for Industrial Renewal* (Evanston, IL: Center for Urban Affairs and Policy Research, Northwestern University, 1985).

5. The Calumet region has been defined in various ways. For the purposes of this study, it refers to the area in the city south of Seventy-ninth Street and the metropolitan area's suburban districts from Blue Island, Illinois, to Gary, Indiana. The most important suburbs in terms of their employment and their contribution to the district's production chain are Blue Island, East Chicago, Gary, Hammond, Harvey, and Whiting. John Appleton, *The Iron and Steel Industry of the Calumet District: A Study in Economic Geography* (Urbana: University of Illinois Press, 1927); Powell Moore, *The Calumet Region: Indiana's Last Frontier* (Indianapolis: Indiana Historical Bureau, 1959), 141–77, 216–56; Alfred Meyer and Nora Mitchell, "Manufactural Geography of Hammond, Indiana: A Study in Geographic Anomaly," *Proceedings of the Indiana Academy of Science* 72 (1962): 190–211. James Walker, "Planning for the Future of East Chicago, Indiana: General Survey of Its Social and Economic Problems" (East Chicago: Chamber of Commerce of East Chicago, 1926), Calumet Regional Archives, East Chicago Collection, box 1, file 1.

6. The median was used rather than the mean because the latter measure is skewed by a few very large firms. The median is the preference in cases where distributions are highly skewed. Regardless, a similar, although larger difference is true for mean size: 522 for the Calumet district and 115 for the metropolitan area.

7. Appleton, *The Iron and Steel Industry*, 1–18, 24–26.

8. For quote, see "Gary's Great Future," *Iron Age* (December 2, 1909): 1715. U.S. Steel had plants in other parts of the metropolis: Rockdale (American Steel and Wire), Joliet (Illinois Steel, American Steel and Wire), and Waukegan (American Steel and Wire), but the firm's core after 1902 was in Calumet. Raymond Mohl and Neil Betten. *Steel City: Urban and Ethnic Patterns in Gary, Indiana, 1906–1950* (New York: Holmes and Meier, 1986); Appleton, *The Iron and Steel Industry*, 29–33; Bate, "The Development of the Iron and Steel Industry," 10–11; "Structural Plant Enlargements in the Chicago District," *Iron Age* (October 21, 1909): 1265; "The American Bridge Company's Gary Shops," *Iron Age* (July 4, 1912): 8–15; Josiah Currey, *Manufacturing and Wholesale Industries of Chicago*, 3 vols. (Chicago: Thomas Poole, 1918), 2:16–18.

9. Appleton, *The Iron and Steel Industry*, 29–33; Bate, "The Development of the Iron and Steel Industry," 14, 63–64; South Chicago Trades and Labor Assembly, "Republic Steel Corp.," *Official Year Book, 1948*, 30, 33–34, Chicago Public Library, Special Collections, South Chicago Community Collection [hereafter SCCC], box 3, file 6; Franklin Reck, *Sand in Their Shoes: The Story of American Steel Foundries* (n.p.: American Steel Foundries, 1952), 19–20, 33–49; "The Inland Steel Company's Works," *Iron Age* (October 20, 1910): 910–12.

10. Bate, "The Development of the Iron and Steel Industry," 14, 64. Also see Appleton, *The Iron and Steel Industry*, 101–5. This point about the diversity of relationships existing between firms along the production chain is made by Howell Harris in his examination of the U.S. foundry industry. Some foundries were "captive" and directly linked through ownership and management to a machine shop. Most were "jobbers" and had a range of competitive and contractual relationships with each other and with their customers, and produced an assortment of products under varying production forms. See "The Rocky Road to Mass Production: Change and Continuity in the U.S. Foundry Industry, 1890–1940," *Enterprise and Society* 1 (2000): 405–6.

11. For quote, see "Chicago Steel Men Hit 'Plus' Price Plan," *Chicago Commerce* (March 25, 1922): 11. Also see Meyer and Mitchell, "Manufactural Geography of Hammond"; Moore, *The Calumet Region*, 226; "Steel Users Rap 'Pittsburgh Plus' Plan," *Chicago Commerce* (April 29, 1922): 19; "Bliss and Laughlin," *Commerce* (February 1941): 37–39; and Alfred Meyer and Ben Lair, "Manufactural Geography of Greater Blue Island, Illinois: A Study in Industrial Saturated Space," *Proceedings of the Indiana Academy of Science* 74 (1964): 237.

12. For quote, see Roger Fiske, "Equipment Machines for Large Works," *Iron Age* (April 1, 1926): 899, 902; Moore, *The Calumet Region*, 241–44.

13. For quote, see Fiske, "Equipment Machines for Large Works," 899, 902. Moore, *The Calumet Region*, 241–44; "New Products from an Old Foundry," *Iron Age* (July 12, 1928): 78–80; "Riverside Iron Works" and "South Chicago Pattern Works" in Currey, *Manufacturing and Wholesale Industries*, 3:379, 118–19; "Valley Mould and Iron Corporation," Newspaper Clippings on the History of South Chicago, ca. 1934, SCCC, box 3, file 18.

14. John Maling, "Indiana's Part in the Development of the Great Calumet Region," *The Calumet Region Today* (Chicago: Calumet Record, 1925), 39, CRCC, box 4, file 18.

15. Quote from "Chicago Is the Center of the Steel Industry," *Chicago Commerce* (September 27, 1924): 7. Also see "Orders Pour in for Chicago-Made Steel," *Chicago Commerce* (December 20, 1924): 9–10; William Hogan, *Economic History of the Iron and Steel Industry in the United States*, 5 vols. (Lexington, MA: D.C. Heath, 1971), 3:982–90; East Side Lions Club, "History and Progress of East Side, 1851–1951," in *East Side Commemoration Pamphlet*, Chicago Public Library, Special Collections, SCCC, box 3, file 21, p. 8; and "Youngstown-Inland Steel Merger," *Iron Age* (February 2, 1928): 342–43, 369.

16. Carl Smith, *The Plan of Chicago: Daniel Burnham and the Remaking of the American City* (Chicago: University of Chicago Press, 2006), 82–83, 86; Christopher Thale, "Calumet River System," in *The Encyclopedia of Chicago*, ed. James Grossman, Ann Durkin Keating, and Janice Reiff (Chicago: University of Chicago Press, 2004), 117–18; Harold Mayer, *The Port of Chicago and the St. Lawrence Seaway* (Chicago: University of Chicago Press, 1957), 1–27.

17. Quotes from "Chicago Is the Center of the Steel Industry," 19; and "Steel Plants Planning Vast Expansions," *Chicago Commerce* (June 20, 1925): 27.

18. Theodore Longabaugh, "Far South End of Chicago" (typescript, November 29, 1937), 3–4, CRCC, box 4, file 19.

19. For quotes, see Van Vlissingen, "Why the Calumet District Draws Manufacturers," 1, 2. I do not attempt to provide a history of Pullman's industrial and residential development. There are several excellent studies, and readers should consult Stanley Buder, *Pullman: An Experiment in Industrial Order and Community Planning, 1880–1930* (New York: Oxford University Press, 1967); Carl Smith, *Urban Disorder and the Shape of Belief: The Great Chicago Fire, the Haymarket Bomb, and the Model Town of Pullman* (Chicago: University of Chicago Press, 1995); Janice Reiff, "'His Statements . . . Will Be Challenged': Ethnicity, Gender, and Class in the Evolution of the Pullman/Roseland Area of Chicago, 1894–1917," *Mid-America* 74 (1992): 231–52; Janice Reiff, "Rethinking Pullman: Urban Space and Working-Class Activism," *Social Science History* 24 (2000): 7–32; Graham Taylor, *Satellite Cities: A Study of Industrial Suburbs* (New York: Appleton, 1915), 28–69; and James Gilbert, *Perfect Cities: Chicago's Utopias of 1893* (Chicago: University of Chicago Press, 1991). Also see Janice Reiff's entries in *The Encyclopedia of Chicago*, ed. Grossman, Keating, and Reiff, 445, 665–66, 722–23. The argument made in this section draws on material taken from Pullman as well as surrounding areas such as Riverdale, Roseland, and Kensington, all of which were intimately tied to Pullman.

20. The point about Pullman's and Calumet's "complex social geography" has been made by Janice Reiff. See Reiff, "His Statements." 238.

21. Mia Gray, Elise Golub, and Ann Markusen, "Big Firms, Long Arms, Wide Shoulders: The 'Hub-and-Spoke' Industrial District in the Seattle Region," *Regional Studies* 30 (1996): 651–66; Ann Markusen, "Sticky Places in Slippery Space: A Typology of Industrial Districts," *Economic Geography* 72 (1996): 293–313.

22. Simon Dekker, "The History of Roseland and Vicinity" (typescript, c. 1935), 92–93, 105, CRCC, box 5, file 7; Reck, *Sand in Their Shoes*, 33–49; "History of West Pullman and Environs, c. 1920" (typescript, c. 1920), CRCC, box 6, file 7; West Pullman Peporter [*sic*], "Pictorial West Pullman" (n.p., 1907), CRCC, box 6, file 7.

23. "Chicago Leads in Rail Appliance Sales," *Chicago Commerce* (January 5, 1924): 15.

24. "The Gary Car Axle Mill," *Iron Age* (August 18, 1910): 381; Reck, *Sand in Their Shoes*, 19–20; "Chicago Leads in Rail Appliance Sales," 15–16.

25. The quote is from A. Schneider, "Chemical and Dye Industry Climbs to High Place in Nation," *Chicago Commerce* (December 7, 1929): 296. For Sherwin-Williams, see Dekker, "The History of Roseland," 107; and *The Story of Sherwin-Williams* (n.p., n.d.), CRCC, box 2, file 26.

26. "History of West Pullman"; West Pullman Peporter [*sic*] "Pictorial West Pullman"; "Replaces Fifty-Year Old Furnace," *Iron Age* (June 12, 1930), 1750–51, 1795; *Directory of the Iron and Steel Works of the United States* (Philadelphia: American Iron and Steel Association, 1908), 209–12; Dekker, "The History of Roseland"; "International Harvester," *Poor's Industrial Section, 1927* (New York: Poor's Publishing, 1927), 1436; Appleton, *The Iron and Steel Industry*, 29–33.

27. *Chicago Commerce* (November 5, 1921): 25; Gilbert Lacher, "Making Radiators under New Conditions," *Iron Age* (March 22, 1923): 805.

28. Thomas Leary and Elizabeth Sholes, *From Fire to Rust: Business, Technology and Work at the Lackawanna Steel Plant, 1899–1983* (Buffalo: Buffalo and Erie County Historical Society, 1987), 10; *Poor's Industrial Section, 1927*, 373–74, 2313–15, 2866–68.

29. *Poor's Industrial Section, 1927*, 2021–23, 2866–68.

30. For quotes, see Plumbe, *Chicago*, 43–44; Kenneth Warren, *The American Steel Industry, 1850–1970: A Geographical Interpretation* (1973; repr., Pittsburgh: University of Pittsburgh Press, 1988), 220, 198–200; and Bate, "The Development of the Iron and Steel Industry," III. Also see Appleton, *The Iron and Steel Industry*, 101–5.

31. For quote, see Louis Cain and Robert Aduddell, "Inland Steel Corporation," in *Iron and Steel in the Twentieth Century*, ed. Bruce Seely (New York: Bruccoli Clark Layman, 1994), 221. Robert Cramer, *Manufacturing Structure of the Cicero District, Metropolitan Chicago* (Chicago: University of Chicago, Department of Geography Research Paper No. 27, 1952), 55–71; Appleton, *The Iron and Steel Industry*, 101–5; "Chicago Is Strong in Iron and Steel," *Chicago Commerce* (November 21, 1925): 39–40; "Chicago Steel Men Hit 'Plus' Price Plan," II.

32. For quote, see "Chicago Leads in Rail Appliance Sales," 15. "Chicago Steel Men Hit 'Plus' Price Plan"; advertisement, *Chicago Commerce* (December 7, 1929): 365.

33. For quote, see A. Backert, "The Chicago Iron Trade in 1906," *Iron Age* (January 3, 1907): 49. T. Wright, "The Chicago Iron Trade in 1908," *Iron Age* (January 7, 1909): 29–30; Walter Isard and Caroline Isard, "The Transport-Building Cycle in Urban Development: Chicago," *Review of Economic Statistics* 25 (1943): 224–26; Helen Monchow, *Seventy Years of Real Estate Subdividing in the Region of Chicago* (Evanston, IL: Northwestern University Press, Studies in the Social Sciences, No. 3, 1939), 69–72.

34. Bruce Seely, "Eugene J. Buffington," Louis Cain, "Joseph L. Block," Louis Cain, "Leopold E. Block," and Cain and Aduddell, "Inland Steel," all in *Iron and Steel*, ed. Seely, 49–54, 66–67, 218–23; "Inland Steel Company" and "The Buda Company," in *Poor's Industrial Section, 1927*, 1320–22, 2304–5; Vincent Carosso, *Investment Banking in America: A History* (Cambridge, MA: Harvard University Press, 1970), 32.

35. Carosso, *Investment Banking*, 29–47, 82–83; Thomas Navins and Marian Sears, "The Rise of a Market for Industrial Securities, 1887–1902," *Business History Review* 29 (1955): 105–38.

36. Stephen Cutcliffe, "Illinois Steel Company," in *Iron and Steel*, ed. Seely, 217–18; Thomas Misa, *A Nation of Steel: The Making of Modern America, 1865–1925* (Baltimore: John Hopkins University Press, 1995), 138–71.

37. Cain and Aduddell, "Inland Steel"; "The Manufacture of Drop Forgings at Chicago," *Chicago Journal of Commerce* (June 20, 1888): 3; Duane Doty, "The Drop-Forge Works," newspaper clippings, December 1892, CRCC, box 2, file 10.

38. Seely, "Eugene J. Buffington"; Robert Casey, "George Lewis Danforth, Jr." in *Iron and Steel*, ed. Seely, 107–8; Cain, "Leopold E. Block"; Cain and Aduddell, "Inland Steel"; Reck, *Sand in Their Shoes*.

39. See descriptions of individual firms in *Poor's Industrial Section, 1927*.

40. Civil Case File 21616, *C. A. Treat vs. The Standard Steel and Iron Company and Joseph T. Torrence*, Bill of Exceptions (December 3, 1890), 80, Record Group 21, Records of the United States District Court, Records of the U.S. Circuit Court for the Northern Districts of Illinois, Eastern Division at Chicago; Civil Records, 1871–1911; Civil Case Files, 1871–1911. Discussion of the court proceedings is taken from this source.

41. A. T. Andreas, *History of Chicago: From the Earliest Period to the Present Time*, 3 vols. (Chicago: Andreas Company, 1884–86), 3:478; "Joseph Thatcher Torrence," in *Dictionary of American Biography*, ed. Dumas Malone (New York: Scribner's, 1936), 19:594.

42. Moore, *The Calumet Region*, 220–26; "Joseph Thatcher Torrence."

43. Moore, *The Calumet Region*, 220–26.

44. Ibid., 223–27; Walker, "Planning for the Future of East Chicago," 6–8; Derek Vaillant, "East Chicago," in *The Encyclopedia of Chicago*, ed. Grossman, Keating, and Reiff, 251–52.

45. *Treat vs. The Standard Steel and Iron Company*, 80, 84, 21–27.

46. Gordon Winder, "Building Trust and Managing Business over Distance: A Geography of Reaper Manufacturer D. S. Morgan's Correspondence, 1867," *Economic Geography* 77 (2001): 95–121.

47. Homer Hoyt, "South Chicago," in Chicago Plan Commission, *Forty-four Cities in the City of Chicago* (Chicago: Chicago Plan Commission, 1942), 50.

48. Henry Lee, "The Calumet Region as an Industrial Center," *Chicago, the Great Market Magazine* (July 1908): 76; "Civic Industrial Committee and Local Industries Committee of City Council Make Eleventh Industrial Excursion," *Chicago Commerce* (July 19, 1912): 17.

CHAPTER SEVEN

1. For quote, see "Experts Tell Why Zoning Is Necessary," *Chicago Commerce* (November 19, 1921): 11. Frances Alexander, "The Making of the Modern Industrial Park: A History of the Central Manufacturing District of Chicago, Illinois" (MA thesis, George Washington University, 1991); 25; "Committee of City Council and Association Make Twelfth Joint Inspection Tour of Manufacturing and Railway Districts," *Chicago Commerce* (June 13, 1913): 27–28.

2. "Committees of City Council and Association Make Twelfth Joint Inspection Tour."

3. For a discussion of the terms, see Robert Boley, *Industrial Districts Restudied: An Analysis of Characteristics* (Washington, DC: Urban Land Institute, 1961). To be consistent, planned industrial districts will be use throughout this chapter to refer to all the variants of corporate-controlled and planned industrial areas. Also see Robert Boley, *Industrial Districts: Principles in Practice* (Washington, DC: Land Institute, 1962); and Theodore Pasma, *Organized Districts: A Tool for Community Development* (Washington, DC: Government Printing Office, 1954).

4. Thomas Bender, *Toward an Urban Vision: Ideas and Institutions in Nineteenth-Century America* (Lexington: University of Kentucky Press, 1975); Stanley Buder, *Pullman: An Experiment in Industrial Order and Community Planning, 1880–1930* (New York: Oxford University Press, 1967); Margaret Crawford, *Building the Workingman's Paradise: The Design of American Company Towns* (London: Verso, 1995); Thomas Dublin, *Women at Work: The Transformation of Work and Community in Lowell, Massachusetts, 1826–1860* (New York: Columbia University Press, 1979), 15–22; John Garner, *The Company Town: Architecture and Society in the Early Industrial*

Age (New York: Oxford University Press, 1992); David Meyer, *The Roots of American Industrialization* (Baltimore: John Hopkins University Press, 2003); Heidi Vernon-Wortzel, *Lowell: The Corporation and the City* (New York: Garland, 1992), 8–20. The American planned industrial district was very similar to what appeared in Europe from the end of the nineteenth century. The one major difference was that many more American districts were established, owned, and run by railroad companies. Railroad-run estates were rare in Europe, which tended to be owned by individual industrial developers or controlled by local and national governments. John Armstrong, "The Development of the Park Royal Industrial Estate in the Interwar Period: A Re-Examination of the Aldcroft/Richardson Thesis," *London Journal* 21 (1996): 64–79; Douglas Farnie, *The Manchester Ship Canal and the Rise of the Port of Manchester, 1894–1975* (Manchester: Manchester University Press, 1980); Peter Scott, "Industrial Estates and British Industrial Development, 1897–1939," *Business History* 43 (2001): 73–98.

5. Alexander, "The Making of the Modern Industrial Park"; Boley, *Industrial Districts* and *Industrial Districts Restudied*; Michael Jucius, "Industrial Districts of the Chicago Region and Their Influence on Plant Location" (MA thesis, University of Chicago, 1932); William Mitchell and Michael Jucius, "Industrial Districts of the Chicago Region and Their Influence on Plant Location," *Journal of Business* 6 (1933): 139–56; Clinton Stockwell, "Central Manufacturing District" and "Clearing," in *The Encyclopedia of Chicago*, ed. James Grossman, Ann Durkin Keating, and Janice Reiff (Chicago: University of Chicago Press, 2004), 124, 173; Robert Wrigley, "Organized Industrial Districts with Special Reference to the Chicago Area," *Journal of Land and Public Utility Economics* 23 (1949): 80–98.

6. Peter Hall, *Cities of Tomorrow* (Oxford: Blackwell, 1996), 122–29; Edward Relph, *The Modern Urban Landscape* (Baltimore: John Hopkins University Press, 1987), 62–69.

7. Hall, *Cities of Tomorrow*; Relph, *Modern Urban Landscape*.

8. Gernot Grabher, "Rediscovering the Social in the Economics of Interfirm Relations," in *The Embedded Firm*, ed. Gernot Grabher (London: Routledge, 1993), 1–31; Robert Lewis, *Manufacturing Montreal: The Making of an Industrial Landscape, 1850 to 1930* (Baltimore: John Hopkins University Press, 2000); Peter Maskell and Anders Malmberg, "The Competitiveness of Firms and Regions: 'Ubiquitification' and the Importance of Localized Learning," *European Urban and Regional Studies* 6 (1999): 9–25.

9. The quote is from Mary Corbin Sies and Christopher Silver, "The History of Planning History," in *Planning the Twentieth-Century American City*, ed. Mary Corbin Sies and Christopher Silver (Baltimore: John Hopkins University Press, 1996), 11. Boley, *Industrial Districts*, 41, 51, 55, 61, 65.

10. Alexander, "The Making of the Modern Industrial Park"; Peter Dicken, Phillip Kelly, Kris Olds, and Henry Yeung, "Chains and Networks, Territories and Scales: Towards a Relational Framework for Analysing the Global Economy," *Global Networks* 1 (2001): 89–112; Peter Dicken and Anders Malmberg, "Firms in Territories: A Relational Approach," *Economic Geography* 77 (2001): 345–63; Maskell and Malmberg, "The Competiveness of Firms."

11. Dicken and Malmberg, "Firms in Territories"; Grabher, "Rediscovering the Social"; Michael Storper and Richard Walker, *The Capitalist Imperative: Territory, Technology, and Industrial Growth* (New York: Blackwell, 1989); Alexander, "The Making of the Modern Industrial Park"; Jucius, "Industrial Districts of the Chicago Region"; Mitchell and Jucius, "Industrial Districts of the Chicago Region."

12. William Cronon, *Nature's Metropolis: Chicago and the Great West* (New York: Norton, 1991); Stephen Adams and Orville Butler, *Manufacturing the Future: A History of Western Electric* (Cambridge: Cambridge University Press, 1999); David Hounshell, *From the American System to Mass Production, 1800–1932: The Development of Manufacturing Technology in the United States* (Baltimore: John Hopkins University Press, 1984); Gordon Winder, "Before the Corporation and Mass Production: The Licensing Regime in the Manufacture of North American Harvesting Machinery, 1830–1910," *Annals of the Association of American Geographers* 85 (1995): 521–52; Sharon Darling, *Chicago Furniture: Art, Craft, and Industry, 1833–1983* (New York: Norton, 1984).

13. Louis Cain, "The Sanitary District of Chicago: A Case Study of Water Use and Conservation" (Ph.D. diss., Northwestern University, 1969); John Lurie, *The Chicago Board of Trade, 1859–1905: The Dynamics of Self-Regulation* (Urbana: University of Illinois Press, 1979); Harold Mayer, "The Railway Pattern of Metropolitan Chicago" (Ph.D. diss., University of Chicago, 1943); Harold Mayer, *The Port of Chicago and the St. Lawrence Seaway* (Chicago: University of Chicago Press, 1957).

14. Hobart Chatfield-Taylor, *Chicago* (Boston: Houghton Mifflin, 1917); Dominic Pacyga and Ellen Skerrett, *Chicago: City of Neighborhoods* (Chicago: Loyola University Press, 1986); John W. Stamper, *Chicago's North Michigan Avenue: Planning and Development, 1900–1930* (Chicago: University of Chicago Press, 1991).

15. For quote, see Francis Stetson Harman, "The Central Manufacturing District of Chicago," in *Manufacturing and Wholesale Industries of Chicago*, 3 vols., ed. Josiah Currey (Chicago: Thomas Poole, 1918), 3:440. Alexander, "The Making of the Modern Industrial Park"; Mitchell and Jucius, "Industrial Districts of the Chicago Region"; Louise Carroll Wade, *Chicago's Pride: The Stockyards, Packingtown, and Environs in the Nineteenth Century* (Urbana: University of Illinois Press, 1987).

16. Alexander, "The Making of the Modern Industrial Park"; Chicago Plan Commission, *Industrial and Commercial Background for Planning Chicago* (Chicago: Chicago Plan Commission, 1942), 26–27. While the contemporary literature provides all sorts of firm and employment estimates, it is extremely difficult to verify their veracity. We must take contemporary claims about the scale of the districts with some skepticism.

17. Harman, "The Central Manufacturing District," 440.

18. "Industries Form Groups as City Grows," *Chicago Commerce* (April 14, 1923): 15–16.

19. "Another Successful Tour of Manufacturing and Railroad Districts by Committees of Association and City Council," *Chicago Commerce* (July 10, 1914): 25–31; Jucius, "Industrial Districts of the Chicago Region"; Mayer, "The Railway Pattern of Metropolitan Chicago."

20. Jucius, "Industrial Districts of the Chicago Region"; Mitchell and Jucius, "Industrial Districts of the Chicago Region"; "Another Successful Tour"; Mayer, "The

Railway Pattern of Metropolitan Chicago"; "Clearing Has Greatest Railroad Yard," *Chicago Commerce* (May 12, 1923): 19–20.

21. Chicago Plan Commission, *Industrial and Commercial Background*, 27–28; Jucius, "Industrial Districts of the Chicago Region"; Mayer, "The Railway Pattern of Metropolitan Chicago," 69–71.

22. J. J. Fagg, "A Re-Examination of the Incubator Hypothesis: A Case Study of Greater Leicester," *Urban Studies* 17 (1980): 35–44; Edgar Hoover and Raymond Vernon, *Anatomy of a Metropolis* (New York: Anchor, 1962); Robert Leone and Raymond Struyk, "The Incubator Hypothesis: Evidence from Five SMSAs," *Urban Studies* 13 (1976): 325–31. This research points to a decided incubator role for the central business district and the adjacent area, although evidence indicates that non-central areas were also the home to small and new firms.

23. Alexander, "The Making of the Modern Industrial Park," 122–23; *Speaking of Ourselves* (Chicago: Central Manufacturing District, c. early 1940s), 19–21; "Industrial Development Plans," *Chicago Commerce* (April 28, 1928): 20; "Industrial Development Plans," *Chicago Commerce* (October 28, 1928): 20; "Plans for Industrial Development," *Chicago Commerce* (August 24, 1929): 26.

24. "Latest News of Chicago Industrial Development Plans," *Chicago Commerce* (December 6, 1924): 32; "Latest Industrial Development News," *Chicago Commerce* (August 29, 1925): 22; "Industrial Development Plans," *Chicago Commerce* (October 6, 1928): 20; L. Thomas, "Hatching New Industries: An Incubator for Fledgling Manufacturers," *Factory and Industrial Management* 77 (1929): 949–51.

25. "Industrial Development Plans," *Chicago Commerce* (October 6, 1928): 20.

26. Ibid.

27. "Industrial Development Plans," *Chicago Commerce* (September 3, 1927): 26; "Industrial Development Plans," *Chicago Commerce* (April 28, 1928): 20.

28. "Industrial Development Plans," *Chicago Commerce* (April 28, 1928): 20.

29. Ibid.

30. Reyner Banham, *A Concrete Atlantis: U.S. Industrial Buildings and European Modern Architecture, 1900–1925* (Cambridge, MA: MIT Press, 1986); Lindy Biggs, *The Rational Factory: Architecture, Technology, and Work in America's Age of Mass Production* (Baltimore: John Hopkins University Press, 1996); Grant Hildebrand, *Designing for Industry: The Architecture of Albert Kahn* (Cambridge, MA: MIT Press, 1974); Robert Lewis, "Redesigning the Workplace: The North American Flexible Factory in the Interwar Period," *Technology and Culture* 42 (2001): 665–84; Terry Smith, *Making the Modern: Industry, Art, and Design in America* (Chicago: University of Chicago Press, 1993).

31. Alexander, "The Making of the Modern Industrial Park," 124–31; Currey, *Manufacturing and Wholesale Industries*, 3:28–30, 138–40; Fred Foltz, "Industry Moves to the First Floor," *Chicago Commerce* (August 1937): 25; Columbia University, "Serving, Saving and Saluting the South Loop" www.lib.colum.edu/archhistory/alschuler/htm, accessed July 19, 2007.

32. For quotes, see H. Phelps, "How 100 Buildings Are Kept in Trim," *Factory Management and Maintenance* 94 (April 1936): S-225; and Foltz, "Industry Moves to the First Floor," 25. Biggs, *The Rational Factory*; Lewis, "Redesigning the Workplace."

33. Lewis, "Redesigning the Workplace"; "Industrial Development Plans," *Chicago Commerce* (March 17, 1928): 24.
34. Scott Joy, "The Central Manufacturing District, Chicago, Ill.," *Architectural Forum* 34 (1921): 177–82; Alexander, "The Making of the Modern Industrial Park," 136–37.
35. Alexander, "The Making of the Modern Industrial Park," 124–37.

CHAPTER EIGHT

1. *Chicago Commerce* (October 24, 1925): 25; (October 31, 1925): 40; and (November 26, 1927): 28; "Industrial Development Plans," *Chicago Commerce* (September 1, 1928): 18.
2. Quotes are from *Chicago Commerce* (November 26, 1927): 28; and (February 16, 1929): 26. Also see "Industrial Development Plans," *Chicago Commerce* (September 1, 1928), 18.
3. "Tin Cans—Another First for Chicago," *Chicago Commerce* (March 16, 1929): 11.
4. Michael Conzen, "The Progress of American Urbanism, 1860–1930," in *North America: The Historical Geography of a Changing Continent*, ed. Robert Mitchell and Paul Groves (Totowa, NJ: Rowman and Little, 1987), 347–70; William Cronon, *Nature's Metropolis: Chicago and the Great West* (New York: Norton, 1991); David Meyer, "Emergence of the American Manufacturing Belt: An Interpretation," *Journal of Historical Geography* 9 (1983): 145–74; David Meyer, *The Roots of American Industrialization* (Baltimore: John Hopkins University Press, 2003); Gordon Winder, "The North American Manufacturing Belt in 1880: A Cluster of Regional Industrial Systems or One Large Industrial District?" *Economic Geography* 75 (1999): 71–91; Gordon Winder, "Building Trust and Managing Business over Distance: A Geography of Reaper Manufacturer D. S. Morgan's Correspondence, 1867," *Economic Geography* 77 (2001): 95–121.
5. Alfred Meyer and Paul Miller, "Manufactural Geography of Chicago Heights, Illinois," *Proceedings of the Indiana Academy of Science* 66 (1956): 209–29; Arthur Longini, *Chicago–Chicago Heights Industrial Economic Blueprint* (Chicago: Chicago and Eastern Illinois Railroad, 1957), 330–46; E. Palma Beaudette, *Chicago Heights, Illinois* (n.p., 1914); *Directory of Illinois Manufacturers, 1928–1929* (Chicago: Illinois Manufacturers' Association, 1928), 555; Illinois, Bureau of Labor Statistics, *Bureau of Labor Statistics for 1910: Part II, Industrial Opportunities* (Springfield: Illinois State Journal, 1911); "New Products from an Old Foundry," *Iron Age* (July 12, 1928): 78–80.
6. The data is taken from various issues of *Iron Age*. For these firms—foundries, steel mills, metalworking shops, machinery makers—along with the machinery and equipment themselves, the name and address of the company where they bought their new products were taken down. In total, 211 separate transactions were involved. In the case of a Chicago firm buying more than one product from a selling firm, the transaction was counted as one transaction, not multiple transactions. Gilbert Lacher, "A Model Pattern Shop and Storage Building," *Iron Age* (May 5, 1921): 1165–68; Gilbert Lacher, "Production Methods at Fittings Plant," *Iron Age* (June 22, 1922): 1729–33; Gilbert Lacher, "Wisconsin Steel Works Adds Open Hearth," *Iron*

Age (May 15, 1924): 1421–25; Roger Fiske, "Blooming Mill Has Heavy Drive," *Iron Age* (October 15, 1925): 1019–23.

7. Gilbert Lacher, "LaSalle Co. Completes Cold Drawing Plant," *Iron Age* (January 5, 1922): 27–31; "New Products from an Old Foundry"; "Plant Makes Upset Forgings Exclusively," *Iron Age* 109 (February 9, 1922): 401–4; Gilbert Lacher, "Making Radiators under New Conditions," *Iron Age* (March 22, 1923): 805–9; Gilbert Lacher, "New Inland Merchant and Billet Mills," *Iron Age* (August 7, 1924): 303–7.

8. *Chicago Commerce* (November 5, 1921): 25; Lacher, "Making Radiators."

9. For a description of the making of steam locomotives and specialty steel products by the Philadelphia firms Baldwin Locomotive and Midvale Steel, see Philip Scranton and Walter Licht, *Work Sights: Industrial Philadelphia, 1890–1950* (Philadelphia: Temple University Press, 1986), 185–213; and John Brown, *The Baldwin Locomotive Works, 1831–1915: A Study in American Industrial Practice* (Baltimore: John Hopkins University Press, 1995).

10. "Steel Treaters to Meet in Chicago," *Iron Age* (September 9, 1926): 680–702; "Foundrymen to Meet in Detroit," *Iron Age* (September 16, 1926): 769–82. These articles listed the name, address, products being shown, and representatives of all the firms with exhibitions. There were 222 and 310 exhibitors at the foundrymen and steel treaters' conventions, respectively. Not surprisingly, almost all of the exhibitors were firms from the Manufacturing Belt.

11. Ann Durkin Keating, *Building Chicago: Suburban Developers and the Creation of a Divided Metropolis* (Columbus: Ohio State University Press, 1988), 75; Alfred Meyer and Marilyn Tschannen, "Manufactural Geography of Harvey, Illinois," *Proceedings of the Indiana Academy of Science 75* (1965): 178–90; Helen Monchow, *Seventy Years of Real Estate Subdividing in the Region of Chicago* (Evanston, IL: Northwestern University Press, Studies in the Social Sciences, No. 3, 1939), 105–7; *Directory of Illinois Manufacturers, 1924–1925* (Chicago: Illinois Manufacturers' Association, 1924).

12. *Thomas' Register of American Manufacturers and First Hands in All Lines: The Buyers' Guide, 1905–1906* (New York: Thomas Publishing, 1905), 313; *Thomas' Register of American Manufacturers and First Hands in All Lines, 1915* (New York: Thomas Publishing, 1915), 1899–1900.

13. In the absence of company records, information about the sales of the firm's products was obtained from *Iron Age*'s weekly column "Machinery Markets and News of the Works" for January through April 1924.

14. Linked means that both the creditors' address (if a Chicago creditor, a street address; if a non-Chicago creditor, a city or town) and product could be identified, thus making it possible to categorize a firm by geographic location and by industrial sector.

15. The number of creditors in the South and the West would be much lower (by about a quarter) if the total excluded the large number found in cities—most notably, St. Louis (21 creditors) and Minneapolis (11)—that were on the margins of the Manufacturing Belt.

16. For discussion of some of the industrial centers that Chicago's bankrupt firms received manufactured inputs from, see Philip Scranton, *Endless Novelty: Specialty Production and American Industrialization, 1865–1925* (Princeton, NJ: Princeton

University Press, 1997); Meyer, "Emergence of the American Manufacturing Belt"; and Conzen, "The Progress of American Urbanism."

17. Scranton, *Endless Novelty*, 139–46, 163–92.

18. Records of the United States District Court, Northern District of Illinois, Eastern Division at Chicago, Bankruptcy Records, Act of 1898, 1898–1972, Bankruptcy Case Files, 1898–1946 [hereafter Act of 1898], Bankruptcy Case File 39549, *In the Matter of Metal Crafts Incorporated*, "Schedule A(3)" (March 24, 1928).

19. The locale effect was measured by comparing (a) the actual number of creditors that were in the same geographic zone as the bankrupt firm that they owed money to (b) the expected number of firms. Chicago was divided into five zones: center, south, north, west, and suburbs.

20. Act of 1898, Bankruptcy Case 28736, *In the Matter of American Pneumatic Chuck Company*, "Schedules A(2) and (A)3" (November 9, 1920); Act of 1898, Bankruptcy Case 28819, *In the Matter of Illinois Motor Company*, "Schedule A(3)" (December 30, 1920).

21. Act of 1898, Bankruptcy Case 28505, *In the Matter of Chicago Ferrotype Company*, "Schedule A(3)" (July 7, 1920) and "Petition: Schedule A and B Thereunder, and Summary of Debts and Assets" (July 7, 1920); Act of 1898, Bankruptcy Case 28795, *In the Matter of Jewell Electrical Instrument Company*, "Schedule A(3)" (December 22, 1920); Prudence Walker, ed., *Directory of Illinois Manufacturers, 1920* (Chicago: Illinois Manufacturers' Association, 1920), 98; *Directory of Illinois Manufacturers, 1924–1925*.

22. Chicago, Department of City Planning, Research Division, *Locational Patterns of Major Manufacturing Industries in the City of Chicago* (Chicago: Department of City Planning, 1960), 24.

23. For an interesting discussion of the regional geography of the relationship between manufacturers and the legal world, see Winder, "Building Trust and Managing Business."

24. Act of 1898, Bankruptcy Case 28505, *In the Matter of Chicago Ferrotype Company*, "Petition for Consolidation of Case 28505 with 28504" (July 7, 1920); *Directory of Illinois Manufacturers, 1924–1925; Directory of Illinois Manufacturers, 1928–1929*; "News of Industrial Development Plans," *Chicago Commerce* (January 20, 1923): 26.

25. "Leads as Maker of Precision Instruments," *Chicago Commerce* (December 8, 1923): 15.

26. Ibid.

27. *Directory of Illinois Manufacturers, 1924–1925; Manufacturers in Chicago and Metropolitan Area* (Chicago: Chicago Association of Commerce, 1940); *Illinois Manufacturers Directory, 1951* (Chicago: Manufacturers' News, 1951); Josiah Currey, *Manufacturing and Wholesale Industries of Chicago*, 3 vols. (Chicago: Thomas Poole, 1918), 2:260–63; Stephen Adams and Orville Butler, *Manufacturing the Future: A History of Western Electric* (Cambridge: Cambridge University Press, 1999), 26–44, 80–84.

28. "Industrial Development Plans," *Chicago Commerce* (January 19, 1929): 54.

29. "Industrial Development Plans," *Chicago Commerce* (January 19, 1929): 54; (October 26, 1929): 24; and (November 9, 1929): 24.

30. "Zenith Radio Corp.," *Commerce* (September 1936): 43.

31. "Zenith Radio Corp."; *Manufacturers in Chicago and Metropolitan Area.*

32. "Chicago—Nation's Radio Capital," *Chicago Commerce* (September 15, 1928): 8. Chicago's first radio broadcast was made by KYW, the station operated by Westinghouse Electric, in 1921. *Polk's Chicago (Illinois) City Directory, 1928–1929* (Chicago: R.L. Polk, 1928).

33. For quotes, see "Chicago—Nation's Radio Capital," 8–10; and B. Grisgby, "Chicago the Radio Hub—A Marvel of Industry," *Chicago Commerce* 25 (December 7, 1929): 189. For further evidence, see the discussion of industrial and movie film production in "Chicago Has a Hollywood, All Its Own," *Chicago Commerce* (December 6, 1924): 17–18.

CHAPTER NINE

1. R. Ardrey, "The Chicago Iron Trade in 1910," *Iron Age* (January 5, 1911): 60; "Chicago Hardware Show," *Iron Age* (July 5, 1906): 52; "Illinois Retail Hardware Association," *Iron Age* (February 21, 1907): 626–32; "The Machinery and Supply Convention," *Iron Age* (May 16, 1907): 1484–93.

2. For quote, see Philip Scranton, *Endless Novelty: Specialty Production and American Industrialization, 1865–1925* (Princeton, NJ: Princeton University Press, 1997), 172. For Chicago, see A. T. Andreas, *History of Chicago: From the Earliest Period to the Present Time*, 3 vols. (Chicago: Andreas Company, 1885), 3:733–42; Sharon Darling, *Chicago Furniture, Art, Craft, and Industry, 1833–1983* (New York: Norton, 1984); Jeremy Kinney, "'We Hold the Merchandising Idea as Paramount': The Virtues of Flexible Mass Production in the 1920s American Furniture Industry," *Business and Economic History* 28 (1999): 83–93; and Civic Industrial Department, "Shows Chicago Furniture Trade Center," *Chicago Commerce* (May 14, 1921): 11.

3. "The Great Central Market for Furniture," *Chicago, the Great Central Market Magazine* (October 1908): 57–61; "Shows Chicago Furniture Trade Center"; George Armstrong, "Chicago as Piano Producing Center," *Chicago, the Great Central Market Magazine* (July 1907): 54–62; Scranton, *Endless Novelty*, 140–41; Craig Roell, *The Piano in America, 1890–1940* (Chapel Hill: University of North Carolina Press, 1989); "Chicago's Output of Furniture Is Huge," *Chicago Commerce* (November 21, 1925), 47; "Study Sales of Furniture in Chicago," *Chicago Commerce* (May 12, 1928), 21.

4. U.S. Department of Commerce, *Fifteenth Census of the United States: Manufactures, 1930* vol. 1 (Washington, DC: Government Printing Office, 1933); Josiah Currey, *Manufacturing and Wholesale Industries of Chicago*, 3 vols. (Chicago: Thomas Poole, 1918), 1:398–99.

5. William Cronon, *Nature's Metropolis: Chicago and the Great West* (New York: Norton, 1991), 148–206; Committee on Business Research, "Lumber Industry Thrives in Chicago," *Chicago Commerce* (August 20, 1927): 15; Theodore Karamanski, "Lumber," in *The Encyclopedia of Chicago*, ed. James Grossman, Ann Durkin Keating, and Janice Reiff (Chicago: University of Chicago Press, 2004), 495–96.

6. Quotes are from George Plumbe, *Chicago, the Great Industrial and Commercial*

Center of the Mississippi Valley (Chicago: Chicago Association of Commerce, 1912), 44; and Edward Hines, "Chicago and Its Lumber Industry," *Chicago Commerce* (May 11, 1929): 13. Also see Committee on Business Research, "Lumber Industry Thrives," 15; and Darling, *Chicago Furniture*, 58–64.

7. Quote from "Chicago's Output of Furniture Is Huge," 47. *The Lakeside Annual Business Directory of the City of Chicago, 1881* (Chicago: Chicago Directory Company, 1881); Darling, *Chicago Furniture*, 47–48; *Directory of Illinois Manufacturers, 1924–1925* (Chicago: Illinois Manufacturers' Association, 1924).

8. "Shows Chicago Furniture Trade Center," 11.

9. Quotes from ibid. *Lakeside Annual Business Directory*; *Directory of Illinois Manufacturers, 1924*; *Directory of Illinois Manufacturers, 1928–1929* (Chicago: Illinois Manufacturers' Association, 1928).

10. Darling, *Chicago Furniture*.

11. Ibid.; "Industrial Development Plans," *Chicago Commerce* (September 21, 1929): 22; U.S. Department of Commerce, *Fourteenth Census of the United States: Manufactures, 1919*, vol. 9 (Washington: Government Printing Office, 1923), and *Fifteenth Census of the United States: Manufactures, 1929*, vol. 3 (Washington, DC: Government Printing Office, 1933); Roell, *The Piano in America*.

12. "Study Sales of Furniture," 21; Chicago Association of Commerce, "Chicago: The Great Central Market," in *Chicago City Directory 1923* (Chicago, 1923), 65–66; American Furniture Mart, *America's Greatest Furniture Mart* (Chicago: Wells, 1925); Mark Wilson, "Mail Order," in *The Encyclopedia of Chicago*, ed. Grossman, Keating, and Reiff, 505–6.

13. Quote from *The Lakeside Annual Directory of the City of Chicago, 1904* (Chicago: Chicago Directory Company, 1904), 2511–13. Also see Darling, *Chicago Furniture*, 58–64.

14. "Industrial Development Plans," *Chicago Commerce* (February 11, 1928), 28; Darling, *Chicago Furniture*, 297–98.

15. For quote, see "The Great Central Market for Furniture," 57. For Chicago, see Darling, *Chicago Furniture*, 53–58, 61–64; and Scranton, *Endless Novelty*, 170–76. For Philadelphia, see Howell John Harris, "Getting It Together: The Metal Manufacturers' Association of Philadelphia, c. 1900–1930," in *Masters to Managers: Historical and Comparative Perspectives on American Employers*, ed. Sanford Jacoby (New York: Columbia University Press, 1991), 111–31; and Philip Scranton, "Webs of Productive Association in American Industrialization: Patterns of Institution-Formation and Their Limits, Philadelphia, 1880–1930," *Journal of Industrial History* 1 (1998): 9–34.

16. Armstrong, "Chicago as Piano Producing Center"; Hines, "Chicago and Its Lumber Industry," 15–16; *The Lakeside Annual Directory of the City of Chicago, 1874–1875* (Chicago: Williams, Donnelley, 1874).

17. Quote from "The Great Central Market for Furniture," 59. Also see Chicago Association of Commerce, "Chicago: The Great Central Market," 65–66; Darling, *Chicago Furniture*, 61–63; and Scranton, *Endless Novelty*, 172–73. Between 1895 and 1908, the FMA was known as the Chicago Furniture Exposition Association.

18. "The Great Central Market for Furniture," 59.

19. Ibid.

20. Quotes from "Furniture Exhibition Building to Cost $5,000,000 Likely to Be Put Up in Chicago," *Chicago Commerce* (February 18, 1922), 32; "American Furniture Mart Building Is Beginning to Rise on North Side," *Chicago Commerce* (October 13, 1923), 42. Also see Darling, *Chicago Furniture*, 292–95.

21. "Furniture Exhibition Building"; John W. Stamper, *Chicago's North Michigan Avenue: Planning and Development, 1900–1930* (Chicago: University of Chicago Press, 1991); Darling, *Chicago Furniture*, 293; American Furniture Mart, *America's Greatest*.

22. "Chicago Building Great Business Marts," *Chicago Commerce* (August 1, 1925): 36; "Latest News of the Chicago Industrial Development Plans," *Chicago Commerce* (December 13, 1924): 42, and (December 20, 1924): 32; "Chicago's Output of Furniture Is Huge," 47; American Furniture Mart, *America's Greatest*; "New Furniture Mart Caps Chicago's Lead," *Chicago Commerce* (June 28, 1924): 9–10.

23. "Tell Why Chicago Leads in Furniture," *Chicago Commerce* (June 28, 1924): 12, 29.

24. Deborah Fulton Rau, "The Making of the Merchandise Mart, 1927–1931: Air Rights and the Plan of Chicago," in *Chicago Architecture and Design, 1923–1993*, ed. John Zukowsky (Munich: Prestel, 1993), 99–117; Stamper, *Chicago's North Michigan Avenue*.

25. "Latest News of Chicago Industrial Development Plans," *Chicago Commerce* (July 21, 1923): 42, (July 12, 1924): 30, and (August 4, 1923): 38.

26. "Latest News of Chicago Industrial Development Plans," *Chicago Commerce* (January 27, 1923): 36, (December 6, 1924): 32, (June 2, 1923): 44, and (November 21, 1923); "Latest Industrial Development Plans," *Chicago Commerce* (February 27, 1926): 21.

27. Quote from "Chicago Building Great Business Marts," 36. "Industrial Development Plans," *Chicago Commerce* (February 18, 1928): 29.

28. "Chicago Building Great Business Marts," 36; "Industrial Development Plans," *Chicago Commerce* (February 18, 1928): 29. The manufacturing data is from the *Directory of Illinois Manufacturers, 1924–1925*.

29. "Clearing House Is Established to Serve the Chicago Candy Industry," *Chicago Commerce* (December 31, 1921): 6; U.S. Department of Commerce, *Fifteenth Census of the United States: Wholesale Distribution, 1930*, vol. 2 (Washington, DC: Government Printing Office, 1933), 447; Stamper, *Chicago's North Michigan Avenue*.

30. "Why I Moved My Factory to Chicago," *Chicago Commerce* (September 15, 1928): 7, 28.

31. Harold Barger, *Distribution's Place in the American Economy since 1869* (Princeton, NJ: Princeton University Press, 1955), 65–76; Alfred Chandler, *The Visible Hand: The Managerial Revolution in American Business* (Cambridge, MA: Belknap Press, 1977), 15–49, 207–326; Glenn Porter and Harold Livesay, *Merchants and Manufacturers: Studies in the Changing Structure of Nineteenth-Century Marketing* (Baltimore: John Hopkins University Press, 1971); Scranton, *Endless Novelty*, 109–13, 244–45, 320–21; Vincent Carosso, *Investment Banking in America: A History* (Cambridge, MA: Harvard University Press, 1970).

32. Barger, *Distribution's Place in the American Economy*; Chandler, *The Visible Hand*; Porter and Livesay, *Merchants and Manufacturers*; Scranton, *Endless Novelty*;

Carosso, *Investment Banking in America*; U.S. Department of Commerce, *Fifteenth Census of the United States: Wholesale Distribution, 1930*, 2:7.

33. Chandler, *The Visible Hand*, 224–35, 285–314.

34. Robert Aduddell, "Edward L. Ryerson, Jr." in *Iron and Steel in the Twentieth Century*, ed. Bruce Seely (New York: Bruccoli Clark Layman, 1994), 374–76; "Chain of Warehouses Standardized," *Iron Age* (August 14, 1924): 386–87.

35. Quote from "The Illinois Steel Company's warehouse," *Iron Age* (August 30, 1906): 53. Aduddell, "Edward L. Ryerson, Jr."; "Chain of Warehouses Standardized."

36. Chandler, *The Visible Hand*, 235–39, 299–302, 391–402.

37. The quotes are from Andreas, *History of Chicago*, 3:338; and Chicago Association of Commerce, "Chicago: The Great Central Market," 105. Also see "Bulwarks of Great Central Market," *Chicago Commerce* (July 18, 1913): 18–19; and E. Ferebee, "The Use of Public Merchandise Warehouses in Chicago," *Journal of Business of the University of Chicago* 6 (1933): 318–26.

38. Chicago Association of Commerce, "Chicago: The Great Central Market," 105–6; "2,500 National Distributors," *Chicago Commerce* (November 21, 1925): 4.

39. U.S. Department of Commerce, *Fifteenth Census of the United States: Wholesale Distribution, 1930*, 2:7.

40. Quote from ibid., 43, also see 19–20, 445–52. William Trout Chambers, "A Geographic Study of Joliet, Illinois: An Urban Center Dominated by Manufacturing" (Ph.D. diss., University of Chicago, 1926), 5.

41. The question of "hand-to-mouth" practices before the Great Depression has been neglected. An exception is the following study of the automotive industry: Michael Schwartz and Andrew Fish, "Just-in-Time Inventories in Old Detroit," *Business History* 40 (1998): 48–71.

42. Quote is from "Mill Competition Hurts Jobbers," *Iron Age* (June 3, 1926): 1575. "The Machinery and Supply Convention," 1484–93; W. Todd, "The Machinery and Supply Jobber," *Iron Age* (May 16, 1912): 1245–46. Also see "Jobbing Trade Sees Up-Grade in Steel," *Iron Age* (June 1, 1922): 1565–66; "Warehouse Distribution Dwindling," *Iron Age* (January 6, 1927): 21–24; George Tegan, "Better Prospects for Steel Jobbers," *Iron Age* (January 5, 1928): 48; and T. Gerken, "Jobbers Improve Their Position," *Iron Age* (January 2, 1930): 28–29.

43. U.S. Steel, *The Distribution of Steel to Major Consuming Industries* (n.p., 1939).

44. Chicago Association of Commerce, *Survey of the Metals Trades and Allied Industries* (Chicago: Chicago Association of Commerce, 1925), 9, 13.

45. Quote from ibid., 9.

46. Chicago Association of Commerce, *Wholesale and Retail Trade of Chicago, Illinois* (Chicago: Chicago Association of Commerce, 1927).

47. Chicago Association of Commerce, *Survey of the Metals Trades*, 9.

48. Quote from "The Ryerson Iron and Steel Warehouses," *Iron Age* (July 9, 1908): 99. Also see Aduddell, "Edward L. Ryerson, Jr."; and Edward L. Ryerson, "Steel Servicing Pioneer Business," *Chicago Commerce* (December 7, 1929): 364–66.

49. Quotes are from "The Scully Steel & Iron Company's Warehouses," *Iron Age* (September 8, 1910): 571; and "A New Perforating Plant in Chicago," *Iron Age* (June 15, 1911): 1485. For another example—A.M. Castle and Company—see "New Steel

Warehouses to Use Chicago River," *Iron Age* 91 (January 23, 1913): 257; "New Chicago Iron and Steel Warehouse," *Iron Age* 92 (August 7, 1913): 297.

50. Ryerson, "Steel Servicing Pioneer Business."

CHAPTER TEN

1. "Chicago Has Big Automotive Industry," *Chicago Commerce* (January 24, 1925): 11.

2. David Farber, *Sloan Rules: Alfred P. Sloan and the Triumph of General Motors* (Chicago: University of Chicago Press, 2002); Allan Nevins and Frank Hill, *Ford: Expansion and Challenge, 1915–1933* (New York: Scribner's, 1957).

3. Reyner Banham, *A Concrete Atlantis: U.S. Industrial Building and European Modern Architecture, 1900–1925* (Cambridge, MA: MIT Press, 1986); Lindy Biggs, *The Rational Factory: Architecture, Technology, and Work in America's Age of Mass Production* (Baltimore: John Hopkins University Press, 1996); Grant Hildebrand, *Designing for Industry: The Architecture of Albert Kahn* (Cambridge, MA: MIT Press, 1974).

4. Clarence Hooker, *Life in the Shadows of the Crystal Palace, 1910–1927: Ford Workers in the Model T Era* (Bowling Green, OH: Bowling Green State University Press, 1997); Nelson Lichtenstein and Stephen Meyer, eds., *On the Line: Essays in the History of Auto Work* (Urbana: University of Illinois Press, 1989); Stephen Meyer, *The Five Dollar Day: Labor Management and Social Control in the Ford Motor Company* (Albany: State University of New York Press, 1981); Richard Feldman and Michael Betzold, eds., *End of the Line: Autoworkers and the American Dream* (Urbana: University of Illinois Press, 1990); Thomas Lewis, *Divided Highways: Building the Interstate Highways, Transforming American Life* (New York: Viking, 1997).

5. David Hounshell, *From the American System to Mass Production, 1800–1932: The Development of Manufacturing Technology in the United States* (Baltimore: John Hopkins University Press, 1984); Nevins and Hill, *Ford*.

6. James Rubenstein, *The Changing US Auto Industry: A Geographical Analysis* (London: Routledge, 1992); Michael Conzen, "The Progress of American Urbanism, 1860–1930," in *North America: The Historical Geography of a Changing Continent*, ed. Robert Mitchell and Paul Groves (Totowa, NJ: Rowman and Littlefield, 1987), 347–70; David Meyer, "The National Integration of Regional Economies, 1860–1920," in *North America*, ed. Mitchell and Groves, 321–46; Philip Scranton, *Endless Novelty: Specialty Production and American Industrialization, 1865–1925* (Princeton, NJ: Princeton University Press, 1997); Gerald Bloomfield, "Shaping the Character of the City: The Automobile Industry and Detroit, 1900–1920," *Michigan Quarterly Review* 25 (1986): 167–81; Olivier Zunz, *The Changing Face of Inequality: Urbanization, Industrial Development, and Immigrants in Detroit, 1880–1920* (Chicago: University of Chicago Press, 1982).

7. Karel Williams, Colin Halsam, and John Williams with Andy Adcroft and S. Joyal, "The Myth of the Line: Ford's Production of the Model T at Highland Park, 1909–1916," *Business History* 35 (1993): 67. Also see Karel Williams, John Williams, Tony Cutler, and Colin Halsam, "The End of Mass Production?" *Economy and Society* 16 (1987): 405–39; and Karel Williams, Colin Halsam, and John Williams, "Ford versus

'Fordism': The Beginning of Mass Production," *Work, Employment and Society* 6 (1992): 517–55.

8. Paul Hirst and Jonathan Zeitlin, "Flexible Specialization versus Post-Fordism: Theory, Evidence and Policy Implications," *Economy and Society* 20 (1991): 1–56; Scranton, *Endless Novelty*; Richard Walker, "The Geographical Organization of Production-Systems," *Environment and Planning D* 6 (1988), 377–408; Jonathan Zeitlin, "Flexibility and Mass Production at War: Aircraft Manufacture in Britain, the United States and Germany, 1929–1945," *Technology and Culture* 26 (1995): 46–79; Stephen Amberg, "The Triumph of Industrial Orthodoxy: The Collapse of Studebaker-Packard," in *On the Line*, ed. Lichtenstein and Meyer, 190–218; Stephen Meyer, "The Persistence of Fordism: Workers and Technology in the American Automobile Industry, 1900–1960" in ibid., 73–99; Michael Schwartz and Andrew Fish, "Just-in-Time Inventories in Old Detroit," *Business History* 40 (1998): 48–71; Williams, Halsam, and Williams, "Ford versus 'Fordism.'"

9. Peter Dicken and Anders Malmberg, "Firms in Territories: A Relational Approach," *Economic Geography* 77 (2001): 345–63; Peter Maskell and Anders Malmberg, "The Competitiveness of Firms and Regions: 'Ubiquitification' and the Importance of Localized Learning," *European Urban and Regional Studies* 6 (1999): 9–25; Meric Gertler, David Wolfe, and David Garkut, "No Place Like Home? The Embeddedness of Innovation in a Regional Economy," *Review of International Political Economy* 7 (2000): 688–718; Ann Markusen, "Sticky Places in Slippery Space: A Typology of Industrial Districts," *Economic Geography* 72 (1996): 293–313; Udo Staber, "The Structure of Networks in Industrial Districts," *International Journal of Urban and Regional Research* 25 (2001): 537–52.

10. Deborah Leslie and Suzanne Reimer, "Spatializing Commodity Chains," *Progress in Human Geography* 23 (1999): 401–20.

11. The extent of Chicago's automotive industry is difficult to assess because the definitions that the manufacturing census used changed from year to year. The totals include automobile bodies and parts, automobiles, vehicle hardware, motorcycles, transmissions, engines, automotive stampings, and electrical equipment. In all years, the numbers underestimate the number of firms making products for the automotive industry. See the various manufacturing reports of the census and the U.S. Department of Commerce, *Biennial Census of Manufactures, 1925* (Washington, DC: Government Printing Office, 1928).

12. "Chicago Has Big Automotive Industry," 11.

13. Chicago Association of Commerce, Ways and Means Committee, "Growth of Automobile Industry and Chicago's Place in Its Phenomenal Development," *Chicago Commerce* (February 13, 1914): 28–30; Rubenstein, *The Changing US Auto Industry*, 25–51.

14. Quote from "Chicago Has Big Automotive Industry," 11. Also see "Latest Industrial Development," *Chicago Commerce* (March 26, 1927): 29; "Industrial Development Plans," *Chicago Commerce* (March 17, 1928): 24, and (November 3, 1928): 24.

15. John Rae, *American Automobile Manufacturers: The First Forty Years* (Philadelphia: Chilton, 1959), 6–44; Rubenstein, *The Changing US Auto Industry*, 28–36; Chicago

Association of Commerce, "Growth of Automobile Industry," 30; William Cronon, *Nature's Metropolis: Chicago and the Great West* (New York: Norton, 1991).

16. F. K. Hendrickson, "Automobile's Influence on Machine Tool Industry," *Automotive Industries* 45 (November 10, 1921): 926; Edward Duggan, "Machines, Markets and Labor: The Carriage and Wagon Industry in Late-Nineteenth-Century Cincinnati," *Business History Review* 51 (1977): 308–25; Hounshell, *From the American System*, 189–215.

17. Josiah Currey, *Manufacturing and Wholesale Industries of Chicago*, 3 vols. (Chicago: Thomas Poole, 1918), 3:7–10.

18. Ibid., 3:282–85; *Directory of Illinois Manufacturers, 1924–1925* (Chicago: Illinois Manufacturers' Association, 1924), 238.

19. Currey, *Manufacturing and Wholesaling Industries*, 3:30–32, 86–88.

20. "24 Dealers Exhibit at Automobile Club," *Automotive Industries* (April 7, 1921): 783; "Chicago World Center in Taxicab Industry," *Chicago Commerce* (September 8, 1923): 15–16, 41; "The Automotive Business in Chicago," *Chicago Commerce* (June 2, 1928): 11.

21. Quote from Currey, *Manufacturing and Wholesaling Industries*, 3:330. "Chicago Has Big Automotive Industry"; "Chicago Active in the Motor Truck Field," *Chicago Commerce* (July 25, 1925): 13–14; "The Automotive Business in Chicago."

22. "Chicago Has Big Automotive Industry."

23. Alfred Chandler, *The Visible Hand: The Managerial Revolution in American Business* (Cambridge, MA: Belknap Press, 1977); Nevins and Hill, *Ford*, 283–93; Rae, *American Automobile Manufacturers*; Rubenstein, *The Changing US Auto Industry*, 101–8; "Show Chicago's Place in Automotive Accessories: A Leader in Manufacture, a Leader in Distribution," *Chicago Commerce* (June 2, 1928): 9–10; Schwartz and Fish, "Just-in-Time Inventories"; Williams, Halsam, and Williams, "Ford Versus 'Fordism.'"

24. Schwartz and Fish, "Just-in-Time Inventories."

25. Ibid.

26. Quote from "Chicago Has Big Automotive Industry," 11. "Industrial Development Plans," *Chicago Commerce* (September 8, 1928): 28.

27. William Hogan, *Economic History of the Iron and Steel Industry in the United States*, 5 vols. (Lexington, MA: Heath, 1971), 3:1000; "On the Highway of the Industry," *Automotive Industries* (February 23, 1929): 260.

28. Quote from "The Corey Steel Co.," *Chicago Commerce* (October 12, 1929): 25. *Polk's Chicago (Illinois) City Directory, 1928–1929* (Chicago: R.L. Polk, 1928).

29. "News of Industrial Development Plans," *Chicago Commerce* (January 19, 1924): 66; "Chicago World Center in Taxicab Industry"; "Cab Company to Make New Ambassador Car," *Automotive Industries* 44 (January 20, 1921): 139; "24 Dealers Exhibit at Automotive Club"; "Show Chicago's Place in Automotive Accessories"; *Directory of Illinois Manufacturers, 1924–1925*; *Directory of Illinois Manufacturers, 1928–1929* (Chicago: Illinois Manufacturers' Association, 1928); "Latest News of Chicago Industrial Development Plans," *Chicago Commerce* (May 12, 1923): 48; "Variety No Bar to Mass Production," *Iron Age* 124 (September 19, 1929): 723–29.

30. Norman Shidle, "Vehicle Builders Buying More Parts from Outside Sources," *Automotive Industries* (October 27, 1928): 578.

31. "Stewart-Warner One of Most Successful Parts and Accessories Companies," *Automotive Industries* (January 21, 1928): 80–82; "Show Chicago's Place in Automotive Accessories"; "Automotive Supply Manufacturers Extending Warehouses System," *Automotive Industries* (August 18, 1928): 217–19, 237.

32. District Court of the United States, Bankruptcy Case Files, Bankruptcy Case 28646, *In the Matter of Chicago Standard Axle Company*, "Schedule A(3)," filed December 10, 1920; Prudence Walker, ed., *Directory of Illinois Manufacturers, 1920* (Chicago: Illinois Manufacturers' Association, 1920).

33. "Latest News of Chicago Industrial Development Plans," *Chicago Commerce* (April 7, 1923): 40.

34. "C.G. Spring Acquires United States Bumper," *Automotive Industries* (January 13, 1921): 97; "Latest News of Chicago Industrial Development Plans," *Chicago Commerce* (July 7, 1923): 34.

35. Quote from "Industrial Development Plans," *Chicago Commerce* (February 2, 1929): 20. "Pines Winterfront Erecting New Plant," *Automotive Industries* (February 2, 1929): 183; "Pines Winterfront Company," *Commerce* (July 1937): 41–45. For Motorola, see Thomas Guifoyle, "Motorola Inc.," in *The Encyclopedia of Chicago*, ed. James Grossman, Ann Durkin Keating, and Janice Reiff (Chicago: University of Chicago Press, 2004), 935–36.

36. "Pullman Steel Body to Appear at Sales," *Automotive Industries* (January 24, 1924): 203; "Buda Manufacturing New Engine for Bus," *Automotive Industries* (February 7, 1924): 309; "New Yellow Cab Weighs 500 Pounds Less than Previous Model," *Automotive Industries* (April 10, 1924): 818.

37. "Latest News of Chicago Industrial Development Plans," *Chicago Commerce* (May 16, 1925): 18. For the other marts, see Deborah Rau Fulton, "The Making of the Merchandise Mart, 1927–1931: Air Rights and the Plan of Chicago," in *Chicago Architecture and Design, 1923–1993*, ed. John Zukowsky (Munich: Prestel, 1993), 99–117; and Sharon Darling, *Chicago Furniture, Art, Craft, and Industry, 1833–1933* (New York: Norton, 1984), 292–95.

38. "Business Done at Auto Show Is Called Excellent Augury for the Future," *Chicago Commerce* (February 4, 1922): 9.

39. "Chicago Has Big Automotive Industry"; "Battery Makers to Form Organization in March," *Automotive Industries* (January 31, 1924): 257; "Manufacturers Form Battery Association," *Automotive Industries* (March 27, 1924): 740.

40. The location quotient (LQ) measures the degree to which an activity is concentrated in a specific area compared to all activities. An LQ of 1 means that the activity is geographically distributed to the same extent as all other activities. A number greater than 1 means that the activity is concentrated in the area, while less than 1 means that it is not.

41. Bankruptcy Case 28646, *In the Matter of Chicago Standard Axle Company*.

42. District Court, Bankruptcy Case 28534, *In the Matter of B&W Manufacturing Company*, "Affidavit of Mailing," December 20, 1920; District Court, Bankruptcy Case 28657, *In the Matter of Sunderland Manufacturing Company*, "Schedule A(3)," 1920.

43. District Court, Bankruptcy Case 39641, *In the Matter of Mohawk Auto Equipment Company*, "Affidavit of Mailing," filed December 3, 1928; District Court, Bankruptcy Case 40632, *In the Matter of Unexcelled Auto Products Company*, "Affidavit of Mailing," filed October 13, 1928.

44. Nevins and Hill, *Ford*, 255–57; Rubenstein, *The Changing US Auto Industry*, 55–69; "Latest News of Chicago Industrial Development Plans," *Chicago Commerce* (September 6, 1924): 24.

45. John Drury, "Hegewisch: The Dream of Early Industrialist," newspaper clipping, *Chicago Daily News* (July 1, 1959), Chicago Public Library, Special Collections, Calumet Region Community Collection, box 4, file 27; William Rowan, "South Chicago's History for First Hundred Years," *Daily Calumet*, June 6, 1936, Chicago Public Library, Special Collections, South Chicago Community Collection, box 4, file 5.

46. Drury, "Hegewisch."

47. Ibid.; Harold Mayer and Richard Wade, *Chicago: Growth of a Metropolis* (Chicago: University of Chicago Press, 1969), 137.

48. An examination of the lists of new buildings, new starts, and moves taken from *Iron Age* and *Chicago Commerce* in the 1920s demonstrates that most automotive firm activity took place within the automotive districts.

49. "Chicago Has Big Automotive Industry," 30.

50. Ibid.

51. Nevins and Hill, *Ford*; Schwartz and Fish, "Just-in-Time Inventories."

52. *Directory of Illinois Manufacturers, 1924–1925*; "News of Industrial Development Plans," *Chicago Commerce* (March 11, 1924): 28.

53. "Latest News of Chicago's Industrial Development Plans," *Chicago Commerce* (September 6, 1924), 24; "Latest Industrial Development Plans," *Chicago Commerce* (December 5, 1925): 22; "Industrial Development Plans," *Chicago Commerce* (December 10, 1927): 26; "Plans for Industrial Development," *Chicago Commerce* (May 18, 1929): 22; Hogan, *Economic History of the Iron and Steel Industry*, 1006.

CONCLUSION

1. "Civic-Industrial Committee and Local Industries Committee Make Eleventh Inspection Excursion," *Chicago Commerce* (July 19, 1912): 15.

2. "Committee of City Council and Association Make Twelfth Joint Inspection Tour of Manufacturing and Railway Districts," *Chicago Commerce* (June 13, 1913): 24.

3. David Gordon, "Capitalist Development and the History of the American Cities," in *Marxism and the Metropolis: New Perspectives in Urban Political Economy*, ed. William Tabb and Larry Sawers (New York: Oxford University Press, 1978), 25–63; Allen Scott, "Locational Patterns and Dynamics of Industrial Activity in the Modern Metropolis," *Urban Studies* 19 (1982): 111–42; and various chapters in Robert Lewis, ed., *Manufacturing Suburbs: Building Work and Home on the Metropolitan Fringe* (Philadelphia: Temple University Press, 2004).

4. For quote, see John F. McDonald, *Employment Location and Industrial Land Use in Metropolitan Chicago* (Champaign, IL: Stipes, 1984), 90. Also see Christine Meisner

Rosen, *The Limits of Power: Great Fires and the Process of City Growth in America* (Cambridge: Cambridge University Press, 1986), 145, 125, 144; Stanley Buder, *Pullman: An Experiment in Industrial Order and Community Planning, 1880–1930* (New York: Oxford University Press, 1967), 38; and Graham Taylor, *Satellite Cities: A Study of Industrial Suburbs* (New York: Appleton, 1915).

5. Taylor, *Satellite Cities.*

6. William Cronon, *Nature's Metropolis: Chicago and the Great West* (New York: Norton, 1991). Also see David Meyer, "Emergence of the American Manufacturing Belt: An Interpretation," *Journal of Historical Geography* 9 (1983): 145–74; Brian Page and Richard Walker, "From Settlement to Fordism: The Agro-Industrial Revolution in the American Midwest," *Economic Geography* 67 (1991): 281–315; and Gordon Winder, "The North American Manufacturing Belt in 1880: A Cluster of Regional Industrial Systems or One Large Industrial District?" *Economic Geography* 75 (1999): 71–91.

7. "Committee of City Council and Association Make Twelfth Joint Inspection Tour," 25.

8. Social scientists have written on local alliances and growth coalitions. For some important statements, see Kevin Cox and Andrew Mair, "Locality and Community in the Politics of Local Economic Development," *Annals of the Association of American Geographers* 78 (1988): 307–25; David Harvey, *The Limits to Capital* (Chicago: University of Chicago Press, 1982), 419–21; and John Logan and Harvey Moloch, *Urban Fortunes: The Political Economy of Place* (Berkeley: University of California Press, 1987). For some historical studies, see Robert Lewis, *Manufacturing Montreal: The Making of an Industrial Landscape, 1850 to 1930* (Baltimore: John Hopkins University Press, 2000); Paul-Andre Linteau, *The Promoter's City: Building the Industrial Town of Maisonneuve, 1883–1918* (Toronto: Lorimer, 1985); and Harold Platt, *The Electric City: Energy and the Growth of the Chicago Area, 1880–1930* (Chicago: University of Chicago Press, 1991).

9. "Committees of City Council and Association Make Twelfth Joint Inspection Tour," 24, 27; Francis Stetson Harman, "The Central Manufacturing District of Chicago," in *Manufacturing and Wholesale Industries of Chicago*, 3 vols., ed. Josiah Currey (Chicago: Thomas Poole, 1918), 1:438.

10. "Committees of City Council and Association Make Twelfth Joint Inspection Tour," 27.

11. Peter Dicken and Anders Malmberg, "Firms in Territories: A Relational Approach," *Economic Geography* 77 (2001): 353–56; Peter Maskell and Anders Malmberg, "Localised Learning and Industrial Competitiveness," *Cambridge Journal of Economics* 23 (1999): 169–70, 173–74; Ron Martin, "The New 'Geographical Turn' in Economics: Some Critical Reflections," *Cambridge Journal of Economics* 23 (1999): 80.

12. Robin Bachin, *Building the South Side: Urban Space and Civic Culture in Chicago, 1890–1919* (Chicago: University of Chicago Press, 2004); James Barrett, *Work and Community in the Jungle: Chicago's Packinghouse Workers, 1894–1922* (Urbana: University of Illinois Press, 1987); Buder, *Pullman*; Michael Ebner, *Creating Chicago's North Shore: A Suburban History* (Chicago: University of Chicago Press, 1988); Dominic Pacyga, *Polish Immigrants and Industrial Chicago: Workers on the South Side, 1880–*

1922 (Chicago: University of Chicago Press, 1991); Thomas Philpott; *The Slum and the Ghetto: Neighborhood Deterioration and Middle-Class Reform, Chicago, 1880–1930* (New York: Oxford University Press, 1978); Louise Carroll Wade, *Chicago's Pride: The Stockyards, Packingtown, and the Environs in the Nineteenth Century* (Urbana: University of Illinois Press, 1987). This work was preceded by the social ecology work of the Chicago school. See Robert Park, Ernest Burgess, and Roderick McKenzie, eds., *The City* (Chicago: University of Chicago Press, 1925); and Harvey Zorbaugh, *The Gold Coast and the Slum: A Sociological Study of Chicago's Near North Side* (Chicago: University of Chicago Press, 1929).

13. United States, Bureau of Commerce, *Thirteenth Census of the United States Taken in the Year 1910: Manufactures, 1909* (Washington, DC: Government Printing Office, 1903), 10:903. Also see Edward Muller, "The Pittsburgh Survey and 'Greater Pittsburgh': A Muddled Metropolitan Geography," in *Pittsburgh Surveyed: Social Science and Social Reform in the Early Twentieth Century*, ed. Maurine Greenwald and Margo Anderson (Pittsburgh: University of Pittsburgh Press, 1996), 75; and Michael Conzen, "The Progress of American Urbanism, 1860–1930," in *North America: The Historical Geography of a Changing Continent*, ed. Robert Mitchell and Paul Groves (Totowa, NJ: Rowman and Littlefield, 1987), 362–63.

14. Meyer, "Emergence of the American Manufacturing Belt"; Cronon, *Nature's Metropolis*; Page and Walker, "From Settlement to Fordism"; Winder, "The North American Manufacturing Belt."

15. Michael Schwartz and Andrew Fish, "Just-in-Time Inventories in Old Detroit," *Business History* 40 (1998): 48–71.

16. Michael Storper and Richard Walker, *The Capitalist Imperative: Territory, Technology, and Industrial Growth* (New York: Blackwell, 1989), 53–54; Philip Scranton, *Endless Novelty: Specialty Production and American Industrialization, 1865–1925* (Princeton, NJ: Princeton University Press, 1997); Bennett Harrison, *Lean and Mean: The Changing Landscape of Corporate Power in the Age of Flexibility* (New York: Basic Books, 1994); Gary Herrigel, "Large Firms, Small Firms, and the Governance of Flexible Specialization: The Case of Baden Wurttemberg and Socialized Risk," in *Country Competitiveness: Technology and the Organizing of Work*, ed. Bruce Kogut (New York: Oxford University Press, 1993), 15–35.

17. Storper and Walker, *The Capitalist Imperative*, 138–42; Harrison, *Lean and Mean*.

18. There are few, if any, histories of these four cities that adequately lay out their industrial geography. However, the following works provide important insights: James Kenyon, *Industrial Localization and Metropolitan Growth: The Paterson-Passaic District* (Chicago: University of Chicago, 1960); Philip Scranton, *Figured Tapestry: Production, Markets, and Power in Philadelphia Textiles, 1885–1941* (Cambridge: Cambridge University Press, 1989); Robert Fogelson, *The Fragmented Metropolis: Los Angeles, 1850–1930* (Cambridge, MA: Harvard University Press, 1967); and Greg Hise, *Magnetic Los Angeles: Planning the Twentieth-Century Metropolis* (Baltimore: John Hopkins University Press, 1997).

19. Pacyga, *Polish Immigrants*, 259.

20. Robert Cramer, *Manufacturing Structure of the Cicero District, Metropolitan Chicago* (Chicago: University of Chicago, Department of Geography Research Paper

No. 27, 1952); Marcel De Meirleir, *Manufactural Occupance in the West Central Area of Chicago* (Chicago: University of Chicago, Department of Geography, Research Paper No. 11, 1950).

21. Chicago Plan Commission, *Industrial and Commercial Background for Planning Chicago* (Chicago: Chicago Plan Commission, 1942), 49.

22. Olivier Zunz, *The Changing Face of Inequality: Urbanization, Industrial Development, and Immigrants in Detroit, 1880–1920* (Chicago: University of Chicago Press, 1982), 178.

23. Edward Muller, "Industrial Suburbs and the Growth of Metropolitan Pittsburgh, 1870–1920," in *Manufacturing Suburbs*, ed. Lewis, 124–42; Margaret Byington, *Homestead: The Households of Mill Town* (New York: Charities Publication, 1910); Anne Mosher, *Capital's Utopia: Vandergrift, Pennsylvania, 1855–1916* (Baltimore: John Hopkins University Press, 2004).

24. William Baker, "The Area Growth of Omaha, Nebraska, with Emphasis on the Westside Area" (Ph.D. diss., University of Nebraska, 1958), 269.

25. Henry McKiven, *Iron and Steel: Class, Race, and Community in Birmingham, Alabama, 1875–1920* (Chapel Hill: University of North Carolina Press, 1995), 133.

26. Greg Hise, "'Nature's Workshop': Industry and Urban Expansion in Southern California, 1900–1950," in *Manufacturing Suburbs*, ed. Lewis, 178–199; Richard Walker, "Industry Builds Out the City: The Suburbanization of Manufacturing in the San Francisco Bay Area, 1850–1940," in ibid., 92–123.

27. Ann Durkin Keating, *Building Chicago: Suburban Developers and the Creation of a Divided Metropolis* (Columbus: Ohio State University Press, 1988); Buder, *Pullman*; Raymond Mohl and Neil Betten, *Steel City: Urban and Ethnic Patterns in Gary, Indiana, 1906–1950* (New York: Holmes and Meier, 1986).

28. Nathan Weinhouse, "Final Report, East Chicago Real Property Inventory and Occupancy Survey" (typescript, May 20, 1935), 5–7, Calumet Regional Archives, East Chicago Collection, box 3, file 17.

29. Josiah Currey, *Manufacturing and Wholesale Industries of Chicago*, 3 vols. (Chicago: Thomas Poole, 1918), 1:395–413; Richard Schneirov, *Labor and Urban Politics: Class Conflict and the Origins of Modern Liberalism in Chicago, 1864–1897* (Urbana: University of Illinois Press, 1998); Barrett, *Work and Community*; Lizabeth Cohen, *Making a New Deal: Industrial Workers in Chicago, 1919–1939* (New York: Cambridge University Press, 1990).

30. Schneirov, *Labor and Urban Politics*; Barrett, *Work and Community*; Cohen, *Making a New Deal*.

APPENDIX

1. Quote from John Cover, *Business and Personal Failure and Readjustment in Chicago* (Chicago: University of Chicago Press, 1933), 5. The bankruptcy cases for metropolitan Chicago are held at the National Archives and Records Administration in Chicago and are archived in Record Group 21. For a history of bankruptcy, see Peter Coleman, *Debtors and Creditors in America: Insolvency, Imprisonment for Debt, and Bankruptcy, 1607–1900* (Madison: State Historical Society of Wisconsin, 1974);

David Skeel, *Debt's Dominion: A History of Bankruptcy Law in America* (Princeton, NJ: Princeton University Press, 2001), 1–70; and Charles Warren, *Bankruptcy in United States History* (Cambridge, MA: Harvard University Press, 1935).

2. Of the 35 firms from which it is possible to gain a sense of firm size, 22 had fewer than 25 workers, while 13 had 25 or more. The largest was a carriage maker with 112 employees, while two automotive parts makers had 96 and 70 workers and a publisher 88. The reason I say "sense" is that employment numbers are hard to come by at the best of times and the number of wage earners given in the bankruptcy records may not cover all of the workers that were employed in the period leading up to the bankruptcy. Despite this, it is clear that the sample does not include any very large firms.

3. William Cronon, *Nature's Metropolis: Chicago and the Great West* (New York: Norton, 1991), 270.

4. William Mollard, "Ties That Bind: Geographical Business Connections in the Corridor in the 1870s," in *The Industrial Revolution in the Upper Illinois Valley*, ed. Michael Conzen, Glenn Richard, and Carl Zimring; Studies on the Illinois and Michigan Canal Corridor (Chicago: Committee on Geographical Studies, University of Chicago, 1993), 34–35; Edward Balleisen, *Navigating Failure: Bankruptcy and Commercial Society in Antebellum America* (Chapel Hill: University of North Carolina Press, 2001), 3; Cover, *Business and Personal Failure*, 17.

5. "Parts Makers Report Better Dealer Trade," *Automotive Industries* 45 (October 27, 1921): 836.

6. Mollard, "Ties That Bind," 34. For an excellent study based on company records, see Gordon Winder, "Building Trust and Managing Business over Distance: A Geography of Reaper Manufacturer D. S. Morgan's Correspondence, 1867," *Economic Geography* 77 (2001): 95–121.

Pages with illustrations are referred to by the page number followed by f; tables are referred to by the page number followed by t.

chemical industry: fixed capital in, 99; and furniture industry, 220; geography, 143*t*; place in the industrial structure of Chicago, 26*t*, 71*t*, 79*t*, 143

Chicago Board of Trade, 73, 176

Chicago and Calumet Stock Yards Company, 80

Chicago Association of Commerce: and capitalist class, 282–83; as Chicago developer, 7, 68; Civic Industrial Committee, 1, 2, 45, 67, 168, 270, 272; furniture survey, 221; on housing, 63; opening of American Furniture Mart, 228

Chicago Ferrotype Company, 206–9, 210*f*

Chicago Harbor, 53

Chicago Harbor Commission, 176

Chicago Heights: attractions, 145, 190–91; development, 135–40; as factory district, 280; and Inland Steel, 146; plant additions in, 97; residential geography of, 42

Chicago Railway Equipment Company, 107, 311n22

Chicago Real Estate Board, 48, 59

Chicago River: and city origins, 21, 43; and factory districts, 6, 69–70, 111, 205–6; as factory location, 32–33, 74, 76, 108, 203, 211; and harbor, 77, 149; and the Near West Side, 62; and older neighborhoods, 83; and plant closing, 54–55; and sewage, 75; and warehousing, 207, 213, 254

Chicago Plan Commission: and business networks, 176; and Charles Wacker, 87, 136; overview of industrial geography, 91

Chicago Zoning Commission, 136, 167

Cicero, IL: conflict in, 282; development of, 36, 85; Diamond T and, 266; immigrants, 31, 269; industrial district, 59, 88, 91, 279; industrial suburb, 23–24; inter-firm linkages,

155–56, 202; land development in, 36, 272–73; population, 40; Western Electric and, 211; working-class suburb, 42

Cincinnati: automotive industry, 245; business links with Chicago, 120, 125, 126, 135, 138, 192, 199; as Chicago competitor, 23; and decentralization, 65, 270; managerial knowledge in, 158; specialized economy, 29, 278; as trade show center, 215–16; workforce, 24

Clearing, IL: advertisement for, 180*f*; aerial view of, 181*f*; factories, 99–102; factory design, 184–88, 186*f*; factory district, 106; fixed capital and, 103; industrial suburb, 16, 59, 91; land development, 62; planned industrial district, 35, 109, 167–88; size of manufacturing, 175*t*; suburb, 1; yards, 168*f*

Cleveland: as automotive center, 247, 252; business links with Chicago, 125–28 passim, 190–208 passim, 255, 258, 259, 276; industrial structure, 29, 278–79

coalitions, local: and annexation, 36; Calumet, 54–58, 135–40, 159; and Calumet Harbor, 149; and city-building, 4*f*, 15, 43, 49, 68, 160, 194, 271–73; conflicts within, 271–72, 284; and factory design, 185; and factory districts, 7; and industrial development, 37, 67, 111; and industrial incubator, 182–83; and land development, 17, 51–57 passim, 62, 103, 272–73; and locational assets, 15–17, 54, 108, 171, 280–81; and marts, 227–28; and neighborhoods, 43; and planned industrial districts, 102, 168–74; and planning thought, 170; and production system, 13–14; and segregation, 171; sentiment, 47, 54, 59, 150; trust, 165; vision 21, 45–68, 141, 177, 244

commodity chain. *See* production chain
Continental Can Company, 184,
189–90
Cook County Real Estate Board, 48
Crane Company, 25, 35, 36, 39; investment by, 82–84, 96–97, 103; moves
of, 82–84
creditors. *See* bankruptcy records

Detroit, 57, 137, 151, 193; automotive
industry, 245–66 passim, 276, 279;
creditors of, 125, 126, 195–96; firm
mobility, 92; as industrial center, 11,
28, 29, 138, 139–40, 192, 276, 277,
279; metropolitan, 13; suburbs of,
5, 280
Diamond T Motor Car Company, 251,
262*f*, 266
disinvestment, 61, 66; factory districts
and, 87, 89; in Packingtown, 78–80

electrical equipment industry, 28; concentration in, 29; as growth industry,
25, 39, 81; inter-firm relations, 155,
206–14, 247, 251; land development,
62–63; West Side and, 37, 85, 282
electricity, 6; and manufacturing, 52;
and metropolitan space, 48, 108–9;
service, 122, 255
Elgin, IL: industrial development, 85;
population, 40; as satellite city, 24,
36, 37, 84
Elgin Metal Novelty Company, 85; and
piano industry, 220
ethnic: conflict, 283; housing markets,
42, 62, 280; institutions, 5, 42, 113,
135, 256; labor markets, 5, 6, 41, 62,
281; neighborhoods, 5, 41, 42–43,
273, 280, 281; segregation, 63
Europe: American firms in, 157; immigrants from, 30–31; industrial
districts in, 11–12, 285, 321n4; technology, knowledge, and capital from,
30, 157, 221; and wholesaling, 235
Evanston, IL, 40–52, 187

external economies, 13, 15, 39, 58, 227,
278; downtown, 113, 182, 207; of the
Furniture Club, 228; labor-intensive
industries, 9, 173, 182; and relational
proximity, 10; and suburbs, 11, 70

factory: additions to, 92–106, 113–14,
153–54; daylight, 185–88, 186*f*; design, 82, 99, 113, 170, 171, 184–88;
and fixed capital, 96–103, 212; moving, 106–14; proprietorial, 8; relocation of, 90–92, 106–12, 160, 163;
sites, 49, 55, 56*f*, 57; starts, 92–106,
113–14; suburban, 5, 36, 66
factory district: city-building process,
4–10, 16, 270–71, 284–86; fixed
capital, 103, 106, 111, 113–14;
geography, 32–39, 69–89, 214; and
industrial inspection tours, 1–3;
inter-firm relations, 14, 16, 117–40,
155, 166, 211; and locational assets,
271–75; and place, 93; and planned
industrial districts, 169–74, 181;
problems of, 66; production chain,
275–79; production system, 14; and
property development, 57–59; and
residential areas, 39–44, 86–88, 102,
279–84; travels through, 59–62;
workers, 4–5, 11
finance: and land development, 48, 52,
54; and manufacturing, 15, 120–21,
128, 145, 276; and railroads, 177;
and the state, 57. *See also* banking
food industry: geography, 71*t*, 79*t*;
packing of, 190; place in the industrial structure of Chicago, 26*t*, 28,
30, 34*t*, 38*t*; processing, 40; sales
organization, 237; small firms, 143;
wholesaling, 238
Furniture Club of America, 228
furniture industry: and the American Furniture Mart, 227–30, 231;
associations, 225–27, 228; batch
producer, 10, 14–15, 118; Chicago as producer, 25, 217–28, 269;

International Harvester Company
(*continued*)
    relations, 153; interlocking director-
    ships, 154; production chain, 159;
    workforce, 281

Jewell Electrical Company, 206–9, 210*f*
J.G. Hoffman Company, 133*f*, 134
jobbers, 122; benefits offered by, 239;
    and bicycle industry, 123; and candy
    industry, 232; changing position
    of, 236; and marts, 230; and steel
    industry, 240–42
Joliet, IL: and Illinois Steel, 85, 237;
    industrial development, 84–85;
    metropolitan economy, 35, 239, 270;
    population, 40; as satellite city 24,
    36, 37, 84, 279
Joseph T. Ryerson and Sons, 241–43;
    acquired by Inland Steel, 236
just-in-time: automotive industry,
    252–55, 266; printing industry, 28;
    and spatial proximity, 122; whole-
    saling, 239–42 passim

Kenwood, IL, 36, 111
knowledge, 175; access to, 267; flows of,
    93–95, 190, 223, 226, 276; as know-
    how, 15, 95–96; local business, 30,
    113, 248; managerial, 158; and new
    starts, 95–96; spillover, 173; spin-
    offs, 85, 153, 173, 277; switching
    lines and, 250–51; tacit, 164

labor: and capital relations, 8–9, 15, 63,
    121, 282–84; and central city, 4; and
    factory districts, 4, 37, 77; force, 10,
    11, 25, 38, 70, 85, 170; know-how, 15
labor markets, 4, 6, 13, 16, 28, 42,
    92–93, 109, 211, 281
land developers: Bogue and Hoyt,
    49–50, 111; Calumet and Chicago
    Canal and Dock Company, 37, 49,
    54–57, 66, 162; Chicago Heights
    Land Association, 136–37; Chicago

Land Company, 49; Gary Land
    Association Company, 37, 49;
    Harvey Land Association, 47, 194;
    Hodge, Nicolson and Porter, 108,
    111; West Pullman Land Associa-
    tion, 57. *See also* industrial real estate
    companies; land development
land development: and the Calumet,
    54–59, 136–38, 145–46, 159,
    162–63; and Hegewisch, 265; for
    industry, 108–12, 141–42; local
    alliances, 16–17; locational assets,
    15, 271–75; and Packingtown,
    73–74; planned industrial districts,
    167–88; property development,
    48–52; suburbanization, 36–37. *See
    also* industrial real estate companies;
    land developers
leather industry: of Boston, 126;
    decline, 25; and furniture industry,
    220; geography, 33, 34*t*, 77, 143;
    inter-firm relations, 119, 123, 128,
    208; Leather Exchange, 119; of
    place in the industrial structure of
    Chicago, 26*t*, 71*t*, 79*t*; resource-
    based, 30; small scale, 29
linkages, business: automotive indus-
    try, 244–67; bankruptcy records,
    287–90; bicycle industry, 126–31;
    of the Calumet, 135–40, 141–66;
    distribution, 28, 230, 235–43; and
    factory districts, 16, 93, 274, 277–78;
    furniture, 216–35; of the Furniture
    Club, 228; industrial district, 11–13,
    16; and the industrial inspection
    tours, 272; institutional, 31, 176;
    of the instrument district, 206–14;
    inter-firm, 6, 11, 12, 16, 117–40,
    152–53, 213, 244–67, 274, 275–79,
    284; intra-metropolitan, 10–16,
    152–53, 200–214, 284–86; and large
    firms, 9; local, 11, 93, 108; and lo-
    cational assets, 274–75; lumber, 73;
    and neighborhoods, 281–83; of the
    planned industrial district, 167–88;

and production chain, 15, 275–79; and promoters, 165; regional, 12–13, 118, 128–29, 134, 139–40, 155–59, 190–93, 195–200, 253; as relational proximity, 10–11; and relocation, 92; and urban economy, 3, 7, 15, 16, 275; and wholesalers, 235–43

locale effect, 132–34, 202–5, 203*t*, 204*f*, 261–62, 263, 326n19

locational assets: and annexation, 36; automotive industry, 265; Bridgeport (Chicago), 70; Calumet, 54, 165; central factory district, 37; and fixed capital, 96–97, 103; and growth coalitions, 108–9, 135, 280–82; and industrialization, 31; instrument district, 209; and metropolitan production system, 15–16, 43, 271–75; Milwaukee Avenue district, 77; place commitments, 93; and planned industrial districts, 171, 175, 176, 182, 183; spatial proximity, 135

Loop, the, 55, 59, 62, 77, 230, 269, 282, 284, 286; business in, 138, 184, 202, 205–6, 213, 230, 213, 233, 238, 242; as factory district, 32–35, 37, 102, 106, 108, 123, 196, 211; manufacturing in, 2, 37, 72–73, 77, 99, 107, 108, 125, 126, 129–31, 132, 134, 206, 207, 224, 260

Los Angeles, 17, 172; as industrial center, 195, 217, 280; metropolitan, 13, 278; suburbs of, 6

Lower West Side (Chicago), 8, 33, 37, 69, 81, 84, 85; development of, 71–73, 77–78; industrial structure, 71*t*, 79*t*

Lumberman's Board of Trade, 72, 226

Lumberman's Exchange, 72, 226

machine layout, 99

machinery industry: associations, 215–16, 230–31; as batch producer, 14, 281; in the Calumet, 143, 144*t*; Chicago as producer of, 28, 71, 269; in Cicero, 85; electrical, 199; furniture industry and, 220, 221; geography of, 97; inter-firm relations, 119, 126, 131, 139, 147, 153, 191–96, 199–211 passim, 256, 258; investment in, 17, 98–99; know-how, 96; metropolitan-building, 6, 30; place in the industrial structure of Chicago, 26*t*, 29, 38*t*, 79*t*; relocating of, 164; scale of, 143; suppliers of, 10, 118, 148, 190–96; wholesaling in 240–43, 274

Madison Furniture Company, 224, 224*t*, 225*t*

magnet firms, 6, 55, 76; Chicago as, 30; and locational assets 273; and marts, 230; and neighborhoods, 33, 36, 62, 93; and satellites, 151

Manufacturing Belt: and automotive industry, 246, 258; and business linkages, 13, 118, 140, 192*f*, 195*f*; and the Calumet, 142, 159; and Chicago firms, 126, 128–29, 134, 191–200 passim, 207–8; and industrial district, 285; and production chain, 275; specialization in, 126, 138

marts, 176, 216, 227–35; Automotive Equipment Mart, 260; Chicago Machinery Mart, 230–31; Merchandise Mart, 230, 260; Millinery Mart, 231; Railway Exchange Building, 230; Textile Exchange Building, 230

meatpacking industry: disinvestment in, 78–80, 81; geography of, 33, 40, 43, 71, 74–76, 78–80; as growth industry, 26–27, 92; and Hammond, 137; and innovation, 175; inter-firm relations, 118, 220; leading industry, 24, 25, 28, 29, 30, 269; and wholesaling, 236, 237; working-class districts, 42, 88, 279

Metal Crafts Inc., 201, 221, 222*t*

metropolitan area. *See* capitalist metropolis

metropolitan production system, 13–17
Michigan, Lake, 21, 23, 46, 49, 53, 55, 57, 84, 162, 200, 266
Milwaukee: automotive industry, 245; as goods supplier, 128, 134, 138, 191, 199
Milwaukee Avenue district: development, 70, 76–78; industrial structure, 71*t*, 79*t*; poor conditions of, 86, 88
Montreal: metropolitan, 13, 32; suburbs, 5–6
Morgan, J. P., 157–58

National Furniture Exchange, 226, 227
National Piano Manufacturers' Association, 226
neighborhood: African American, 41, 269; and business linkages, 96, 132–33, 200–207 passim, 276–77; and the Calumet, 153–54, 192; and Clarence Stein, 171; and conflict, 292–84; dense, 4–5; different types, 15, 41, 52; disinvestment in, 87–88; ethnic, 43; everyday activities, 282; and factories, 62–67, 77, 83, 92, 113, 263; institutions, 42–43; and local alliances, 43, 59, 103–4; and metropolitan areas, 8, 123, 273, 284–85; and occupation, 281; production chain, 275
networks, business. *See* linkages, business
New York: control of business decisions, 283–84; "economic machinery" of, 107; factory districts, 277; finance, 154, 157; and furniture industry, 216, 217; industrial scale, 21, 23, 28, 269; industrial structure, 25, 29; industrial suburbanization, 107; inter-firm relations, 125–38 passim, 191–208 passim, 221, 233, 258; production chain, 276; rival to Chicago, 64, 278; and Western Electric, 211; and wholesaling, 238
New York Regional Plan, 65
New York Regional Survey, 65

Oak Park, IL, 40, 42, 109

Packingtown: annexation, 40; and the Central Manufacturing District, 176, 179; development of, 36, 73–76, 77–80; and fixed capital, 106, 107; housing, 41–42, 86–89, 282, 284; immigrants in, 269; industrial structure, 71*t*, 79*t*; land development, 272–73; and meatpacking, 33; specialization, 37
paper industry: bags and boxes, 208, 256; fixed capital, 99; inter-firm relations, 119, 123, 125, 134, 208, 212, 256; as new industry, 26; place in the industrial structure of Chicago, 26*t*, 79*t*; and wholesalers, 233*t*
P.A. Starck Piano Company, 218*f*
Philadelphia, 30, 191, 194, 239; creditors of, 120, 125, 126, 138, 199, 258; as financial center, 157; as industrial center, 23, 25, 28, 29, 125, 126, 139, 277; metropolitan, 13, 139–40, 278; suburbs, 5–6, 10
piano industry, 73, 218*f*, 220, 223, 243; growth of, 216–27
Pines Winterfront Company, 99, 101*f*, 259, 262*f*
Pittsburgh: Chicago competitor, 23, 141, 145; decline of, 91; freight rates, 149; industrial structure, 28, 29, 278; inter-firm relations, 126–38 passim, 190–99 passim, 241, 276; know-how, 158; and metropolitan scale, 13, 32; satellite towns, 11; steel producer, 27–28, 37, 245; steel towns, 5; working-class districts, 280
Pittsburgh-plus pricing, 97, 149
planning: by business leaders, 2–3; Chicago, 64–68, 271; City Planning Commission, 64; corporate planning, 145–46; Gary, IN, 49; in industrial districts, 142; land development and, 111, 272–73; in planned industrial